AP

GALILEO, COURTIER

Science and Its Conceptual Foundations

David Hull, Editor

GALILEO, COURTIER

The Practice of Science
in the Culture of Absolutism

MARIO BIAGIOLI

THE UNIVERSITY OF CHICAGO PRESS
Chicago & London

Mario Biagioli is associate professor in the Department of History at the University of California, Los Angeles.

This book has been brought to publication with the support of a grant from the National Endowment for the Humanities, an independent federal agency, and both author and publisher acknowledge this support with thanks.

The University of Chicago Press, Chicago 60637
The University of Chicago Press, Ltd., London
© 1993 by The University of Chicago
All rights reserved. Published 1993
Printed in the United States of America

02 01 00 99 98 97 96 95 94 93   1 2 3 4 5 6

ISBN (cloth): 0-226-04559-5

Library of Congress Cataloging-in-Publication Data

Biagioli, Mario.
   Galileo, courtier: the practice of science in the culture of absolutism/Mario Biagioli.
      p. cm.--(Science and its conceptual foundations)
      Includes bibliographical references and index.
   1. Galilei, Galileo, 1564-1642--Knowledge--Science. 2. Science-
-History. 3. Despotism. I. Title. II. Series.
   QB36.G2B54 1993
   509.4'09'032--dc20   92-33736

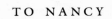

TO NANCY

CONTENTS

ILLUSTRATIONS

ACKNOWLEDGMENTS

This project has followed me (or I have followed it) around for several years and, under its spell, I have talked about it with almost everybody who had the patience to listen or the interest to discuss it. I apologize for not being able to acknowledge in a synthetic form all the comments, criticisms, and support I have received along the way.

David Harden, John Heilbron, David Hull, Nancy Salzer, Albert Van Helden, Richard Westfall, and an anonymous (but very perceptive) reviewer have read the entire book manuscript, provided detailed comments and criticism, and corrected several embarrassing mistakes. Bill Ashworth, Paul Feyerabend, Maurice Finocchiaro, Anthony Grafton, William Shea, Randy Starn, two anonymous reviewers, and especially Roger Hahn had the patience to do the same with earlier versions of the manuscript. Pnina Abir-Am, Ugo Baldini, Peter Barker, Peter Dear, Owen Gingerich, Carlo Ginzburg, Richard Goldthwaite, Keith Hutchison, Nicholas Jardine, Thomas Kuhn, Tim Lenoir, Lauro Martines, Laurie Nussdorfer, Trevor Pinch, Roy Porter, Robert Westman, and Norton Wise have read one or more versions of some of these chapters and provided comments important for their further development. Pierre Bourdieu, Ian Burney, Ernest Coumet, Olivier Darrigol, Mary Foertsch, Carl Ipsen, Tom Laqueur, Daniel Milo, Pietro Redondi, Jacques Revel, and Barbara Shapiro discussed with me what they thought I was trying to do when that was not yet quite clear to me. "Frank criticism" from graduate students at UCLA and Stanford has provided further material for useful doubts and reflections. This book owes much to all that I have received from all of these friends and colleagues.

In different and complementary ways, John Heilbron and Randy Starn have played a particularly important role in shaping this work. Their influence cannot be as clear to the reader as it is to me, and I want to thank them for the ongoing support and detailed criticism they have offered through the years. This book would have not been quite the same without the continuing dialogue I had with them.

This research topic emerged early during my graduate career. Graduate school is an inherently bittersweet memory, and I want to thank my fellow students first at the Visual Studies Workshop and then at Berkeley for making it exciting or, more frequently, just bearable. Ted Brown, Henry Kyburg, Donald Kelley, and Nathan Lyons sustained my initial interest in history and philosophy of science and encouraged me to pursue it at a time when becoming a scholar was still a concept alien to me. Although at times I have cursed them (without complete conviction) for their advice, I am ultimately very grateful for what they did. As Koyré taught us, mistakes can have fortuitously rewarding outcomes.

Special thanks to Susan Abrams, friend and editor, for all her advice and support, to Skuta Helgason for having introduced me to Psychic TV (*et al.*), and to Sande Cohen for helping make the southlands an intellectually lively place.

Earlier versions of some material in chapters one through four appeared in *Isis, Studies in History and Philosophy of Science,* and *History of Science,* and I thank the publishers of these journals for allowing me to reproduce here some of this previously published material. Grants from the National Science Foundation, the UCLA Academic Senate, and especially from the Biagioli Foundation have made this work materially possible.

This book is dedicated to Nancy Salzer. Our extensive and intensive discussions about all relevant aspects of this work have been essential to their articulation in this book. Moreover, her editorial advice has been crucial to turning this text into something that might convey the impression of being a coherent piece of work. Only the personal and intellectual companionship she has provided all along has made it possible for me to think and work through this project. While a dedication can be nothing but a gesture, I hope it will convey some of my deeper appreciation for all I have received from her over the years.

If fleas had rituals they would be about dogs.
Ludwig Wittgenstein

Is it you, is it me, or is it history?
Psychic TV

Court Culture and the Legitimation of Science

THAT GALILEO SPENT MOST OF HIS MATURE LIFE at the Medici court as mathematician and philosopher of the grand duke of Tuscany is certainly no news. However, that Galileo's courtly role was integral to his science is something that has not attracted the attention of historians and philosophers of science. The tendency to distinguish between Galileo the scientist and Galileo the courtier has not been limited to scholarly studies. Although in his insightful *Life of Galileo* Bertolt Brecht did not present him as a disinterested or "pure" scientist, he attributed to Galileo the ethos and culture of the artisan rather than of the courtier. To Brecht, the artisans represented progressive forces while the courtiers epitomized the *ancien régime*—a culture Brecht saw as at odds with the potentially positive and modern values he attributed to science.[1]

The common representation of Galileo's identity and science as alien to court values does not stem only from some authors' belief in the sharp distinction between "science" and "society." It is also maintained by those who are willing to link the development of modern science to social change but do not see the court as embodying the "good" forces of modernity.

While I am quite sympathetic to Brecht's view of the relationship between the artisanal tradition and the development of modern science (especially as that relationship has been articulated by Edgar Zilsel and Paolo Rossi)[2] and believe that it points to important aspects of the scien-

1. Bertolt Brecht, *Galileo* (New York: Grove Press, 1966). I am aware of the anachronism entailed by my use of "science" in relation to Galileo's activities. Here and elsewhere in the book I could have used more cumbersome terms such as "natural philosophy" or "mathematical natural philosophy" which would have been more appropriate to Galileo's historical context. However, I have decided to adopt "science" as a shorthand label for these activities—a label I am consciously using "under erasure."

2. Edgar Zilsel, "The Genesis of the Concept of Scientific Progress" and "Origins of Gilbert's Scientific Method," in Philip Wiener and Aaron Noland, eds., *Roots of Scientific*

tific revolution, I also think that we now have much evidence that aristo-cratic culture (not just individual aristocratic patrons and practitioners) played a crucial role in that process.[3] In particular, the court contributed to the cognitive legitimation of the new science by providing venues for the social legitimation of its practitioners, and this, in turn, boosted the epistemological status of their discipline.

The shift from the artisan's workshop to the princely court as a crucial site for the new science reflects, I think, the increasing interest (noticeable throughout the humanities and social sciences) in the workings of ritual, representation, and discourse. Also, it reflects a more complex approach to the relationship between power and knowledge. In *Galileo, Courtier,* power is treated neither as limited to its more material forms nor as a "thing" external to the process of knowledge-making.[4] Readers familiar with the Renaissance and Baroque court, its culture, and its etiquette know how intimately power was ingrained in manners, discipline, and po-lite discourse and that, despite its apparently "soft" appearance, power was extremely effective in fashioning individual identities, behaviors, and thoughts. At the same time, power did not exist outside these practices (as their independent cause); rather it was constituted by them.

This view of the relationship between power, knowledge, and self-fashioning applies well to an analysis of Galileo's scientific career. In many ways, I am presenting a study of a scientist's self-fashioning. Although self-fashioning is a category not commonly employed by historians of sci-ence, it has been effectively used in historical and literary studies of early modern Europe.[5] My choice of this approach to Galileo's scientific career reflects the character of his own social trajectory. Galileo began his career as a member of a specific socioprofessional culture—that of the mathe-maticians. However, in the process of moving to court, he successfully re-

Knowledge (New York: Basic Books, 1957), pp. 219–50, 251–75; Paolo Rossi, *I filosofi e le macchine, 1400–1700* (Milan: Feltrinelli, 1984).

3. These relationships were initially indicated by, among others, R. J. W. Evans, *Rudolph II and His World* (Oxford: Oxford University Press, 1973); Dario Franchini et al., eds., *La scienza a corte* (Rome: Bulzoni, 1979); Robert S. Westman, "The Astronomer's Role in the Sixteenth Century: A Preliminary Study," *History of Science* 18 (1980): 105–47; and Owen Hannaway, "Laboratory Design and the Aim of Science: Andreas Libavius versus Tycho Brahe," *Isis* 77 (1986): 585–610.

4. For a succinct statement of this view, see Michel Foucault, "Truth and Power," in Colin Gordon, ed., *Power/Knowledge* (New York: Pantheon, 1980), pp. 109–33.

5. See, for instance, Stephen Greenblatt, *Renaissance Self-Fashioning* (Chicago: University of Chicago Press, 1980); and Randolph Starn, "Seeing Culture in a Room for a Renais-sance Prince," in Lynn Hunt, ed., *The New Cultural History* (Berkeley: University of California Press, 1988), pp. 205–32. The exception to the neglect of individual self-fashioning by historians of science is the recent work by Steven Shapin, especially his "The House of Experiment in Seventeenth-Century England," *Isis* 79 (1988): 373–404; and idem, "A Scholar and a Gentleman," *History of Science* 29 (1991): 279–327.

fashioned himself as an unusual type of philosopher, a type of identity for which there were no well-established social roles or images. In a sense, Galileo reinvented himself around 1610 by becoming the grand duke's philosopher and mathematician. Although in doing so he borrowed from and renegotiated existing social roles and cultural codes, the socioprofessional identity he constructed for himself was definitely original. Galileo was a *bricoleur*.

This book traces Galileo's court-based articulation of the new socioprofessional identity of the "new philosopher" or "philosophical astronomer" and analyzes the relationship between this identity and Galileo's work. It does so by reconstructing the culture and codes of courtly behavior that framed Galileo's everyday practices, his texts, his presentation of himself and his discoveries, and his interaction with other courtiers, patrons, mathematicians, and philosophers. *Galileo, Courtier* is neither a biography nor a social history of Galileo's career. Although I follow Galileo through several nonconsecutive years and scientific disputes, and analyze several of his texts, my chief aim is to provide a detailed, sometimes microscopic, study of the structures of his daily activities and concerns and to show how these framed his scientific activities. In fact, while the volume's organization tends to mirror the chronological development of Galileo's career, its analysis is not primarily diachronic. There are a number of important strains that run chronologically through the chapters, but I do not present Galileo's scientific career as moving from intellectual childhood to maturity, and do not privilege his later (and allegedly more significant) works such as the *Dialogue on the Two Chief World Systems* and the *Discourse on the Two New Sciences*. I am more interested in identifying and studying the synchronic processes, conditions, resources, and constraints that shaped his everyday life and scientific activity and that—over several decades—produced the historical artifact that we now call Galileo's career. The analysis concludes with some reflections about the possible relationship between court culture, political absolutism, the legitimation of science, and the development of early scientific institutions.

Important aspects and episodes of Galileo's career are left out of this study. The bibliography of works on Galileo comprises several volumes already and its growth rate is not yet declining. Consequently, coverage has not been a priority of this work. Rather, I have tried to provide a new interpretive framework through a few case studies based on scientific disputes and episodes from Galileo's career. The richness and complexity of Galileo's work has forced me to be quite selective and leave out some of his texts that I think could have been successfully interpreted within the proposed framework.[6]

6. In particular, I am thinking of the 1612–13 dispute on sunspots with the Jesuit mathematician Christopher Scheiner (which is treated only briefly here), the "Letter to the Grand

Also, I do not wish to claim that the point of view adopted here can make sense of Galileo's entire career and texts. Galileo did not begin his professional life as a courtier but became one in 1610 when he was forty-six years old. While I would argue that a move to court was something Galileo may have considered much earlier than 1610, only some of his earlier work was framed by court culture or targeted at a court audience. Similarly, court culture did not equally affect all his many scientific interests. As we will see, while there was a close symbiosis between Galileo's work in astronomy and his court career, his interest in mechanics did not fit the court environment particularly well. Similarly, I do not want to argue that the strategies of social and cognitive legitimation I analyze in the case of Galileo were the only ones available for the legitimation of the new science and cosmology. While throughout the book I suggest homologies between Galileo's experience and that of other scientific practitioners who happened to choose similar court- or patronage-based strategies of legitimation, I do so not to present this interpretive framework as something that can cover the entire scientific revolution, but only to probe the boundaries of its applicability and to identify areas for further research. For instance, recent historiography has shown that different disciplines developed different patterns of social and epistemological legitimation depending on different national, religious, and political contexts. In several instances, the credibility of the mathematical sciences resulted from their success in solving technical problems. That these scenarios are not discussed in this book does not mean that I underestimate their importance, but simply that I have decided to concentrate on those processes of legitimation that relied on the representation of the new science as something fitting the culture of princes, patrons, and courtiers.

Patronage is one of the recurring themes of this study. I will provide evidence that concerns for patronage and social climbing were not external to Galileo's work. Court patronage was not simply a "resource" to be used by shrewd, clearheaded operators (like Brecht's Galileo). Patronage was part and parcel of the process of self-fashioning of all courtiers. As I argue in Chapter 1, patronage was an institution without walls, an elaborate and comprehensive system that constituted the social world of Galileo's science. In short, Galileo is presented not only as a rational manipulator of the patronage machinery but also as somebody whose discourse, motivations, and intellectual choices were informed by the patronage culture in which he operated throughout his life. Not only was Galileo's style embedded in court culture, but, as I hope to make clear by

---

Duchess," and the 1632 *Dialogue on the Two Chief World Systems*—a text that I consider only in relationship to Galileo's trial of 1633. I would have liked to extend this analysis to cover in more detail Galileo's involvement, beginning in 1611, with the Accademia dei Lincei.

the end of the book, his increasing commitment to Copernicanism and his self-fashioning as a successful court client fed on each other.

If Galileo's science was not external to court culture and patronage concerns, neither was it determined by these concerns. The portrait I am proposing is not that of a "slave of the system"—of somebody who fit himself within received roles and expectations in order to receive legitimation. Power does not censor or legitimate some body of knowledge that exists independently of it. By emphasizing the process of self-fashioning, I do not assume either an already existing "Galileo" who deploys different tactics in different environments and yet remains always "true to himself," nor a Galileo who is passively shaped by the context that envelops him. Rather, I want to emphasize how he used the resources he perceived in the surrounding environment to *construct* a new socioprofessional identity for himself, to put forward a new natural philosophy, and to develop a courtly audience for it. As indicated by Prince Leopold de' Medici's convening the Accademia del Cimento between 1657 and 1667, Galileo had an impact on Florentine court culture well after his death in 1642.

Obviously, any process of self-fashioning is not without tensions. If the court made possible Galileo's legitimation of his new socioprofessional identity, it also constrained it in ways that, at times, may have collided with Galileo's specific desires. While in certain contexts the fit between Galileo's work and court discourse was remarkable, on other occasions we find irresolvable tensions between patronage strategies and scientific authorship, or between Galileo's attempts to draw the patron on his side to legitimize his scientific claims and the prince's interest in preserving his power and image by not tying them to possibly problematic claims.

These tensions are an ongoing theme of the book and are eventually brought to bear on a reinterpretation of Galileo's trial. Without denying the obvious cosmological and theological dimensions of the events of 1633, I suggest that an understanding of court patronage and culture (and the tensions inherent in them) throws new light on this much-examined event. The very processes that allowed Galileo to fashion himself as a successful philosopher and courtier may have informed the dynamics of his trial.

The relationship between patronage, court culture, and Galileo's career is not just a matter of the history or sociology of scientific professions. Copernicus and some of his followers faced a crucial obstacle when they tried to legitimize their work as not only a *mathematical* computational model but also a *physical* representation of the cosmos. The received hierarchy among the liberal arts was such an obstacle.[7] According to this

7. Robert S. Westman, "The Astronomer's Role in the Sixteenth Century: A Preliminary Study," *History of Science* 18 (1980): 105–47.

hierarchy (one that was justified by scholastic views on the differences between the disciplines and their methodologies), mathematics was subordinated to philosophy and theology. The mathematicians were not
expected (or supposed) to deal with the physical dimensions of natural
phenomena, which (together with the causes of change and motion) were
considered to be the philosophers' domain. Consequently, the philosophers perceived Copernicus not just as putting forward a new planetary
theory, but as "invading" their own disciplinary and professional domain.
In general, this invasion was unacceptable to them and, having higher disciplinary status than the mathematicians, the philosophers had resources
to control such an invasion. The usual tactic (one that worked quite well in
institutions that accepted this disciplinary hierarchy) was to delegitimize
the mathematicians' claims by presenting them as coming from a lower
discipline.

The so-called Copernican revolution was two revolutions in one. The
acceptance of dramatic cosmological changes required drastic modifica-
tions in the organization of the disciplines that studied the cosmos. As we
know, this process was a very long one. The legitimation of Copernican
astronomy implied a restructuring of the hierarchies among the liberal
arts which, in turn, involved an increase in the mathematicians' social status. Such changes did not simply result from the strength of the new theories but from an institutional migration as well. Although the traditional
disciplinary hierarchy was quite entrenched in the university, it was not so
at court. There, one's status was determined by the prince's favor rather
than by the discipline one belonged to.

The court, then, was a social space in which mathematicians could
gain higher social status and credibility, thereby offsetting the disciplinary
gap traditionally existing between them and the philosophers. This increased socio-disciplinary status would in turn contribute to the legitimation of the new worldview they were proposing. If we look at the so-called
scientific revolution from the point of view of its sites of activity we may
notice (at least on the Continent) a trajectory that leads from the university, to the court, and, eventually, to the scientific academy. To a large extent, Galileo's career exemplifies this trajectory of social and cognitive
legitimation. After being a university mathematician, he became a natural
philosopher at a court and then a member of what is often considered the
first scientific academy—the Accademia dei Lincei. This pattern of institutional migration (one that mathematicians shared with the visual artists and, to some extent, the writers) is another of the ongoing themes of
Galileo, Courtier.

The almost exclusive focus on Galileo adopted in this book may create
the impression that his career was radically different from that of other

mathematicians. As I have indicated elsewhere, this is both true and un-true.[8] Although his courtly career and title of philosopher was exceptional for a mathematician, there are many other ways in which Galileo fit well the mathematician's traditional social role. For instance, in terms of train-ing, social status, and career pattern it would not have been easy to tell Galileo from other leading Italian mathematicians before 1610. Although Galileo's father, Vincenzio (a well-known musician and music theorist), had sent him to the University of Pisa to study medicine (and to help re-lieve the financial difficulties of his family), Galileo eventually left Pisa without a degree in 1585.[9] Like many other mathematicians, Galileo did not study mathematics at the university. In Florence he studied mathe-matics under Ostilio Ricci, an applied mathematician and military en-gineer who instructed the Florentine court pages and lectured on perspective to the painters, sculptors, and architects of the Accademia del Disegno—the Medici-sponsored academy of fine arts.[10] It was in this pro-fessional culture of applied mathematics at the crossroads of architecture, mechanics, fortifications, and the visual arts that Galileo spent his early years in Florence.

After 1588, Galileo taught mathematics, astronomy, mechanics, and fortification techniques at Siena, Pisa, and Padua—both inside and out-side the universities in those cities. That he could teach in a university without a university degree suggests that mathematics was perceived as a technical rather than philosophical discipline, one that was learned through apprenticeship rather than formal university training. The posi-tion of the teachers of mathematics in the university was marginal. In ad-dition to the disciplinary and epistemological gap that we have seen between mathematics and philosophy, mathematics was often considered as a mechanical art (because of its role in surveying, mechanics, and book-keeping). This did not confer much status on its professors, who in fact made between one-sixth and one-eighth the salary of philosophers.[11] Fi-

8. In Mario Biagioli, "The Social Status of Italian Mathematicians, 1450–1600," *History of Science* 27 (1989): 41–95, I have provided some background for Galileo's strategies by sketching out the culture, social status, career patterns, institutional affiliations, and intel-lectual traditions of Italian mathematicians who operated before and around him.

9. For a biography of Galileo's early years, see Stillman Drake, *Galileo at Work* (Chicago: University of Chicago Press, 1978).

10. Thomas B. Settle, "Ostilio Ricci, a Bridge between Alberti and Galileo," *Actes du XIIe Congrès International d'Histoire des Sciences* (Paris, 1971): pp. 229–38. On the Florentine culture of mathematician-artists, see idem, "Egnazio Danti and Mathematical Education in Late Sixteenth-Century Florence," in John Henry and Sarah Hutton, eds., *New Perspectives on Renaissance Thought* (London: Duckworth, 1990), pp. 24–37.

11. Biagioli, "Social Status of Italian Mathematicians," 53.

nally, the marginality of mathematicians in the university was reflected by the few chairs dedicated to their discipline (one at Padua and Pisa and two at Bologna) and by its subordinate role in the curriculum.

At Padua (where he taught from 1592 to 1610) Galileo dabbled successfully in mechanical inventions (he obtained a patent for a machine to lift water) and acted as a consultant to the Venetian Arsenal.[12] Beside his activities as a university professor, Galileo supplemented his relatively low income by teaching mathematics, mechanics, and especially fortifications to private students—some of whom he boarded in his own house. In 1599, he took an artisan, Marcantonio Mazzoleni, into his house as an instrument maker. Mazzoleni's chief task was to construct geometrical and military compasses, computation devices that Galileo would then sell mostly to his private students.

Up to this point, Galileo's career was typical of a competent and enterprising mathematician. Things changed abruptly in 1610 when, after improving the telescope (which had initially been developed in the Netherlands) and making several remarkable astronomical discoveries, Galileo left the University of Padua for the Medici court to become the *philosopher* (not just the *mathematician*) of the grand duke. It is at this point that Galileo's trajectory began to deviate from that of his fellow mathematicians. However, if the courtly skills and tactics that characterized much of his later career were exceptional for a mathematician, his desire for social and epistemological legitimation was rooted in the professional culture he shared with many leading mathematicians of his time.

Although this book begins with an analysis of the emergence of Galileo's patronage networks and his strategies in the early university-based phase of his career, its main emphasis is on the 1610–33 period, that is, from his arrival at the Medici court to the trial that followed the publication of his *Dialogue on the Two Chief World Systems*. The geographical focus of the narrative changes in time. The first part of the book is about Galileo at the Florentine court, while the second looks at his interaction with the Roman court and with the papal prince.

Chapter 2 explores Galileo's successful move to court not as the result of what we today would call the scientific value of his astronomical work, but as the outcome of his presentation of his newly discovered Jovian satellites as fitting the discourse of the court and the Medicis' dynastic mythologies. Chapter 3 continues the analysis of Galileo's scientific activities at the Florentine court. There, the focus is on a little-studied dispute on

12. Antonio Favaro, "Galileo e Venezia," *Galileo Galilei e lo Studio di Padova* (Florence, 1883; reprint, Padua: Antenore, 1966), vol. 2: 69–102; idem, "Intorno ai servigi straordinari prestati da Galileo Galilei alla Repubblica Veneta," *Atti del Reale Istituto Veneto di Scienze, Lettere e Arti*, series 7, 1 (1889–90): 91–109. On the Arsenal, see Ennio Concina, *L'Arsenale della Repubblica di Venezia* (Milan: Electa, 1984).

buoyancy that took place in Florence in 1611–13. This was the first sustained confrontation between Galileo and Aristotelian philosophers, his previous disciplinary superiors. It was on this occasion that he tried to make use of the disciplinary privileges deriving from his newly acquired title of philosopher. Equally interesting is the light this case study casts on the place of science at court, its "etiquette," and its relation to spectacle. As we will see, the high point of this event was a dispute between Galileo and the philosopher Papazzoni at the grand duke's table; a dispute that involved Cardinals Barberini and Gonzaga as well. Moving from the analysis of this dispute, Chapter 4 shows that the dynamics of self-fashioning and of the social and cognitive legitimation of science can throw new light on a crucial problem of the philosophy of science: the so-called incommensurability between scientific paradigms.

At this point, there is a jump in the narrative, and *Galileo, Courtier* moves to Rome. The next aspect of Galileo's career to be analyzed is the 1619–26 debate on comets with the Jesuit astronomer Orazio Grassi. The setting in which Galileo was operating in 1619 was quite different from that of 1611. After his 1615 "Letter to the Grand Duchess Cristina," the legitimation of the Copernican hypothesis became increasingly important in Galileo's career and work. In that letter Galileo tried to fend off the theologians' scriptural objections to Copernicus' heliostatic cosmos. His attempt was not successful. In the spring of 1616 Copernicus's *De revolutionibus* was put on the Index until corrected (that is, until all nonhypothetical references to earthly motions were removed) while a recent work by Father Paolo Foscarini on the concordance of Copernican astronomy and the Scriptures was condemned outright.[13] Although Galileo's name was not mentioned in the condemnation, he was ordered by Cardinal Bellarmine not to treat the Copernican doctrine as absolutely (i.e., physically) true but only as a hypothesis (i.e., mathematically). With Galileo's growing commitment to an increasingly difficult legitimation of Copernican astronomy, the Roman court became his most important theater of operation.

The Intermezzo that follows Chapter 4 provides a schematic picture of the cultural and academic environment of Rome and of its relationship to the court. In particular, the Intermezzo identifies some of the specific dimensions of Roman academic and courtly cultures that would play a major role in Galileo's subsequent tactics and troubles.

Chapter 5 offers a contextual analysis of the dispute on comets between Galileo and the Jesuit mathematician Orazio Grassi. During this

13. Most of the texts involved in the events of 1616 are translated, with a commentary, in Richard J. Blackwell, *Galileo, Bellarmine, and the Bible* (Notre Dame: University of Notre Dame Press, 1991).

increasingly Rome-centered debate, Galileo published the *Assayer*, a text that has since puzzled many Galileo scholars because of its problematic scientific content. This chapter tries to do with Rome what Chapter 2 does for Florence; it analyzes Galileo's use of court discourse to legitimize his views on comets and to present himself as a sophisticated and courtly natural philosopher.

Chapter 6 proposes a reinterpretation of Galileo's trial. After setting the context by analyzing the peculiar patronage dynamics and generational cycles of the Roman court, this chapter looks at a typical courtly event—the fall of the courtier—as described in contemporary court treatises. By bringing some aspects of the fall of the courtier to bear on Galileo's trial, I suggest that the events of 1633 were as much the result of a clash between the dynamics and tensions of baroque court society and culture as they were caused by a clash between Thomistic theology and modern cosmology. In short, Galileo's career and his attempts to legitimate Copernican astronomy were terminated by those same processes that made it possible to begin with.

The work of Richard Westfall on Galileo's patronage strategies and of Robert Westman on disciplinary hierarchies and their institutional implications have provided important starting points for *Galileo, Courtier*.[14] There are other authors whose work has influenced this analysis in ways that can hardly be registered in footnotes. Norbert Elias's study of court society has been an ongoing reference, as has Pierre Bourdieu's work on the processes of cultural distinction.[15] Michel Foucault's and Louis Marin's analyses of the discourse of power of the *ancien régime*, and Paul Feyerabend's discussion of Galileo's "opportunism," have provoked a number of reflections that, in variously mediated forms, have found their ways into this text.

14. Richard S. Westfall, "Science and Patronage: Galileo and the Telescope," *Isis* 76 (1985): 11–30; Westman, "The Astronomer's Role."

15. Norbert Elias, *The Court Society* (New York: Pantheon, 1983); idem, *The History of Manners* (New York: Pantheon, 1982); idem, *Power and Civility* (New York: Pantheon, 1982); Pierre Bourdieu, *Distinction: A Social Critique of the Judgment of Taste* (Cambridge, Mass.: Harvard University Press, 1984).

ONE

Galileo's Self-fashioning

Disembodied Minds and Chaotically Interacting Bodies

WRITING TO BELISARIO VINTA in May 1610 about the possibility of a position at the Medici court, Galileo apologized for taking up the time of such a high-ranking official for a matter that might seem of little consequence to him.[1] But, Galileo continued, such a decision was of the utmost importance, since it concerned the change "of the whole of my status and being."[2] Galileo's statement suggests that patronage was not something that influenced only the external conditions of his work, such as freedom from teaching duties, financial comfort, and an appropriate title. As he put it, patronage shaped his status and identity—not just his career.[3]

What I propose in this chapter is a systematic analysis of some of the patronage-related processes that structured Galileo's self-fashioning as a court philosopher. This analysis of patronage as the social system of

---

1. "Primo Segretario," Vinta's title, does not translate well into any modern political role. "Secretary of State" may be its least distorting analogue. Vinta obtained that post in December 1609.

2. Galileo Galilei, *Opere,* ed. Antonio Favaro (Florence: 1890–1909), vol. 10, no. 307, p. 353. Henceforth cited as *GO.* Unless otherwise noted, all translations from the Italian are mine.

3. On Galileo's patronage, see Paolo Galluzzi, "Il mecenatismo mediceo e le scienze," in Cesare Vasoli, ed., *Idee, istituzioni, scienza, ed arti nella Firenze dei Medici* (Florence: Giunti-Martello, 1980), pp. 189–215; Richard Westfall, "Science and Patronage: Galileo and the Telescope," *Isis* 76 (1985): 11–30; idem, "Galileo and the Accademia dei Lincei," in Paolo Galluzzi, ed., *Novità celesti e crisi del sapere* (Florence: Giunti Barbèra, 1984), pp. 189–200; idem, "Galileo and the Jesuits," and "Patronage and the Publication of the *Dialogue*," in *Essays on the Trial of Galileo* (Vatican City: Vatican Observatory, 1989); and Michael Segre, "Galileo as a Politician," *Sudhoffs Archiv* 72 (1988): 69–82. Because studies of early modern scientific patronage have become increasingly common in the last few years, I will not list them here but include them in the Bibliography.

CHAPTER ONE

Galileo's science provides a framework for much of what is discussed in the later chapters. It also tries to show that focus on processes of self-fashioning may help bypass some of the deadlocks of the so-called externalists-versus-internalists debate that has characterized much of recent and not so recent science studies.

Genealogically connected to the arguments over the demarcation between "science" and "nonscience" and, ultimately, between *nomos* and *physis,* this debate has taken a remarkable range of shapes, colors, and forms. The case of the scientific revolution tends to be even more complicated because (as suggested by the term "revolution") what happened in the sixteenth and seventeenth centuries is usually represented as marking a sharp distinction between modern science and something different that took place before it. In many ways, much of the historiography of the scientific revolution has been implicated in the construction of what anthropologists would call a "myth of origins"—a narrative by which we Western "moderns" differentiate ourselves from "the rest." Without trying to map out the various ways in which different interpretations of the scientific revolution have tried to draw the line between the "before" and "after" and, within that distinction, between the "scientific" and "social" dimensions of scientific change, I would select only two interpretations that are among the most helpful in illustrating the debate from which this analysis departs.

Reproducing the mind/body dualism, idealistic readings of the scientific revolution have tended to introduce a distinction between science (or the "scientific mind") and flesh-and-blood scientists. Little attention has been paid by this historiography to the social dimensions of science, which were occasionally called upon only to account for the scientists' apparent deviations from assumed rational norms.[4] Although a number of fascinating studies were produced by historians operating within this framework, the demarcation it introduced between theory-based, internal, rational, essential features of science, and its society-influenced, external, irrational, accidental dimensions reflected a problematic inversion: the *explanandum* was turned into the *explanans.* In a fashion that bears a striking resemblance to Aristotelian essentialism, the emergence of a scientist's belief in a given theory was not explained contextually but was taken for granted as the "natural" result of the "force of that theory." Then, the "natural" belief in a theory was assumed to produce an equally "natural" commitment to it.[5] When seen through these lenses, patronage

4. Richard Westfall's studies of Galileo's patronage strategies reflect this type of historiography, one that reaches back to the work of Alexandre Koyré and Edwin Burtt.

5. It is an interesting paradox that such an essentialistic (in the Aristotelian sense) view of scientific rationality as the one held by this historiographical tradition turns out to be in patent contradiction with the notion of scientific rationality it claims to defend.

became a means toward the fullfilment of rational goals (like securing support for a scientist's rational research program) or, alternatively, as some sort of "thing of the flesh" which causes the scientist's body to deviate from the correct path established by his mind's rational commitment to a good theory.[6]

More recent historians, broadly influenced by Mertonian sociology of science and the Kuhnian notion of "paradigm" and related categories such as "scientific community" and "professionalization," have seen the development of scientific societies (and the related establishment of scientific communities) toward the end of the scientific revolution as marking the beginning of "paradigmatic" science. Although I find the distinction between paradigmatic and pre-paradigmatic less heuristically crippling than the one between modern rationality and whatever preceded it, it is still problematic in that it represents much of earlier science in terms of what it is *not*. Science before 1660 *lacked* a well-structured social system, real scientific institutions, organized professions, and forms of professional communication. In short, early science, in this view, was still represented as the "other" of modern science; what had changed were only the parameters by which it was constructed as such.

In such a view a historian would find it difficult to use "paradigm" to link scientific change to the structure and dynamics of the social system of science simply because there is no real social system of science to speak of. Also, one would tend *not* to perceive patronage as the social system of early modern science because such a perception would go against the fundamental assumption that scientific paradigms are supposed to be connected to well-structured scientific *communities*.

To summarize, if idealistic historiography discouraged historians from analyzing the social dimensions of scientific change (or limited the extent of their contextualizations) because of the distinction it drew between the scientist's mind and body, more recent historiographical categories introduced to integrate social and intellectual dimensions ("paradigm," "scientific community," "scientific institutions") do not seem to apply to most of early science.[7] In short, the idealists present this period as

6. On this issue, see Mario Biagioli, "Galileo's System of Patronage," *History of Science* 28 (1990): 42–45.

7. Institutions are reassuring entities for historians because they usually come with conspicuous buildings (and pictures of them), statutes, archives, journals, and records of discussions and prize competitions. However, we should not forget that the success of institutions as historiographical categories reflects not only their actual historical relevance but also their conspicuous archival "presence." The spell that institutions have cast on recent historiography also has a fetishistic character. Institutions allow historians to "touch" the past in very tangible ways. In contrast, patronage—an institution without walls, its reality made of

populated by disembodied minds while the institution-minded historians may see it as a chaotic pattern of interacting bodies.

A finely articulated notion of patronage may allow for a better integration of the social and conceptual dimensions of early modern science.[8] The first step of this project involves the rejection of the notion of patronage as a mere set of rational strategies and relations through which a scientist makes a career (acquires money, power, and free time to do research). By perceiving patronage only in terms of its economic dimensions, we may end up believing clients to be rational individuals fully committed to some sort of research program in favor of which they try to manipulate the patronage system. However, "ends" and "means" are not categories that exist outside of the processes of self-fashioning that shape them. Consequently, it is by linking patronage to the social process of self-fashioning of the clients and patrons (rather than simply to their economic subsistence) that we can relate cultural production and social context.[9] Rather than looking for paradigms, we may focus on the study of the client's identity in all its sociocultural dimensions, as well as on a scrutiny of the processes through which such an identity is shaped.

In fact, the process of identity formation need not take place in well-circumscribed professional groups such as scientific communities or institutions. The sociological and conceptual dimensions of modern science that the historiography informed by Kuhn or Merton attributes to the professional identity one develops by being socialized into a scientific community or social group must be sought for in the process of self-fashioning that early modern scientists underwent by entering into patronage relationships and networks. I am not claiming that patronage is the early analogue of scientific community. I am suggesting that patronage is the key to understanding processes of identity and status formation that are the keys to understanding *both* the scientists' cognitive attitudes *and* career strategies.

---

etiquette-bound rituals rather than of "things" such as buildings and statutes—has eluded many historians of early modern science.

8. Such a notion of patronage could be more than a premodern version of the social system of science, for it could also be used to uncover the nonmodern dimensions of the modern social system of science and to revise some of the claims of the institution-based historiography of modern science. For example, Dorinda Outram, *Georges Cuvier* (Manchester: Manchester University Press, 1984), views scientific institutions as frames within which networks of patronage were developed, suggesting a continuity between the social systems of early modern and modern science.

9. On self-fashioning in this period, Greenblatt's *Renaissance Self-Fashioning* is the *locus classicus*.

Patronage, Status, and Credibility

The institution of patronage one finds represented in Galileo's texts and correspondence is not structurally different from the one that can be reconstructed from the autobiographies and correspondence of Italian baroque artists, poets, and courtiers.[10] Patronage was a social institution widespread in early modern Europe and one which is still very powerful in today's Mediterranean basin. Cicero thought that the origins of Roman *clientela* were so ancient that it must had been brought to Rome by Romulus himself.[11]

Recent historiography of early modern Europe presents patronage as a fundamental form of social binding and hierarchical organization. Also, the domain of patronage dynamics has been shown to comprehend what today we call the public and private spheres. For instance, patronage bondings could be so deeply ingrained in one's identity that the difference between family and patronage ties, or between friendship and clientelism, was often blurred.[12] In the case of Renaissance Florence, patronage has become a standard historiographical category for the analysis of ritual interaction in civic life, sense of heritage, kinship and friendship bonds, and political and economical activity. Similarly, the study of patronage dynamics in their symbiotic relationship with courtly etiquette and self-fashioning has become central to the understanding of the culture, politics, and structure of early modern courts. By now, patronage studies have successfully broadened their scope well beyond basic considerations about who paid the bills into the study of processes through which sociocultural identities and hierarchies were developed and maintained in early modern Europe.[13]

10. See, for instance, Benvenuto Cellini, *The Autobiography of Benvenuto Cellini*, trans. John Addington Symonds (New York: Doubleday, 1961); Giambattista Marino, *Lettere* (Turin: Einaudi, 1966); Giorgio Vasari, *Vita di Michelangelo*, ed. Paola Barocchi (Milan-Naples: Ricciardi, 1962), vol. 1.

11. Cicero, *De re publica*, ii.16, quoted in Ronald Weissman, "Taking Patronage Seriously," in F. W. Kent, Patricia Simons, and J. C. Eade, eds., *Patronage, Art and Society in Renaissance Italy* (Oxford: Oxford University Press, 1987), p. 33.

12. See Weissman, "Taking Patronage Seriously," esp. pp. 27–30.

13. Examples of works in the history of Renaissance Florence that have treated patronage as a complex social institution are Richard Trexler, *Public Life in Renaissance Florence* (New York: Academic Press, 1980); F. W. Kent, *Household and Lineage in Renaissance Florence* (Princeton: Princeton University Press, 1977); Ronald Weissman, *Ritual Brotherhood in Renaissance Florence* (New York: Academic Press, 1982); and Christiane Klapisch-Zuber, "Kin, Friends, and Neighbors," in *Women, Family, and Ritual in Renaissance Italy* (Chicago: University of Chicago Press, 1985), pp. 68–93. Relevant essays on patronage in early modern Italy are found in Kent, Simons, and Eade, *Patronage, Art and Society in Renaissance Italy*. More traditional views on patronage in early modern Europe are found in Guy Fitch Lytle and

Moreover, we should not think of patronage as an "option." Unless one was engaged in a complex network of patronage relationships, a career and social mobility were impossible, especially if one belonged or wanted to belong to the upper classes. Patronage was a voluntary activity only in the narrow sense that by not engaging in it one would commit social suicide. That sometimes both clients and patrons tended to deny the mutual interests underlying their patronage relationship and to present the relationship as a disinterested, voluntary choice is a fact that should not be taken at face value. That clients represented themselves as "loving" to serve and patrons as "loving" to be served may be viewed as something akin to the claims of disinterestedness that anthropologists have encountered in the context of gift-exchange in so-called traditional societies. Marcel Mauss has shown that although those who give or return gifts present themselves as behaving voluntarily, they actually act under strong social obligations.[14] Giving and returning gifts is a process through which members of a community try to gain or maintain status and power. Nobody wants to admit to being forced to give or return gifts because that would amount to an admission of the limits to one's power and autonomy. Therefore, the representation of gift-exchange (or patronage relationships) as a free and disinterested activity is a gesture aimed at legitimizing those who engage in that game.

If patronage provides a key for interpreting the behavior of early modern historical actors in general, it becomes an even more powerful tool if what we are analyzing is a scientist's life. Close connection between

Stephen Orgel, eds., *Patronage in the Renaissance* (Princeton: Princeton University Press, 1981), and Yves Durand, ed., *Hommage à Roland Mousnier: Clienteles et fidélités en Europe à l'époque moderne* (Paris: PUF, 1981). I have found useful a few works on early modern French patronage and aristocratic culture, such as Sharon Kettering, *Patrons, Brokers and Clients in Seventeenth-Century France* (Oxford: Oxford University Press, 1986); idem, "Gift-Giving and Patronage in Early Modern France," *French History* 2 (1988): 131–51; idem, "The Patronage Power of Early Modern French Noblewomen," *The Historical Journal* 4 (1989): 817–41; idem, "The Historical Development of Political Clientelism," *Journal of Interdisciplinary History* 3 (1988): 419–47; Mark Greengrass, "Noble Affinities in Early Modern France: The Case of Henri I de Montmorency, Constable of France," *European History Quarterly* 16 (1986): 275–311; and Kristen B. Neuschel, *Word of Honor* (Ithaca: Cornell University Press, 1989). Renata Ago, *Carriere e clientele nella Roma barocca* (Bari: Laterza, 1990), provides a remarkable study of how patronage shaped identities and careers in late seventeenth-century Rome. Patronage as a form of social organization in the Mediterranean basin is studied in J. Pitt-Rivers, *Mediterranean Countrymen* (Paris: Mouton, 1963); and Ernest Gellner and John Waterbury eds., *Patrons and Clients in Mediterranean Societies* (London: Duckworth, 1977). For socio-anthropological views of patronage, see J. Boissevain, *Friends of Friends* (Oxford: Oxford University Press, 1974); S. N. Eisenstadt and L. Roniger, *Patrons, Clients and Friends* (Cambridge: Cambridge University Press, 1984); and S. W. Schmidt, L. Guasti, C. H. Lande, and J. C. Scott, *Friends, Followers and Factions* (Berkeley: California University Press, 1977).

14. Marcel Mauss, *The Gift* (New York: Norton, 1967), pp. 6–16.

social status and epistemological credibility characterized early modern Europe. Peter Dear has argued that the early Royal Society of London had parameters for the evaluation of evidence that were sensitive to the social status of the observer. Nobility and credibility were perceived as related; by having many churchmen and aristocrats among its members, the Society gained "social prestige, which could itself be turned to evidential advantage."[15] Steven Shapin and Simon Schaffer have traced a similar relationship between the social class of reporters or witnesses and the degree of trust accorded to their reports.[16] Social status not only regulated trust but affected the very possibility of communication. For instance, the parish priests in charge of data collection for the seventeenth-century Venetian censuses went around mapping the parish's households in company with a nobleman in case noble residents refused to answer questions asked by a mere clergyman.[17]

Social taxonomies of status and credibility are reflected by the hierarchical order of the liberal disciplines which projected contemporary assumptions about social distinction and status on the disciplines, their subject matter, and their methodology. Given the social hierarchies in Galileo's time, it is not surprising to find theology represented as "Queen of the disciplines." Because of its marginal role within dominant Aristotelian philosophy and because of its use in low-class arts like mechanics, mathematics was accorded a relatively low social and cognitive status.[18] As initially suggested by Robert Westman, the low disciplinary status of mixed mathematics like astronomy, optics, and mechanics (especially in relation to philosophy and theology) was probably the most important obstacle in the way of the epistemological legitimation of Copernican astronomy.[19] Unless these disciplinary hierarchies were redrawn, the mathematicians' new worldviews could be dismissed almost a priori by

15. Peter Dear, "*Totius in Verba:* Rhetoric and Authority in the Early Royal Society," *Isis* 76 (1985): 156.

16. Steven Shapin and Simon Schaffer, *Leviathan and the Air Pump* (Princeton: Princeton University Press, 1985), esp. pp. 58–59, 66. The relationship between social status and credibility is also the underlying theme of Steve Shapin, "The House of Experiment ...1 Seventeenth-Century England," *Isis* 79 (1988): 373–404. See also Steven Shapin, "The Invisible Technician," *American Scientist* 77 (November-December 1989): 554–63.

17. Peter Burke, "Classifying the People: The Census as Collective Representation," *The Anthropology of Early Modern Italy* (Cambridge: Cambridge University Press, 1987), p. 29.

18. Biagioli, "The Social Status of Italian Mathematicians, 1450–1600," pp. 41–95.

19. Westman, "The Astronomer's Role in the Sixteenth-Century," 105–47. Westman has developed different elements of his initial thesis in "The Copernicans and the Churches," in David C. Lindberg and Ronald L. Numbers, eds., *God and Nature* (Berkeley: University of California Press, 1986), pp. 76–113; and idem, "Proof, Poetics, and Patronage: Copernicus's Preface to *De revolutionibus*," in David C. Lindberg and Robert S. Westman, eds., *Reappraisals of the Scientific Revolution* (Cambridge: Cambridge University Press, 1990), pp. 167–205. On the Italian case, see Biagioli, "Social Status of Italian Mathematicians, 1450–1600."

philosophers and theologians, who could simply rely on commonly accepted distinctions of subject matter, methodology, and disciplinary sociocognitive status.

In short, the legitimation of the new science involved much more than an epistemological debate. The acceptance of the new worldview depended also on the sociocognitive legitimation of the disciplines and practitioners upholding it. Mixed mathematics had to gain the epistemological status of philosophy. Given the nexus between social status and credibility, high social status was the password to cognitive legitimation, patronage was the institution through which social status and credibility could be gained, and the court was the space in which the most powerful patronage relationships could be established.[20] As Torquato Tasso put it in *Il Malpiglio*, "the arts—however ignoble—gain quality and gentility from the court."[21] Galileo's obtaining the title of "philosopher" of the grand duke at the Medici court in 1610 can be seen as emblematic of this patronage-based trajectory of social and cognitive legitimation.

To understand how Galileo and other mathematicians interacted, communicated, disputed, presented their arguments, and tried to legitimate their claims, we may concentrate on the issues of status and honor and how they were developed and sometimes ruined through patronage dynamics. Because we do not have a *Book of the Courtier* in which Galileo's courtly behavior and tactics for status enhancement are discussed in detail, I propose to study them as they are inscribed in his correspondence from the year 1589 to 1613. The choice of this period is not arbitrary. Judging from the changing titles attributed to Galileo in his correspondence, he experienced a remarkable social acceleration in these years.[22] The letters from the 1580s and 1590s tend to refer to him as *Molto Magnifico* (*Magnifico*

20. The enlistment of noble practitioners in order to legitimize a new discipline was a conscious strategy adopted by Cesi for his Lincei (*GO,* vol. 11, no. 874, p. 507). The relationship between social status and cognitive legitimacy indicates the sociological importance of authors like Guidobaldo del Monte, Tycho Brahe, or Robert Boyle. Through their undisputed aristocratic status as well as through their scientific achievements, they legitimized the new scientific practices, methodologies, and worldviews. The Jesuits' Collegio Romano played a similar role. Although the members of the Society of Jesus did not necessarily come from the aristocracy, nor were they in general brilliant mathematicians, their order brought them a degree of nobility. As shown by the reception of members of religious orders at court, they were perceived as sharing in the "sacredness" of the Church in the same way ambassadors participated in the sacredness of the state they represented (*ASF,* "Miscellanea medicea 447" ["Cerimoniale della Real Corte di Toscana"], pp. 443–44). On the high social status that could be achieved by scientists belonging to religious orders, see Bernard de Fontenelle, "Eloge de Monsieur Cassini," *Eloges des académiciens* (La Haye: Kloot, 1740), 1: 287.

21. Torquato Tasso, *Il malpiglio, o vero de la corte,* reprinted in Cesare Guasti, eds., *I dialoghi di Torquato Tasso* (Florence: Le Monnier, 1901), 3: 10.

22. Unless differently specified, I have relied on the taxonomy of titles provided by Panfilo Persico, *Del segretario libri quattro* (Venice: Damian Zenaro, 1629), pp. 163–65; this was a standard handbook for secretaries of princes.

indicating a private individual below the rank of gentlemen),[23] while those from the 1600s refer to him as *Illustre* and then *Molto Illustre* (*Illustre* being a title for gentlemen and *Molto Illustre* for distinguished gentlemen and *cavalieri*).[24] After he became mathematician and philosopher to the grand duke in 1610, he was usually addressed as *Molto Illustre et Molto Eccellente* (or *Eccellentissimo; Eccellente* being similar to our "Dr." and *Molto Eccellente* or *Eccellentissimo* referring to particularly accomplished doctors of law, medicine, and philosophy).[25]

The following reading of the apparatus of epistolary rituals characteristic of Galileo's correspondence treats them as homologous to the forms of interaction that clients and patrons assumed when dealing with each other in person. Through this sort of "epistolary anthropology," I will try to reconstruct the etiquette of patronage interactions.[26]

The Microphysics of Patronage

Power is not a *thing* but a *process,* and a patron is somebody who can *do* things for the client.[27] A patron has power in the measure in which he or she can make it circulate and be productive. Therefore, an early modern patron was often a broker, a *trait d'union* between the client and a higher source of power such as a greater patron. A client's access to a patron was not just a technical matter of getting to know the right ways to contact the patron. Access to patronage was not a matter of information one could get from the yellow pages. Hierarchies of patrons and clients reflected hier-

23. Stefano Guazzo, in his *Dialoghi piacevoli* (Venice: Bertano, 1585), pp. 94–95, claims that *Molto Magnifico* was a title for merchants and that *Magnifico* could be used for surgeons and notaries.

24. The difference between these titles was much more important than expected. Agostino Mascardi, a client of Cardinal d'Este, realized that he was falling out of grace because his patron had begun to address him as *Molto Magnifico* instead of *Illustre* as he had before (Francesco Luigi Mannucci, "La vita e le opere di Agostino Mascardi," *Atti della Società Ligure di Storia Patria* 42 [1908], p. 89).

25. The escalation of titles in Benedetto Castelli's letters to Galileo exemplifies this pattern. In 1607 he calls Galileo *Eccellentissimo Signor mio;* in 1610 (after his astronomical discoveries) Castelli adds a gentlemanly touch to Galileo's title that becomes *Illustre et Eccellentissimo Signore;* in 1613 he becomes a *Molto Illustre et Eccellentissimo Signore.* In the late 1610s, Castelli also began to add *Padrone Colendissimo,* a title that he kept using routinely in the 1620s and 1630s. However, because titles depended on the status of both the addresser and the addressee, one can find a range of exceptions to this pattern. In fact, as shown by the titles used by the grand duke, princes were rather more conservative with the titles they conferred on their addressee (*ASF,* "Miscellanea medicea 415").

26. For a similar approach, see Neuschel, *Word of Honor,* esp. pp. 72–78; and Kettering, "Gift-Giving and Patronage in Early Modern France," pp. 138–43.

27. This notion of power is broadly derived from Foucault's analysis of the structure of power mechanisms. For a concise statement of his view on this issue, see Foucault, "Truth and Power," pp. 109–33.

archies of status: they mirrored social structures. Therefore, not all clients were eligible for all types of patronage. A lowly client could not approach a very important patron directly.

For instance, during the early phases of his career, Galileo's status was not sufficient to allow him to address Prince Cosimo de' Medici directly. When he eventually did so in 1605, he was aware of crossing an important status boundary and tried to defuse a potential breach of patronage etiquette:

> I have waited until now to write to Your Most Serene Highness, being held back by a respectful concern of not wanting to present myself as presumptuous or arrogant. In fact, I made sure to send you the necessary signs of reverence through my closest friends and patrons, because I did not think it appropriate—leaving the darkness of the night—to appear in front of you at once and stare in the eyes of the most serene light of the rising sun without having reassured and fortified myself with their secondary and reflected rays.[28]

This letter indicates quite clearly that brokers were more than mere distributors of power and privileges. They also preserved social structures and boundaries that might have been otherwise violated by improper contacts resulting from attempts to establish patronage relationships.[29]

The role of brokers was also tied to other rituals of power detectable in Galileo's correspondence. A great patron like the grand duke avoided formalized patronage connections outside his court. Stressing the instability of extra-court patronage and forcing clients to "reapply" regularly for patronage was a strategy through which the prince displayed and reinforced his power. As Matteo Pellegrini put it in his 1624 treatise on the court, "it is in the royal interest to keep everybody suspended between fear and hope".[30]

For instance, Galileo's position as the mathematics tutor of the young prince Cosimo, which he first obtained in 1605, was neither permanent nor official. There was no contract requiring him to come down from Padua, where he was teaching, to Florence each summer to instruct the prince.

28. *GO,* vol. 10, no. 131, pp. 153–54.

29. I am thinking of an interpretation of the role of brokers as mediators between hierarchically organized social castes along the lines of Mary Douglas's anthropological analysis of the threat of pollution related to the maintenance of social boundaries in her *Purity and Danger* (London: Routledge, 1966). In a sense, early modern brokers were also pollution controllers.

30. Matteo Pellegrini, *Che al savio è convenevole il corteggiare libri IIII* (Bologna: Tebaldini, 1624), p. 57. The book was reissued one year later as *Il savio in corte* (Bologna: Mascheroni, 1625).

Instead, every year, Galileo had to write anxiously to check his standing at the Florentine court to make sure of the Medici's continuing interest in his services.[31] And he had to do that through brokers.

Conversely, an important patron could not ask a low-status client directly for services because a refusal on the client's part would stain the patron's image. In such a case, too, brokers played a crucial role in the patronage machinery by communicating the patrons' desires while protecting their image. For instance, several cardinals did not ask Galileo directly for telescopes but had mutual acquaintances advise Galileo to send them good ones.[32] Similarly, when Grand Duke Ferdinand I supported Galileo's request for a higher salary at the University of Padua in 1605, he did not press the Venetian authorities directly but only through his *residente* in Venice, Asdrubale da Montauto.[33] As Galileo was told, the grand duke thought that the republican Venetians disliked to be pressured by princes. Through this nice fiction, Ferdinand let Galileo know that he did not want to put himself on the line for an issue as marginal as Galileo's stipend; at the same time, Ferdinand was sheltering himself from the disturbing consequences of a possibly uncooperative response from Venice.

Galileo's initial career goal was to gain a university post, which he eventually obtained at Pisa in 1589 and at Padua in 1592. The second phase in his pursuit of patronage, which began around 1600 and grew more systematic after 1604, was aimed at obtaining a position at either the Medici or Gonzaga court.[34] Although both phases were largely dependent on Medici power networks, they were structured around two different sets of brokers and patrons. The first phase of Galileo's career materialized primarily through the patronage/brokerage relationship he had established with Guidobaldo del Monte. In contrast, Galileo's nomination to a court position in 1610 can be seen as the fortunate result of a range of long-term patronage strategies centered on young Cosimo de' Medici. These strategies made it possible for Galileo and Cosimo to "grow together" (Galileo as a client and Cosimo as a patron) and were developed through a range of brokers less powerful than Guidobaldo but better situated for the task.

The development of the necessary brokerage connections with courtiers such as Girolamo Mercuriale, Cipriano and Ferdinando Saracinelli (uncle and nephew), Vincenzio and Niccolò Giugni (father and son), Cosimo Concini, Giovambattista Strozzi, Alessandro d'Este, Baccio Val-

31. *GO*, vol. 10, no. 120, p. 144; no. 138, p. 160; no. 190, pp. 210–13; no. 192, pp. 214–15.

32. Ibid. no. 232, pp. 254–55; no. 309, p. 354; no. 320, p. 361; no. 349, p. 388; no. 373, pp. 420–21; vol. 11, no. 831, pp. 463–64.

33. *GO*, vol. 10, no. 126, p. 148.

34. Ibid., no. 97, pp. 106–7; no. 99, p. 109; no. 131, pp. 154–55; no. 190, pp. 210–13; no. 209, pp. 231–34; no. 211, p. 235.

ori, Antonio de' Medici, and Silvio and Enea Piccolomini was no small feat. A number of these connections were family assets—they had been passed on to Galileo by his father, Vincenzio, and they would be passed on to Galileo's son.[35]

Galileo's strategies to gain the support of young Cosimo were not original. Becoming a young prince's tutor was one of the standard avenues through which clients could enter into a patronage relationship with powerful patrons. It was a strategy based on the hope that ties between pupil and tutor would grow stronger as the pupil progressed toward maturity and power.[36] Galileo was introduced to this strategy in the spring of 1601, when Cosimo was eleven years old. It was Mercuriale—a professor of medicine at Pisa and the grand duke's *Protomedico*—who gave Galileo the hint. Mercuriale was a perfect broker for Galileo. A courtier very close to the royal family, he could appreciate Galileo's intellectual assets and see how they could be turned into something appealing to court patronage.

Writing to Galileo in Padua in 1601, Mercuriale told him that the following year the prince would be old enough to begin to study mathematics "and I believe you will have a chance to demonstrate your talent and—who knows—*that may bring you some good luck.*"[37] In the same

35. That patronage connections were a family capital transmitted to its male members can be seen from the father/son or uncle/nephew pairs who appear among Galileo's patrons and brokers (Saracinelli, Giugni, Piccolomini). Galileo too developed his own "patronage clan" connected to the Medici court by placing members of his family in the Medici administration, and by having his son legitimized, given a sinecure by Urban VIII, and later married to the daughter of Geri Bocchineri—a member of a family quickly emerging through the ranks of the Medici bureaucracy. Consequently, although there is no, or very little, continuity between Galileo, his father, his son Vincenzio, and his brother Michelangelo in their specific professional activities, there is a strong continuity in their social roles: they were all connected to courts or worked as civil servants. Also, Galileo's responsibility for the dowries of his sisters suggests that he was never an "individual" but the head of a clan. His role as the head of an impoverished clan may tell something about his not marrying Marina Gamba (although he had three children by her) and his later decision to lock up the two daughters in a convent. By the early 1610s Galileo was no longer poor, but perhaps he was not rich enough to marry the two daughters to people of social status comparable to the one he had recently acquired. Although these are well-known facts, it may be interesting to put them together and view Galileo not just as a mathematician who dedicated his discoveries to the Medici and made a brilliant career at their court, but also as the head of a Florentine clan with some court connections who tried to maximize his patronage assets not only for his sake but also for that of his patrilinear clan. The early phases of Galileo's family-related strategies can be traced in *GO,* vol. 10, no. 65, p. 74; no. 163, pp. 180–81; no. 202, p. 225; no. 206, pp. 227–28; no. 290, pp. 312–14; vol. 11, no. 497, p. 71; no. 522, pp. 95–97. These strategies of familial advancement are discussed (in the case of visual artists) in Peter Burke, *Culture and Society in Renaissance Italy 1420–1540* (New York: Scribner's, 1972), pp. 247–49.

36. For a famous example, see Count Olivares's strategies to win the favor of Prince Philip (the future Philip IV, King of Spain), in John H. Elliott, *Richelieu and Olivares* (Cambridge: Cambridge University Press, 1984), p. 36.

37. *GO,* vol. 10, no. 73, p. 84 (emphasis mine).

letter, he also advised Galileo to perfect the military compass and bring it to Florence, volunteering to act as a *trait d'union* between Galileo and the Medici in case Galileo wanted to show or offer the compass to them. Mercuriale's scheme did not materialize as quickly as he expected, but his lead was a good one, for it was through the dedication of his tract on the military compass that Galileo strengthened his patronage relationship with young Cosimo and with the prince's mother, the powerful Grand Duchess Cristina.

Galileo did not dedicate his compass to Cosimo through the intercession of Mercuriale (who died in 1606) but reached the Medici through other court brokers. It was through Giugni, the Saracinelli, and the Piccolomini that Galileo proposed the dedication of the compass, inquired about his standing with the Medici, assessed the possibility of a position at court, asked for Medici support for a salary raise at Padua, inquired about the development of the suit brought against him by his brother-in-law, and checked the grand duke's will to have Galileo continue to teach mathematics to young Cosimo during the summer.[38]

As far as we know, it was only at the end of 1605 that Galileo bypassed his brokers and wrote directly to young Cosimo for the first time. On that occasion, Galileo did not communicate or ask anything specific, but presented Cosimo with an all-encompassing declaration of his desire to serve his prince in whatever fashion might please him.[39] That letter functioned as a rite of passage. With it, Galileo presented himself as having gained sufficient intimacy with the prince to address him directly, while recognizing that Cosimo had grown old enough to be dealt with no longer as a boy but as an independent young man. They both had become patronage grown-ups.

Although Galileo's early brokers continued to play an important role in his later career, their function became increasingly that of informers after Galileo had hooked up with a Medici *secretario*—Belisario Vinta.[40]

---

38. Ibid., no. 120, p. 144; no. 126, p. 148; no. 129, pp. 150–51; no. 133, pp. 155–56; no. 134, pp. 156–57; no. 136, pp. 158–59; no. 138, p. 160.

39. Ibid., no. 208, pp. 230–31. He seemed to have planned it for some time and consulted with his brokers just before he wrote Cosimo (ibid., no. 129, p. 151).

40. Galileo's correspondence indicates that he also used the services of a number of broker-like individuals (such as Sertini in Florence or Cigoli and Faber in Rome) who might be called "informers." These were not occasional brokers, for they did not handle privileges but rather information. They seem to have had a specific sociological character. Neither Sertini nor Cigoli was powerful, but both had access to the powerful. They could see and hear without being able to manipulate. But they were important in giving Galileo information he could not have received from more powerful patrons or brokers who did not have, or could not display, the perspective of the marginal. Cigoli's information must have been very important if, after Cigoli's death, Galileo used Guiducci as his "spy" in Rome during the crisis over the comets, even though he had plenty of powerful patrons and brokers there. Faber seemed to have played the same role for Welser in Rome (see Giuseppe Gabrieli, "Vita romana del

This was a crucial turning point in Galileo's career. It was during the nego-tiations with the Medici secretary concerning the purchase of Sagredo's lodestone on behalf of Prince Cosimo that Galileo tested the specific patronage tactics that would later secure for him a position at court. Through Vinta, Galileo was able to consolidate his hold on Cosimo. That done, it was only a matter of waiting for the ugly princeling to meta-morphose into a powerful grand duke.

Behind the apparent fragmentariness of Galileo's early tactics to gain Medici patronage we can detect a systematic pattern. Galileo invested in a patron who was destined to become powerful and then carefully improved his connection with him by relying on a series of powerful brokers located closer and closer to him. The most powerful of these brokers was Cosimo's mother, Grand Duchess Cristina. Galileo's other court brokers (Mercu-riale, Saracinelli, and Piccolomini) were also closely connected to young Cosimo, all of them being involved in the planning and supervision of his education.[41]

Having been connected only marginally to the court networks of power during his youth, and possessing mostly un-courtly skills such as mathematics, Galileo would become a top court client only after his astro-nomical discoveries of 1609–10. His later friend and patron Giovanni Ciampoli was probably better connected at twenty than Galileo was at forty. Before 1610, there was a major gap between Galileo's own percep-tion of his worth and his actual value on the courtly market. Probably, that gap could have been bridged only by adhering to (and training the tastes of) a future grand duke. An established courtier would have aimed for the patronage of Cosimo's father, without bothering with the young prince, whose patronage he would have obtained naturally in due course. Gal-ileo's patronage options were much narrower, but he played them right. His career following his discovery of the four satellites of Jupiter which he called "Medicean Stars" was not the fruit of chance but of an earlier sys-tematic weaving of patronage relationships according to typical patterns and tactics. Without those carefully forged relationships, the Medicean Stars would not have projected him into prominence.

But the ascent of a client toward a great patron was not a single-handed feat. Brokers were like talent scouts looking for potentially up-wardly mobile clients in which to invest their connections. While clients like Galileo were betting on young patrons like Cosimo, brokers were bet-

---

600 nel carteggio inedito di un medico tedesco in Roma," *Atti del primo Congresso Nazionale di Studi Romani* (Rome: Istituto di Studi Romani, 1929), 1: 813–27).

41. *GO*, vol. 10, no. 129, pp. 150–51; no. 133, pp. 155–56; no. 136, pp. 158–59; no. 143, pp. 161–62; no. 164, p. 181; no. 223, pp. 246–47; no. 240, pp. 258–59; no. 281, p. 305. Also, Cipriano Saracinelli was the *pedagogo* of Prince Cosimo; see Gaetano Pieraccini, *La stirpe dei Medici di Cafaggiolo* (Florence: Nardini, 1986), 2: 327.

ting on clients like Galileo. Power being a process, it had to be exercised in order for brokers to increase or maintain it. Patrons and brokers were also like bankers, who want and need to loan money in order to make more money. For instance, Vinta was quick to notice Galileo's increased standing at court and did his best to become Galileo's sole broker. When Galileo left the Medici court in the fall of 1608 to go back to Padua, Vinta told him: "contact me and nobody else for your needs."[42]

Brokers *pushed* the clients they considered worthy. In 1601 Mercuriale did not simply inform Galileo about his options at court but explicitly encouraged him to complete his work on the military compass and dedicate it to the Medici. The pushiness of Galileo's early Florentine brokers does not seem much different from the pressure exercised by the Lincei on Galileo for the writing and publication of the *Assayer* after 1620, an episode of Galileo's career analyzed in Chapter 5. Patronage did not simply reward clients *a posteriori,* but also stimulated and accelerated them toward not always happy endings.

The frequent statements of friendship (*amicizia*) found in letters to Galileo in which patronage is formally offered to him should be read in this context. Brokers and patrons saying "I beg you to order me," or Galileo's ritual confirmation of his willingness to "serve" his patrons in return must not be viewed as empty formalities characteristic of the baroque era. They were, instead, ritualized forms of advertisement. The interdependence of the welfare of clients and patrons emerges most explicitly in Galileo's letters of congratulation to patrons on occasions of their promotion. These letters were, at the same time, celebrations of the patron's increased power and reminders of Galileo's requests for patronage. They represented an epistolary genre so common as to be treated in a separate chapter in handbooks about courtly letter-writing like Panfilo Persico's *Del Segretario.* As Persico saw it, "all men run where they see prosperity go, and, in doing so, they signify their happiness [by writing letters of congratulation]".[43] Although Galileo did not have Persico's text in his library, he owned a comparable book entitled *Samples of Various Letters Employed in the Secretariat of Every Prince.*[44]

The letters Galileo sent to great patrons—like the one to Guidobaldo on the occasion of his brother Francesco Maria's elevation to the cardinalate, or the one to Cosimo on the occasion of his father's death and Cosimo's accession[45]—were appreciated but did not trigger any explicit

42. *GO,* vol. 10, no. 277, p. 301.

43. Persico, *Del segretario libri quattro,* p. 316.

44. *Idea di varie lettere usate nella segreteria di ogni principe,* listed in Antonio Favaro, "La libreria di Galileo Galilei," *Bullettino di bibliografia e storia delle scienze matematiche e fisiche* 19 (1886): 273–75.

45. *GO,* vol. 10, no. 23, p. 39; no. 208, pp. 230–31.

CHAPTER ONE

patronage commitment from the patron's side. Great patrons were not supposed to admit that they needed clients. Smaller patrons tended to be more explicit about their desire to put newly acquired power to work through patronage relations. Francesco Morosini, a friend of Sagredo's whom Galileo congratulated on his election to the *Saviato di Terra Ferma*—a Venetian magistracy in charge of the University of Padua among other things—expressed the desire to patronize Galileo "on every major occasion,"[46] "major" referring to Morosini's increased power as well as to the rank of privileges he could now secure for Galileo. Sebastiano Venier, who was elected to the *Saviato* with Morosini, answered another of Galileo's flattering letters of congratulation by saying that the promotion was much welcome for it allowed Venier to "support the desires of my friends."[47] Similar if less explicit messages were returned to Galileo by Alessandro d'Este and Lorenzo Magallotti, whom Galileo congratulated on reaching the cardinalate.[48]

Persico—himself a former secretary to Cardinal Orsini—understood well how these rituals benefited both great patrons and lowly clients:

On the promotions of Cardinals, any occasion and pretext offers people of any type and quality a sufficient excuse to write and congratulate them and their relatives, clients, and friends, hoping that some benefit will come out of this. However, even patrons of great fortune who aim at having a great following of clients and courtiers do not lose the opportunity to congratulate even their inferiors because all friendships are useful at some point, especially if they have been cultivated.[49]

As noted by Persico, brokers and patrons needed clients. If nothing else, clients were necessary to keep the patron's connections active and to test his power. Guidobaldo del Monte did not expect anything back from Galileo for having secured him two university chairs, but he mentioned more than once his desire to see his will (to find Galileo a job) respected. This suggests that, by patronizing Galileo, Guidobaldo was also testing his own power in Florence, Padua, and Venice. His success in seeing his desires taken seriously was not merely rewarding to Guidobaldo's ego but was also an empirical indicator of his power. The use of a client as a probe of one's patronage network was not a practice restricted to great patrons only. Galileo himself made use of such a strategy as early as 1606 when he proposed Johannes Fabricius of Acquapendente to replace the recently deceased Mercuriale as the grand duke's *Protomedico*. In writing to

46. Ibid., no. 90, pp. 101–2.   47. Ibid., no. 91, p. 102.
48. Ibid., no. 62, pp. 72–73; vol. 13, no. 1685, p. 231.
49. Persico, *Del segretario libri quattro*, p. 317.

Cristina about Fabricius, Galileo checked his standing with the grand duchess and tested the Medici's willingness to provide patronage for a socioprofessional role quite similar to the one he would request for himself later on. By the time he wrote the letter about Fabricius, Galileo had already expressed his desire for a position at court, but without success.[50]

He says that Fabricius,

> having reached here all that he could in terms of wealth and reputation, and finding it tedious to endure the continuous duties he has accepted to satisfy his many friends and patrons, because of his age, he is very much looking forward to some leisure both for the preservation of his health and for the completion of some of his works. And all that he needs to satisfy his ambition is to attain those titles and positions reached by others in his profession, and those can be given him only by some great, absolute prince. Consequently, I believe he would be very pleased to serve Your Most Serene Highness.[51]

The striking analogies between Galileo's presentation of Fabricius' state and desires to Cristina and of his own in his 1609 letter to an anonymous courtier referred to as "S. Vesp" and in another letter, to Vinta in 1610, suggest that, while acting as Fabricius' broker, Galileo was testing both his standing with his own patron and the potential success of the strategies he had developed for his own career.[52]

The etiquette of letter-writing offered patrons and clients ways of testing each other's interest in establishing or continuing a patronage relationship. The *cirimonie* mentioned in Galileo's correspondence served this purpose. When a client like Galileo wanted to test a patron or broker's availability, he would write extravagantly flattering letters and then study the response. If the answer was a friendly rebuff of Galileo's *cirimonie*, it meant that he had been accepted as an "intimate" client—a client with whom all those ceremonies would be out of place. But Galileo—as all clients—had to keep offering those ceremonies until he was told to drop them. The refusal of ceremonies was itself a ritual; it meant that Galileo had been granted enough status to enter into a patronage relationship. The delivery and rebuff of ceremonies constituted an epistolary rite of passage.[53] As shown by Richard Trexler's analysis of the rhetorical offers of *cortesia* and *amicizia* in the correspondence of the fourteenth-century

50. *GO*, vol. 10, no. 131, pp. 153–54; no. 146, pp. 164–66.

51. Ibid., no. 146, p. 165.

52. Ibid., no. 209, pp. 231–34; no. 307, pp. 348–53.

53. A good example of the delivery and rebuff of *cirimonie* is in a series of letters between Galileo and Guidobaldo del Monte. See ibid., no. 10, pp. 25–26; no. 27, p. 41.

merchant Francesco Datini with the Florentine notary Lapo Mazzei, the currency of these patronage rituals was by no means limited to early-seventeenth-century Florentine court culture.[54]

Galileo used ceremonies not only to establish new patronage connections but also to test older ones that had been left inactive for some time. He used them with his close friend Sagredo after the two had been out of touch for a long time. He received the rebuff: "I am not going to reply to the ceremonies Your Most Excellent Lordship wrote to me . . . both for lack of time and also to suggest that in the future you should not indulge in these superfluities."[55] For similar reasons, he employed the same "superfluities" with Mercuriale, who wrote back, "I did not know that mathematicians—who deal with certainty only—would then apply themselves to mislead men with eloquence, but I had to change my mind after I received your letter. . . ."[56] Similarly, Galileo flooded the Saracinelli—two of his important brokers at the Medici court—with ceremonies in order to test their continuing availability as patrons.[57] The link between the absence or ritualistic rejection of ceremonies and the establishment of a patronage relationship is made most explicit by Sertini—a Florentine friend and a later client of Galileo—who wrote to him in Padua that "*If we did not know each other,* I would go out of my way to make plenty of excuses and ceremonies to Your Lordship. . . ."[58]

These considerations indicate that patronage was not a chaotic set of personal and voluntary relationships, but that it had specific structural features and a logic that bound patrons, brokers, and clients through their need to circulate power in order to obtain or maintain it. Appropriate boundaries and related rites of passage controlled that circulation while specific rituals allowed patrons, clients, and brokers to test in nondisruptive ways the perceptions of each other's status and interest in entering into a patronage relationship.

But if patronage represented the most important means for social mobility, not every patronage relationship offered the same possibilities of

54. Richard Trexler, *Public Life in Renaissance Florence* (New York: Academic Press, 1980), p. 135.

55. *GO,* vol. 10, no. 246, p. 261.

56. Ibid., no. 46, p. 54.

57. Ibid., no. 133, pp. 155–56: "My uncle Signor Cipriano is by nature very sincere, and with his friends (among whom I am certain he keeps Your Lordship) proceeds with simplicity and frankness and without any sort of ceremony—as I suppose he did not use with Your Lordship when he responded to the very nice letter he had received from you. Whence, His Lordship (having seen what you wrote to me) was surprised that you thought his letter required thanks from you." Nevertheless, Galileo seemed to use more ceremonies later on (ibid., no. 155, p. 178).

58. Ibid., no. 229, p. 251 (emphasis mine).

social ascent and legitimation. Major leaps in social status could not be achieved by compounding the patronage of many small patrons but could result only from the patronage of a single great patron. This explains why the princely court was such a powerful institution in legitimizing socio-professional identities: it was the space of the absolute prince, of the greatest patron.

Galileo understood well how different ranks of patrons could provide different levels of social legitimation. As he told Vinta, a republic like Venice (a sort of patrician corporation) could not offer the type of legitimation he sought. Galileo's frequently expressed desire to serve only one princely patron rather than many smaller ones has usually been read as an attempt to obtain an excellent salary and plenty of free time for research.[59] This interpretation is correct as far as it goes. What it misses is an important dimension of Galileo's desire to move to court—a dimension linked to the social status one could obtain by serving a single princely patron.[60]

Writing early in 1609 to a Florentine courtier, Galileo expressed his desire to obtain a court position.[61] He claimed that his interest in such a position was not related to the *amount,* but to the *type* of work he would have to face at court:

> Regarding the everyday duties, I shun only that type of prostitution consisting of having to expose my labor to the arbitrary prices set by every customer. Instead, I will never look down on serving a prince or a great lord or those who may depend on him, but, to the contrary, I will always desire such a position.[62]

Galileo understood that a patronage relationship with a great patron brought "purity" (i.e., high status) because it was "monogamous," exclusive, and paid through a regular stipend. To serve many low patrons and be paid piecemeal was a sort of prostitution (*servitù meretricia*). By developing an exclusive and full-time relationship with a great patron one participated in "nobility"—a high status that could be transferred from one's social identity to one's discipline or activity. And, as mentioned earlier, high social status was instrumental in securing the epistemological

59. Ibid., no. 146, p. 165; no. 209, p. 233; no. 307, p. 351.

60. Of course, this does not mean that Galileo did not have other lesser patrons who helped and supported him even as, after 1610, he became the philosopher and mathematician of the grand duke. However, he was not *paid* by these other patrons (Salviati, Cesi, Leopold of Austria, Marsili, etc.), and he interacted with or was introduced to them as a client of the Medici grand duke.

61. The content and gist of this letter remind us of the letters Galileo wrote to Vinta after his discoveries to negotiate his appointment at the Florentine court (ibid., no. 307, pp. 348–53, esp. pp. 350–51).

62. Ibid., no. 209, p. 233.

status of a discipline and method like Galileo's, whose legitimacy was undermined by the existing disciplinary hierarchy.[63]

Finally, by going up in the social ladder through connections to an increasingly *smaller* number of patrons of *higher* power, a successful client like Galileo built a pyramid of clients below him. Patronage requests addressed to Galileo took a sudden jump around 1610.[64] It would be wrong to assign that jump to Galileo's popularity as a discoverer. Discoverers get power from their discoveries only through the institutions that legitimize them. Galileo's power as a courtly star in Florence was much higher than that he had or could have had as a university professor of the Republic of Venice. While in Padua, Galileo received a few requests for patronage, but he seems not to have fulfilled any of them. After 1610 he was able to place a range of mathematicians and philosophers (Castelli, Cavalieri, Aggiunti, and Papazzoni) as professors at the universities of Pisa, Rome, and Bologna. A more complex reading of patronage dynamics shows that Galileo was seeking much more than free time at the Medici court.

Marvelous Conjunctures and Providential Deaths

Recurrent features of clients' lives in early modern Europe were the discontinuities and disruptions produced by the termination of patronage relationships, usually as a result of the patron's death. The trajectories of early modern clients were not smooth curves but tortuous and discontinuous paths punctuated by patronage crises. The careers of Commandino, Leibniz, Dee, Kepler, Tycho, and Galileo exemplify these dynamics.

Biographies of Galileo present the turning points of his career as related to discoveries and controversies like the Medicean Stars, the "Letter to the Grand Duchess," the dispute on comets and the *Assayer,* the trial of 1633, and the last dispute with the philosopher Liceti. Another pattern emerges when we compare the chronology of these turning points and those of his patronage relations.

The first phase of Galileo's career was much indebted to Guidobaldo del Monte, the patron/broker through whom Galileo gained the university positions at Pisa in 1589 and at Padua in 1591. Actually, this segment of Galileo's career was an indirect result of the accession of Ferdinand I de' Medici following the death of his older brother, Francesco I in 1587. At the time of Francesco's death, Ferdinand was a cardinal, and he maintained that title until the end of 1588 because there was no other member of the Medici family who could take over the post. In December 1588 Ferdinand

63. Westman, "Astronomer's Role in the Sixteenth Century"; and Biagioli, "Social Status of Italian Mathematicians."

64. We find only four requests for patronage from three clients before 1610 (*GO*, vol. 10, nos. 98, 100, 179, 229), and eleven requests from nine clients between 1610 and 1612 (ibid., nos. 386, 441, 444, 445, 448; vol. 11, nos. 469, 473, 474, 488, 577, 726).

put a client of his—Francesco Maria del Monte—in his place as the "Medici cardinal."[65] Up to this point, although Galileo had been a client of Guidobaldo for almost a year, all his patron's attempts to secure him a post through his brother Francesco Maria—then only a *monsignore*—failed.[66] Things changed rapidly after Francesco Maria became a cardinal at the end of 1588; the very next fall Galileo was a professor at Pisa. Galileo seemed to understand the new possibilities opened by the much increased power of Francesco Maria; in December 1588, he wrote to Guidobaldo congratulating him on his brother's promotion.[67] He was right in his expectations.

In a letter dated December 23, 1588, Guidobaldo wrote to the grand duke to convey (in very uncompromising terms) his joy and gratitude for having secured the cardinalate for his brother Francesco Maria.[68] Guidobaldo concluded by telling the grand duke of his desire to go to Florence to thank him in person, something he was planning to do so as soon as Francesco Maria returned from Rome. It was probably during the joint visit of the Del Monte brothers (which is likely to have coincided with Ferdinand and Cristina's wedding in May) that Galileo's appointment materialized. Quite likely, it was a little present that Ferdinand gave to his two clients to celebrate the wedding festivities. Toward the end of July, Galileo was already writing to Guidobaldo to thank him for the lectureship at Pisa.[69]

It was still through Guidobaldo and other members of the del Monte family that Galileo was introduced to Padua's most important patron, Vincenzio Pinelli. Pinelli died in 1601, but by that time he had in turn introduced Galileo to the patronage of Venetian patricians. In Venice, Sagredo became Galileo's most important patron during the first few years of the seventeenth century. When Guidobaldo died in 1607, Galileo did not need his protection and support any more.[70] Similarly, when

---

65. Ugo Barberi, *I Marchesi Bourbon del Monte Santa Maria di Petrella e di Sorbello* (Città di Castello: Tipografia Unione Arti Grafiche, 1943), pp. 64–65.

66. Guidobaldo began to press his brother for a position for Galileo in May 1588. Guidobaldo's targets were the chair at Pisa and the public lectureship in Florence, which had once been Egnazio Danti's (*GO,* vol. 10, no. 17, pp. 33–34; no. 18, pp. 34–35; no. 20, pp. 36–37; no. 21, pp. 37–38). In July 1588 Galileo wrote Guidobaldo saying that the chair at Pisa was already filled and that the only possibility still open was the public lectureship in Florence connected with the Accademia del Disegno (ibid., no. 19, p. 36). As it turned out, Galileo obtained the chair at Pisa—the one he thought was already filled.

67. *GO,* vol. 10, no. 23, p. 39.

68. *ASF,* "Mediceo principato 802," fol. 500 (the folio number is incorrectly marked as "50"). As far as I know, this letter had not been noticed by historians of science.

69. *GO,* vol. 10, no. 27, p. 41.

70. This also can be detected by the infrequent letters exchanged by the two men after 1600. Although there are probably gaps in the correspondence as it came to us, we do not

Sagredo was sent to Syria in 1608 as the Venetian ambassador, his departure did not damage Galileo's position in Venice because by then he had
already established relationships with a number of powerful Venetians
and Paduans, such as Duodo, Morosini, Priuli, Gualdo, and Venier.
Galileo's patronage mobility in this period was also a result of the *type* of
patron he was relying on.

As a disappointed Sagredo would write to Galileo after the latter's return to Florence, the position he had left behind in Venice was a *safe* one.
His formerly young patrician patrons were growing older and more powerful. Galileo could have risen with them as a matter of course. In Venice
he would not have had to depend on the fragile favor of a young and possibly changeable prince. As Sagredo put it, "who knows what [can be
caused by] the infinite and incomprehensible accidents of the world made
more complicated by the impostures of bad and envious men who, by
planting and nurturing false and injurious concepts in the mind of the
prince, can use his justice and virtue to ruin a gentleman?"[71]

Sagredo's portrait of the dangers of courtly life and patronage was basically accurate. What Sagredo glossed over was what Galileo could not
have found in Venice. True, since the patronage networks in republican
Venice were not focused on a single and absolute princely patron, a client
like Galileo was not so heavily subjected to disruptions caused by patrons'
deaths, departures, or changing tastes as he would be in a principality. As
Galileo himself knew well, so long as he taught at Padua he would be a
subject of an "immortal and immutable prince" (i.e., the republic itself).[72]
However, while making the clients' lives safer, the absence of a great patron reduced the clients' opportunities for great jumps in social (and cognitive) status.

This second phase of Galileo's career that began with his development
of the telescope in 1609 was also characterized by the emergence of a new
patron and the rejection of old ones. Galileo's revolutionary astronomical
discoveries of 1609–10 went hand in hand with a revolution in his patronage relationships. When he gained the patronage of Cosimo II, he broke
the contract of *amicizia* with his Venetian patrons right after they had confirmed and strengthened it, causing legitimate surprise and bitterness.
And, as in the case of the patronage crisis of 1587, a major improvement in
Galileo's status came about as a result of a death-related shift in patronage
possibilities.

In 1587 it was Francesco de' Medici's death that turned Francesco
Maria del Monte (via Ferdinand and Guidobaldo) into a powerful patron

have letters from Guidobaldo after 1597, and the last letter from Galileo dates from December
1602.

71. *GO*, vol. 11, no. 569, p. 171.
72. *GO*, vol. 10, no. 350, p. 350.

of Galileo. In 1609, Ferdinand's death turned Galileo's young princely student Cosimo into a grand duke, just a few months before Galileo's astronomical discoveries. Neither Francesco nor Ferdinand had been direct patrons of Galileo. Grand Duchess Cristina and Belisario Vinta were the main brokers through whom Galileo maintained Medici patronage. Yet it was the deaths of the two Medici grand dukes that proved crucially beneficial for his career.

Cosimo's enthronement in 1609 was particularly advantageous for Galileo's career because new and controversial ideas are better supported by young patrons seeking an image for themselves. Still, it would be simplistic to say that Galileo happened to find the right patron at the right time. Although chance played a great role in bringing about such a remarkable synchronization between his discoveries and the constructive crisis in his patronage networks, we should not forget that such a conjuncture was made possible by patronage tactics he had been pursuing since the early 1600s.[73] And this conjuncture did not come early in his career. By the time Galileo moved to Florence as philosopher and mathematician to the grand duke, he was already forty-six.

The last phase of Galileo's career—one gravitating toward the Roman court—was similarly characterized by the emergence of new patrons: Prince Federico Cesi (and some members of his Accademia dei Lincei), and Pope Urban VIII. The increasing importance of Roman patronage in this phase of Galileo's career was significantly paralleled by the declining power of his Florentine patronage resources.[74] Both Cosimo II and Antonio de' Medici—the latter an old friend and patron of Galileo—died in 1621, while Filippo Salviati and Belisario Vinta had died, respectively, in 1614 and 1613. Cosimo left an adolescent son who would not be able to reign until 1628.[75]

We find another remarkable instance of synchronization between

73. Sometimes, patronage crises were not just synchronized with the turning points of Galileo's scientific or literary activity, but they rather caused them. For instance, the "Letter to the Grand Duchess" of 1615 was largely the result of patronage dynamics. In fact, Galileo had to meet the attempt to disrupt his patronage relationship with the Medici by insinuating doubts of his religious orthodoxy in the very pious ears of the very powerful Cristina.

74. By resources I do not mean financial ones, for Galileo drew his salary from 1610 until his death in 1642 from the grand duke. What I mean here is that, while the acceptance of his astronomical discoveries of 1610 relied strongly on the grand duke's endorsement (as will been shown in chapter 2), Galileo's later attempted legitimation of Copernicanism hinged on the pope's power.

75. Galileo's shift from Florence to Rome as the locus of his patronage networks can be associated with his need to maintain a relationship with a very strong patron. Only with such arrangements could he maintain and increase his status and legitimation. And Cosimo II—a grey, frequently sick, and almost always debilitated prince—was not a real presence. Moreover, after Cosimo's death in 1621, Galileo found himself as a court philosopher and mathematician with a very good salary but without a flesh-and-blood patron. Cosimo's son

Galileo's scientific production and continuing patronage crises: the death of Pope Gregory XV and the election of Maffeo Barberini to the pontificate as Urban VIII precisely when Galileo's *Assayer* (then dedicated to Urban) was in press.[76] Pope Gregory's was the third providential death in Galileo's career. As with Cosimo II thirteen years before, a new patron (one Galileo had been cultivating for a number of years, as he had Cosimo) suddenly reaches an important status and is willing to support Galileo's provocative views in order to develop a necessarily new image for himself. In terms of patronage, the *Sidereus nuncius* was to Cosimo II's reign what the *Assayer* was to Urban VIII's. Galileo was quite right in calling this remarkable synchronization "a marvelous conjuncture" (*una mirabil congiuntura*).[77]

There is no perverse pleasure on my part in stressing the "providentiality" of these deaths for Galileo's career. As we will see, deaths of great patrons—especially of monarchs like popes, who were not members of a hereditary dynasty—were perceived by contemporaries as major patronage crises. Careers were suddenly made and destroyed on those occasions. As shown by Virginio Cesarini and Giovanni Ciampoli (two clients of Urban VIII and supporters of Galileo), who found themselves suddenly lifted to the top of the Roman court, brand-new cadres would come to power with a new pope.[78] In a true pilgrimage toward the center of power, non-Romans too would try to exploit the new patronage system by traveling to Rome to pay homage to the pope. Galileo himself was a participant in the patronage pilgrimage of 1623–24. Alluding to the comets that had appeared a few years before Urban's election, the Florentine poet Jacopo Soldani compared the clients' pilgrimage to the trip undertaken by the Three Magi to pay homage to the baby Jesus.[79] The gift which Galileo presented to the new pope was the *Assayer*.

Ferdinand turned eighteen (and became grand duke) only in 1628. By then, Galileo's strategies had been long focused on Rome.

76. The printing of the *Assayer* began in May 1623, Maffeo Barberini was elected Pope Urban VIII on August 6, and the *Assayer* was dedicated to him as it came off the press in October. Before Urban's election, Ciampoli was the *Secretario Apostolico* (or *Secretario dei Brevi*), while Cesarini was the pope's *Cameriere Secreto*.

77. "I am contemplating things of some importance for the literary republic which, if they do not take place in this marvelous conjuncture, one should not (as far as I am concerned) hope to find another opportunity like this again," Galileo wrote to Cesi on 9 October 1623 (*GO,* vol. 13, no. 1581, p. 135). Cesi agreed with Galileo that the scenario was indeed a "congiuntura sì buona" (ibid., no. 1588, p. 140). On this period, see Pietro Redondi, "The 'Marvelous Conjuncture,'" *Galileo Heretic* (Princeton: Princeton University Press, 1987), pp. 68–103.

78. After the election of Urban, Cesarini became his *Maestro di Camera,* while Ciampoli added to his title of *Secretario dei Brevi* that of *Cameriere Secreto* (*GO,* vol. 13, no. 1564, p. 121).

79. "Others make their way to Rome/ to see in the great throne the new Urban/ loaded with grave and rich burdens/ and try with the hand/ to seize the waving forelock/ of she who

The end of Galileo's Roman successes, marked by the trial of 1633, can also be seen as framed by patronage dynamics. It was well known that patronage quakes could clear the grounds for the construction of brilliant careers at the Roman court but could also shatter them, as had been the case with the mathematician Federico Commandino.[80] Similarly, there are indications that Cosimo II's death in 1621 put Galileo in an unstable patronage situation, one which gave Galileo cause to worry about the security of his position at court.[81]

The pattern was so familiar that *The Courtier's Philosophy* (a court game not unlike today's "Monopoly") published in Madrid in 1587 prescribed that those who landed on square 43 ("Your patron dies") had to go back to the start.[82] Analogously, Galileo's condemnation cannot be separated from the patronage crisis that affected his (and his friend Ciampoli's) relationship with Urban VIII.[83] Galileo's troubles of 1633 were also preceded by the deaths of two of his major patrons—Cardinal del Monte (1626) and Prince Cesi (1630). With Cesi's death—and the earlier one of Cesarini in 1624—Galileo was left with very little support within the Roman court. Therefore, although Galileo's condemnation was triggered by the specific theological implications of his *Dialogue on the Two Chief World Systems,* it was, at the same time, an instance of a more general pattern. It offers a typical example of a patronage-related termination of a client's career.

The last dispute that punctuated Galileo's career (with the philosopher Liceti) marks Galileo's final attempt to regain Medici support through his young admirer Prince Leopold, the future founder of the Accademia del Cimento. This attempt, reminiscent of Galileo's more successful strategy with Cosimo, was mediated by Jacopo Soldani, a friend and broker of Galileo. Soldani's role at the Medici court was not unlike that of the brokers through whom Galileo had gained Cosimo's patron-

---

flees, and then wait for her again in vain" (Jacopo Soldani, "Contro i Peripatetici," reprinted in Nunzio Vaccalluzzo, *Galileo Galilei nella poesia del suo secolo* [Milan: Sandron, 1910], p. 20).

80. Bernardino Baldi, "Vita di Federico Commandino," in Filippo Ugolini and Filippo Polidori, eds., *Versi e prose scelte di Bernardino Baldi* (Florence: Le Monnier, 1859), pp. 513–37; Paul L. Rose, *The Italian Renaissance of Mathematics* (Geneva: Droz, 1975), pp. 185–221; and C. Bianca, "Federico Commandino," *Dizionario biografico degli italiani* (Rome: Istituto della Enciclopedia Italiana, 1982), 26: 602–6.

81. In fact, in April 1621 he told Archduke Leopold: "It will be of most grace and favor to me if the Most Serene Archduchess your sister [Cosimo's widow] will hear at some point from Your Most Serene Highness how you maintain affection for me—an affection that may secure for me her grace in a way that my low merit would never do" (*GO,* vol. 13, pp. 60–61). On this maneuver of Galileo, see also ibid., pp. 64, 70.

82. Alonso de Barros, *The Courtier's Philosophy* (Madrid, 1587); quoted in Geoffrey Parker, *Philip II* (London: Hutchinson, 1979), p. 170.

83. On this point, see also Westfall, "Patronage and the Publication of the *Dialogue.*"

age. In fact, Soldani was Leopold's *Aio*—his primary advisor and tutor, a position identical to Piccolomini's and comparable to that of Saracinelli, of Mercuriale, and, in a sense, of Cristina.[84] Galileo, though, did not live long enough to test the potential of this new patronage connection.

The chronological synchronization of Galileo's publications and the crisis of his patronage relationships indicates more than a series of remarkable coincidences. Although patronage was not a fully predictable process, it was far from being chaotic. It had its logic, etiquette, and periodical crises tuned to generational cycles that could be expected and intelligently bet upon. Successful careers were those of clients who extended their patronage networks and tuned their cultural production to patronage cycles so as to turn the play of chance into "marvelous conjunctures."[85]

Gift-Exchange as the Logic of Patronage

Galileo's correspondence shows that gifts and other economically non-quantifiable services and privileges were the medium through which patronage relationships were articulated and maintained. Even when much cash entered the scene—as with Galileo's remarkable thousand-scudi stipend at the Medici court—we should not view it only in our capitalistic perspective and reduce the significance of a thousand-scudi salary to its buying power alone. Its symbolic dimension was also important: income was both a sign and a material cause of status.

Galileo did not formally negotiate his stipend. All his efforts were aimed at securing Medici patronage in some stable fashion, and when his wishes eventually materialized, he simply told Vinta how much he made at Padua and left the final figure to the generosity of the grand duke. Galileo's stipend was a result of both Galileo's value and of the grand duke's *noblesse oblige*. It was an emolument, a sign of Cosimo's generosity

---

84. Jacopo Soldani is listed as "Aio del Serenissimo Principe" in the *ruoli* of the Medici court since 1630 (*ASF*, "Manoscritti 321," p. 522). His very high salary (600 scudi per year) indicates that he was much more than a tutor, perhaps something of a "big brother" (Prince Leopold was an orphan). Soldani was made a senator in 1637 (*ASF*, "Manoscritti 320," p. 255) and *Maestro di Camera* (one of the highest roles at court) in 1639 (*ASF*, "Miscellanea medicea 438," fol. 212v.). Soldani's role in cultivating the scientific interests of the future founder of the Cimento can be found in a few letters from 1640 between Leopoldo and Soldani, related to the dispute between Galileo and Liceti, that were not included by Favaro in the Edizione Nazionale (*ASF*, "Mediceo principato 5550," fols. 261, 271, 272, 274, 278, 291, 310). These letters are published in Mario Biagioli, "New Documents on Galileo," *Nuncius* 6 (1991): 157–69.

85. The term "conjuncture" was not Galileo's invention. It recurred frequently in court literature (and not only in Italy) in relation to the workings of fortune. José Antonio Maravall, in his *Culture of the Baroque* (Minneapolis: University of Minnesota Press, 1986), claims that "this word [conjuncture] was also very frequently used in the seventeenth century; in Gracian or Cespedes it would not be difficult to find fifty examples of its use" (p. 191).

and power in reciprocating in due proportion Galileo's generous gift of the dedication of his astronomical discoveries to the house of Medici. Galileo's salary was not just a number in the Medici payroll but an item of public information. A Florentine chronicle of the time says that to reward his dedication of the Medicean Stars the grand duke called Galileo back to Florence and gave him a good stipend.[86] Similarly, in describing a public demonstration of Galileo's telescope, held in Rome in April 1611, an *Avviso* (a sort of proto-newspaper, reporting Roman political news and gossip) managed to include the information that Galileo "is at present retained by the Grand Duke with a stipend of 1,000 scudi."[87] Therefore, the stipend was a *public gesture*.[88] If the Medici had been stingy with Galileo, they would have automatically belittled the significance of the Medicean Stars in the public eye. Galileo's stipend was not the product of the dynamics of offer and demand operating in a free-market economy but rather of the economy of honor characteristic of the exchange of status-carrying gifts.

Even when patronage relationships were represented through monetary exchanges, they reflected gift-exchange dynamics. As with the potlatch of the Indians of the American Northwest analyzed by Marcel Mauss, gift-exchange was an exercise of power.[89] In early modern Europe,

86. *ASF,* "Manoscritti 132" ("Diario fiorentino del Settimanni," vol. 7, 1608–20), fol. 39r: "On July 1610, Signor Galileo Galilei having dedicated to the Most Serene House of Medici the four stars newly observed by him revolving around the planet Jupiter [and] having named them Medicean Stars and Planets, the Most Serene Grand Duke with his own letter in sign of gratitude has recalled him from Padua (where he was a public lecturer) to his service with the title of Primary and Extraordinary Mathematician of the University of Pisa without obligation of lecturing or residing there, and of Primary Philosopher and Mathematician to His Most Serene Highness, assigning to him a good stipend."

87. J. A. F. Orbaan, *Documenti sul barocco in Roma* (Rome: Societá Romana di Storia Patria, 1920), 2: 283.

88. On Cosimo II's cultivated image as supporter of artists and scientists, see Pietro Accolti's funeral oration for Cosimo, "Delle lodi di Cosimo II, Granduca di Toscana," in Carlo Dati, ed., *Raccolta di prose fiorentine* (Florence: Stamperia di SAR, 1731), vol. 6, part 1, p. 119.

89. Mauss, *The Gift,* 31–45. The bibliography on gift-exchange is by now quite extensive. The works that I have found most useful are Pierre Bourdieu, *Outline of a Theory of Practice* (Cambridge: Cambridge University Press, 1977); idem, *The Logic of Practice* (Stanford: Stanford University Press, 1990), pp. 98–101; Marilyn Strathern, *The Gender of the Gift* (Berkeley: University of California Press, 1988); Bronislaw Malinowski, "Kula: The Circulating Exchange of Valuables in the Archipelagoes of Eastern Guinea," *Man,* series 1, 19–20 (1920): 97–105; Natalie Zemon Davis, "Beyond the Market: Books as Gifts in Sixteenth-Century France," *Transactions of the Royal Historical Society* 33 (1983): 69–88; Marshall Sahlins, *Stone-Age Economics* (New York: Aldine de Gruyter, 1972), esp. pp. 149–275; Sharon Kettering, "Gift-Giving and Patronage in Early Modern France," *French History* 2 (1988): 131–51; Georges Bataille, *The Accursed Share* (New York: Zone Books, 1988), vol. 1, esp. pp. 63–77; Carlo Zaccagnini, *Lo scambio dei doni nel Vicino Oriente durante i secoli XVIII–XV* (Rome: Centro per le Antichità e la Storia dell'Arte del Vicino Oriente, 1973); "Le don perdu et retrouvé," special issue of *La revue de Mauss* 12 (1991); and Claude Lévi-Strauss, *The Elementary Structures of Kinship* (Boston: Beacon Press, 1969), pp. 52–68.

CHAPTER ONE

patronage was a competitive process of social distinction effected through the important patrons' challenge of each other's power to spend. For instance, Louis XIV controlled the rebellious French aristocracy in part by impoverishing it through a challenge of conspicuous expenditure at court.[90] A similar, if less marked, pattern can be found in late sixteenth- and early seventeenth-century England where the conspicuous expenditure triggered by the court life-style of an absolutist ruler was responsible for the ruin of several aristocratic houses.[91] The Medici had successfully adopted similar strategies to turn former political competitors into a docile court aristocracy. When the ambassadors of Lucca visited Florence in 1619, they found that the Florentine nobility had declined greatly because of the severe costs of court life.[92]

In a sense, court life and etiquette may be seen as a body of rituals aimed at disciplining the potentially subversive challenges to conspicuous expenditure (the potlatches) between the prince and the aristocracy so as to guarantee the prince's victory. Also, through this "civilizing process" (to use Norbert Elias's term) the aristocrats did not lose everything in an afternoon as did some of the Tlingit studied by Mauss. The aristocrats' wealth and power burned down more slowly in a controlled "chain reaction" so as to illuminate the absolute prince's *magnificentia* over a long period of time.

Although gift-giving was not always represented as a form of challenge, the competitive subtext was always present. The relation between a

90. Norbert Elias, "The Sociogenesis of French Court Society," *The Court Society* (New York: Pantheon, 1983), pp. 146–213. For a different use of gift-exchange in relation to the social system of science, see Warren Hagstrom, *The Scientific Community* (New York: Basic Books, 1965), pp. 12–23; and idem, "Gift Giving as an Organizing Principle in Science," in Barry Barnes and David Edge, eds., *Science in Context* (Cambridge, Mass.: MIT Press, 1982), pp. 21–34. Hagstrom limits his use of the notion of gift-giving to discussing the authors' donating discoveries (in the form of research papers) to the scientific community in exchange for recognition. More precisely, he sees gift-exchange as a necessary island of moral economy in the ocean of "rational" economic action characteristic of the social system of science. Gift-exchange is crucial in that it creates particularistic obligations by reducing the rationality of economic action. Hagstrom seems to suggest that, without this softening of economic rationality, scientists would be perceived not as acting according to "higher" values but rather as following more mundane interests. Instead, by "donating" their discoveries to the community, scientists act as if they were "pure" and are rewarded with "pure" recognition. The role of gift-exchange in early modern Europe is also discussed in Paula Findlen, "The Economy of Scientific Exchange in Early Modern Italy," in Bruce Moran, ed., *Patronage and Institutions* (Rochester, NY: Boydell, 1991), pp. 5–24; and idem, *Possessing Nature: Museums, Collecting and Scientific Culture in Early Modern Italy* (Berkeley: University of California Press, forthcoming).

91. Lawrence Stone, *The Crisis of the Aristocracy, 1558–1641* (Oxford: Oxford University Press, 1967), pp. 86–88, 249–67.

92. Amedeo Pellegrini, ed., *Relazioni inedite di ambasciatori lucchesi alle corti di Firenze, Genova, Milano, Modena, Parma, Torino* (Lucca: Marchi, 1901), p. 141.

client and a patron was characterized by the patron's power to give the client more than the client could return. The disparity in gift-giving power was stressed as a way of confirming a situation of dependency. By accepting the gift from the patron without returning it (i.e., without "challenging back"), the client admitted "defeat" and, consequently, dependency.

I am not saying that clients never gave gifts to their patrons. Quite the contrary, a client who could not offer interesting gifts would not go far. In his reflections on the Roman court, Galileo's friend Ciampoli claimed that "There is something that holds true everywhere in the world: you need to give gifts [*donare*] to those in power. . . . Blessed are those who can accelerate their success by giving gifts!"[93] Another Roman courtier, Agostino Mascardi, saw the courtiers' need to give gifts to their patrons as one of the fundamental paradoxes of court society. The courtier was bound to "impoverish himself in order to become rich; to give in order to receive."[94] Another court treatise by Matteo Pellegrini commented—this time quite critically—on the paradoxes underlying gift-exchange practices at court.[95] Similar practices have been uncovered by Carlo Zaccagnini in his remarkable study of gift-exchange at the courts of ancient Near East kingdoms (fifteenth to thirteenth centuries B.C.). There, "among subjects of different rank, the initiative [of gift-giving] belonged naturally to the subject of inferior rank."[96]

Galileo's brokers often reminded him to bring "new things" every time he visited Florence. Gift-giving was the best investment for a client because the patron was bound by his status to reciprocate gifts in proportion to his own (rather than the client's) status. Often, a high patron perceived gifts as challenges. If he accepted them, he was bound to behave as if he had accepted a duel, that is, "heroically." In 1604 Galileo received from the Gonzaga a gold medal and chain and two silver dishes worth 1,340 lire—this for a military compass and the training in its use for which Galileo usually charged his Paduan students about 200 lire.[97] This multiplication of the value of the client's gift also appears in the very generous stipend Galileo obtained at the Medici court in exchange for the dedication of the Medicean Stars.

Instances of patrons who refused material gifts from the client without putting an end to the patronage relationship do not refute my claim of

93. Giovanni Ciampoli, "Discorso di monsignor Ciampoli sopra la corte di Roma," in Marziano Guglielminetti and Mariarosa Masoero, "Lettere e prose inedite (o parzialmente edite) di Giovanni Ciampoli," *Studi secenteschi* 19 (1978): 232.

94. Agostino Mascardi, *Prose vulgari* (Venice: Baba, 1653), p. 20.

95. Pellegrini, *Che al savio è convenevole il corteggiare libri IIII*, p. 38.

96. Zaccagnini, *Lo scambio dei doni nel Vicino Oriente durante i secoli XVIII-XV*, p. 40.

97. *GO*, vol. 19, pp. 147–58.

gift-giving as the medium of patronage. For instance, after having helped Galileo obtain the chair at Padua, Guidobaldo told him that "I do not want you to feel obliged to me about the chair at Padua because I did not have anything to do with it."[98] Guidobaldo lifted the patronage relationship with Galileo above the material level so that he could accept "intellectual gifts" (the only ones Galileo could offer at that stage of his career) as a legitimate return for his patronage. Guidobaldo's apparent "disinterestedness" in the materiality of the gift was a ritualistic representation of his aristocratic identity. First, it presented him as having a lofty (i.e., aristocratic) mind that could appreciate the "immateriality" of Galileo's gifts. Second, it showed that he did not need to be paid for the privileges he distributed. Guidobaldo gave out of *noblesse oblige*.

However, Guidobaldo did get a little something out of Galileo's hiring at Padua. He asked Galileo "what stipend do they give you, because I would like to see you treated according to *my desires* and your merits."[99] This same statement is also found, in almost identical form, in an earlier letter: "I wished to know whether Your Lordship has ever received a rise in salary, because I would like this to reflect *my desire* and your merits."[100] What these references to Guidobaldo's "desire" indicate is that he was testing his power through his client. Guidobaldo was clear in separating his own desires and power from Galileo's worth. To him, Galileo's stipend was a "gift" he was giving him through Guidobaldo's own patronage networks. Therefore, while matching Galileo's worth, the "gift" had to reflect Guidobaldo's power. Too low a stipend to Galileo in Padua would have damaged Guidobaldo's image in the same way too low a stipend to Galileo as philosopher of the grand duke would have been perceived as infamous for the Medici's image.

Both among the Tlingit and early modern European courtiers, conspicuous gift-giving was a sign of the giver's disregard of value. It represented one's "heroicness." Gifts and duels reflected a similar logic. The patron who gave important gifts, or the originator of a daring dispute, was perceived as a challenger, that is, as an aristocrat. The gift acted as a probe. If the receiver was a person of high status, he or she would perceive the gift as an honorable challenge. Conversely, if the receiver was unable to reciprocate, the gift acted as a paternalistic gesture that emphasized the power of the giver-patron. In this case, the gift would become a sort of monument to the patron—a *fetish*. This, I suggest, may explain why many of the official gifts—like the gold medals Galileo received first from the Gonzaga and then from the Medici—carried the patron's effigy. They were signs of the

98. *GO*, vol. 10, no. 45, p. 54.
99. Ibid. (emphasis mine).
100. Ibid., no. 33, p. 45 (emphasis mine).

client's "belonging" to the patron.[101] They were conveniently removable brands. Consequently, patronage and gift-giving represented more than economic exchange: they produced status, identity, and credibility.

Galileo's correspondence offers much evidence of private gift-exchanging rituals among clients and patrons involving presents such as instruments, books, hospitality, letters of introduction, wine, dogs, paintings, wild game, seeds of exotic plants, invitations to parties and ceremonies, access to important circles, services, and privileges. Other gifts—especially those given out by princes—were often part of a standardized etiquette of reception and reward at court.[102] It seems that by the end of the sixteenth century, the symbolic and economic dimensions of gifts had became more distinct and their economic value increasingly quantified. For instance, gold medals and chains became a standardized gift, with the prince expecting his client to retain the gold medal carrying the prince's image ("to remember me" [*per ricordo mio*]) but expecting the client to sell the gold but nonpersonalized chain "according to your needs" (*per i bisogni vostri*).[103] Gold chains came in sizes related to standard cash value, such as two, three, or four hundred scudi.

The exchange of gifts structured both Galileo's official contacts and private friendships. For example, it does not seem that there were mone-

101. In discussing patronage dynamics among various levels of the British aristocracy, Marvin B. Becker reports: "In addition to the tenants of magnates, there were lesser lords who had bound themselves to the potentates of the realm and from whom they frequently received a type of uniform (livery) and a badge which they wore as a sign of their clientage" (*Civility and Society in Western Europe, 1300–1600* [Bloomington: Indiana University Press, 1988], p. 85. I would say that the customs of displaying one's patron's coat of arms on one's house, and of collecting portraits of patrons, are—like the displaying of medals or badges—elements of one's self-fashioning as a client. One may also try to apply these considerations to the early modern collectors' displaying often-unimpressive specimens they had received from their patrons in their natural history museums.

102. Marcello Fantoni, "Feticci di prestigio: Il dono alla corte medicea," in Sergio Bertelli and Giuliano Crifò, eds., *Rituale, cerimoniale, etichetta* (Milan: Bompiani, 1985), pp. 141–61. Hospitality gifts are one of the central topics in Zaccagnini, *Lo scambio dei doni nel Vicino Oriente durante i secoli XVIII–XV*. The ritualistic exchange of gifts is encountered quite frequently in the *Diari di etichetta* of the Medici court (*ASF*, "Diari di etichetta di guardaroba," nos. 1–7). It plays a fundamental role in the ritual of reception of foreign dignitaries. See also *ASF*, "Carte strozziane," series 1, 30," fols. 127–44 ("Donativi").

103. Fantoni, "Feticci di prestigio," 143. That gold chains represented a sort of transitional stage of the gift toward a more quantified and cashlike retribution is shown not only by the various examples listed by Fantoni, but also by the specifically stated value of the two gold chains Galileo received from the Gonzaga and the Medici, by the references found in Galileo's correspondence about court clients being given gold chains and medals (*GO*, vol. II, no. 838, p. 473), and by the example of the chain given by the Medici to Fabricius of Acquapendente on the occasion of his visit to Florence. Similarly, the etiquette diary of the Medici court records that in 1615 "l'ambasciatore del Giappone fu regalato di una collana di 400 scudi" (*ASF*, "Miscellanea medicea 447," p. 328).

tary transactions between Galileo and his close friend Sagredo, with the exception of loans that were considered as something outside their friendship. Galileo was not shy in pressing Sagredo to lobby for his salary increase at Padua. Nor was Sagredo backward in asking to have his instruments fixed or manufactured gratis by Galileo's artisans, or in enlisting Galileo to fetch water from the spring of the Virgin of Monte Artone and send it to him in Venice, to find a bottle of the "Sicilian's oil for wounds" he wanted to take to Syria, to send him a declinatorium, and, after Galileo's move to Florence, to find and send him (again for free) a pair of exotic dogs.[104] On top of these requests, Galileo quite frequently sent small gifts like wine, game, and truffles that were reciprocated by Sagredo with more wine or with seeds of "special" melons from Syria.

The frequency and informality of the exchanges of gifts between Sagredo and Galileo may mislead us into regarding the entire process as just an expression of the friendship between the two. But friendship had a very specific meaning in Galileo's time. As much recent historiography has shown, *amicizia* was a contractual bond ritualistically represented as informal.[105] Sagredo spelled out the contractual dimensions of his *amicizia* with Galileo on several occasions. In August 1602 he told Galileo that

> Our Signor Venier and myself want to take a short trip in Cadore this October . . . but because this trip through wonderful places may turn boring without the company of Your Most Excellent Lordship, I decided to let you know ahead of time so that, to please the two of us, you can arrange things to do us this favor. *And whatever effort you take, we promise to do the same for you on the occasion of the confirmation of your post.*[106]

This directness in indicating the reciprocity of gift-exchange is found elsewhere in Galileo's correspondence. In June 1604, Antonio de' Medici wrote to Galileo, saying "I understand that Your Lordship has a ball that, once thrown in the water, remains between the two waters. *I really beg you to send me one* . . . and you can be sure I will consider this a very special favor, *and that your courtesy will be reciprocated.*"[107] Several years later, An-

---

104. *GO*, vol. 10, no. 75, p. 86; no. 82, p. 90; no. 85, p. 95; no. 87, p. 96; no. 89, p. 100; no. 187, p. 208; For more gift-exchanges between Galileo and Sagredo, see *GO*, vol. 12, no. 1188, p. 246; no. 1198, p. 258; no. 1219, p. 273; no. 1224, p. 278; no. 1230, p. 286; no. 1255, p. 317; no. 1275, pp. 343–44; no. 1281, p. 349; no. 1287, p. 355; no. 1310, p. 376; no. 1341, p. 407.

105. For instance, Trexler, *Public Life in Renaissance Florence*, pp. 131–59.

106. *GO*, vol. 10, no. 82, p. 91 (emphasis mine).

107. Ibid., no. 101, pp. 110–11 (emphasis mine). This experience consisted of a sphere reaching its equilibrium in between two immiscible liquids having specific weights respectively higher and lower than that of the sphere.

tonio de' Medici would use a similar expression to ask Galileo for one of his telescopes:

> I beg you . . . to make one and send it to me. It will be regarded by me as a most special favor, and I will demonstrate to Your Lordship how much I appreciate this demonstration of your love . . . *because, beside giving you back the necessary equivalent favor, I will feel eternally obliged to find the occasion to serve you.*[108]

As in the case of Antonio de' Medici, no important European aristocrat or upper clergy paid for the telescopes they received from Galileo; instead, they reminded Galileo of the favors (the countergift) he would receive. Andrea Labia, acting as a broker to Cardinal Borghese, who did not know Galileo personally, wrote to him that "You should know that such an instrument is so dear to him [Borghese] that if he receives one from you . . . he will not only acknowledge the favor in writing *but you will soon know how much such a gesture may be rewarding to you.*"[109] Similarly, Duke Paolo Giordano Orsini told Galileo that "Needing a telescope for my own pleasure . . . I desire one coming from the hands of Your Lordship; and I beg you to do me the favor of sending it to me as soon as possible. *But with equal promptness, I offer myself to Your Lordship on all occasions you may wish or need me.*"[110]

A few years later, Cardinal Alessandro d'Este sent Galileo his birth date and asked him to cast a horoscope. He concluded by reminding Galileo how the confidentiality of the matter testified to his trust in his virtue and that, when the time would come for him to "return [*contracambiarle*] this favor with anything you may desire," Galileo would realize how much the cardinal felt obliged to him.[111]

That these patrons did not pay Galileo for the telescopes they received from him should not be seen as a sign of their cheapness. Galileo himself did not want to be paid, as this would have demoted him from the rank of gentleman to that of artisan. When Philip IV, the king of Spain, requested (through an appropriately long chain of brokers) a telescope from Galileo and wanted to pay for it, Galileo sent one to Madrid but told the broker that "I never sold any of my instruments and I do not intend to do that, neither in the present nor in the future."[112] Given the status Galileo

---

108. Ibid., no. 238, p. 257 (emphasis mine).

109. Ibid., no. 320, p. 361 (emphasis mine).

110. *GO*, vol. 13, no. 1526, p. 91 (emphasis mine). See also no. 1527, p. 92.

111. *GO*, vol. 12, no. 1308, p. 375. For another case of *contracambio* between Galileo and Cardinal Farnese, see *GO*, vol. 10, no. 371, p. 411.

112. *GO*, vol. 14, no. 1967, pp. 52–53. This exchange was noted by Westfall in his "Galileo and the Jesuits," p. 35.

sought to achieve, it was better for him to look forward to a countergift than to have a possible patronage connection liquidated through a cash deal. Therefore, the *contracambio* mentioned by Sagredo, Don Antonio de' Medici, Labia, Duke Orsini, and Cardinal Farnese was neither a formality nor an exception but represented a fundamental feature of patronage—one that was quite independent from *what* was being exchanged.[113]

Sagredo knew, and Galileo frequently reminded him, of the duties he had contracted by entering into their friendship. Among the countergifts Galileo requested from him was an aggressive lobbying for the confirmation of Galileo's position at Padua and for recurrent salary increases. In 1599, a tired Sagredo, coming back from one of the many discussions he had with Venetian officials on behalf of Galileo, wrote that he had complied with the rules of friendship: "Since I have already satisfied abundantly enough the friendship I hold for you, *the obligations to you which I acknowledge,* and the favor and help that true gentlemen try to extend to the virtuosi who deserve it . . . I should finally desist and make sure that Your Most Excellent Lordship desist as well. . . ."[114] Therefore, even a friendship as informal and intimate as Sagredo and Galileo's was maintained through rituals of gift-exchange.

The Venetians' outrage at Galileo's departure from Padua can also be understood in terms of gift-exchange rituals or, more precisely, of Galileo's violation of a gift-related taboo. In August 1609 Galileo had presented the telescope *as a gift* to the Venetian Senate. In the letter of presentation to Doge Donato, he even relinquished his right to produce telescopes. But in the same letter Galileo also indicated that tenure at Padua would have been an appropriate countergift.[115] After obtaining the countergift he had asked for (and much more), Galileo was in a sense bound by his honor to remain a client of the Republic of Venice.[116]

113. For instance, Sharon Kettering has described comparable dynamics of gifts and counter-gifts in her study of seventeenth-century French political patronage ("Gift-Giving and Patronage in Early Modern France," *French History* 2 [1988]: 131–51), while Paula Findlen in "The Economy of Scientific Exchange in Early Modern Italy" documents patterns that are remarkably similar to this.

114. *GO,* vol. 10, no. 68, pp. 77–78 (emphasis mine). The ritual nature of this exchange has been noted by Richard Westfall in "Science and Patronage," p. 13.

115. ". . . if it will please God and Your Lordship that he, according to your desire, passed the rest of his life in the service of Your Lordship" (*GO,* vol. 10, no. 228, pp. 77–78).

116. The one-thousand-ducat salary the senate had promised to Galileo was remarkably high when we consider that in September 1614 the total budget of the studio (covering all the salaries and other expenses) was about 15,542 ducats (*Relazioni dei rettori veneti di terraferma,* vol. 4, *Podestaria e capitanato di Padova* [Milan: Giuffré, 1975]). However, Galileo's salary was not unique. His friend-enemy, the philosopher Cesare Cremonini, was the best-paid professor at Padua with two thousand ducats. Also, Galileo's new salary came with the clause that it could not be increased further.

By dedicating the satellites of Jupiter (something he had discovered with the instrument he had previously donated to the Senate) to the Medici and by leaving his generously rewarded and tenured position at Padua, Galileo broke the code of honor. He insulted the Venetians by sending back the generous gift he had asked for, *after having accepted it*. There is some evidence that he did something worse. He did not, as was required by the customary etiquette, provide a formal renunciation of the gift he had received—the privilege of life tenure at the University of Padua.[117] Therefore, it is not surprising that at the end of 1612 Sagredo could still write that "it is impossible to believe the disgust provoked by your departure, and in particular by the way in which people say you departed."[118] Formerly good friends of Galileo's like Venier felt so insulted as to threaten to break off relations with Sagredo if he kept on corresponding with Galileo.[119]

But if Galileo's violation of the ethics of gift-exchange marked the end of his patronage relationship with the Venetian republic, it also set the beginning of a new one, with Cosimo II. This new relationship was also developed around gift-giving rituals. Galileo paid all the expenses for the printing of the *Sidereus nuncius* and for the construction of several telescopes ("produced with great expense and labor") he gave to Cosimo and to various European princes and cardinals so that they could view the Medicean Stars.[120] It seems that Galileo tried to put Cosimo in his debt by distributing telescopes and copies of the *Sidereus* as gifts through the channels of Medici diplomacy, that is, by presenting them as coming from the Medici. Within his own limits, Galileo was trying to engage Cosimo in a potlatch.

Galileo mentioned monetary contributions in relation to a second and more luxurious edition of the *Sidereus nuncius* that was supposed to be

---

117. Not only was tenure a very rare privilege, but it also had an important symbolic meaning that went beyond that of a safe job. By giving it to Galileo, the Venetian Republic transformed him (a foreigner) into a lifelong client, almost a kinsman.

118. *GO*, vol. ii, no. 813, p. 447.

119. Ibid., no. 569, p. 172. When, more than a year later, Venier resumed his connection with Galileo, he wrote that Galileo's departure offended many people in Venice and Padua who thought that he should have acknowledged the unusual gift by accepting it or by "some other gesture," and that "those in power—who are very wise—do not talk about the matter [Galileo's departure] as if it were some insignificant event that happened in some remote country" (ibid., no. 591, pp. 215–16).

120. *GO*, vol. 10, no. 277, p. 298. Although the Medici gave Galileo two hundred scudi toward the expenses encountered in building additional telescopes and printing a new edition of the *Sidereus nuncius,* this contribution came *after* Galileo had been told of the grand duke's intention to offer him a position at court. Actually, Vinta tells Galileo of the two hundred scudi in the same letter in which he confirms the grand duke's intention to invite Galileo back to Florence with a suitable title and no teaching duties. In short, Galileo accepted money only after the potlatch was over (ibid., no. 311, p. 356).

in the Florentine vernacular rather than in Latin. He told Vinta: "I really think this second edition should reflect the greatness of the Patron rather than the weakness of the client."[121] Therefore, the Medicean Stars and the *Sidereus nuncius* were a gift to Cosimo in the most literal sense: initially Galileo did not get a penny for it. Moreover, he extended himself by giving presents to European royalty and upper clergy in the name of the Medici. The *Sidereus nuncius* was not a commissioned artwork. And precisely because it was a "pure" gift, it needed to be rewarded by the Medici with an adequate countergift.

Another gift marks the beginning of the patronage relationship between Galileo and Cosimo: the military compass and the related *Istruzioni*. As we have seen, it was the gift of the compass that allowed Galileo to get a hold on the prince by becoming his summer mathematics tutor. The Medici did not reciprocate Galileo's gift with cash, but with other gifts like black taffeta, a gold medal with a four-hundred-scudi gold chain, hospitality at court, gifts of food during the summer visits to teach Cosimo, help in improving Galileo's stipend at Padua, and a job for his brother-in-law, Benedetto Landucci.[122] Actually—as with the telescope for Philip IV—it was important for Galileo *not to be paid* during those summers. That allowed him to present himself as a gentleman. His patronage relationship with the Medici could be presented as voluntary, as based on a reciprocal exchange of gifts.

The two gift-exchanges crucial to the development of Galileo's patronage relationship with the Medici display a similar pattern. The Medici did not give a valuable object as a gift to Galileo in return for the compass and the dedication of its *Istruzioni*. Some black taffeta for a cloak could not match the value of Galileo's gift to the Medici. But they gave him a more prized gift, that of instructing the prince. Galileo's gift was reciprocated by letting him into a more personal patronage relationship. Something quite similar happened in 1610. Galileo offered the dedication of the

121. Ibid., no. 277, p. 299.

122. Ibid., no. 142, p. 161, and no. 295, p. 318. When Galileo was in Florence during the summer to teach mathematics to Prince Cosimo, he was not paid but hosted at court. When the Medici Maggiordomo, Giovanni del Maestro, invited Galileo to the Villa di Pratolino—one of the court's summer residences—in August 1605, he told him that Cristina offered him "a good room, a modest table, a good bed, and free candles," but he did not mention any monetary reward (ibid., no. 122, p. 146). That the Medici thought of hospitality as a gift—and therefore as a reward for Galileo's services—is confirmed by the fact that they would send him gifts of food when he was not staying at court. For instance, when in July 1605 the court was still in Florence and Galileo was staying with his brother-in-law, Cristina had Giovanni del Maestro send "to Signor Galilei at the house of Signor Benedetto Landucci, his brother-in-law, . . . 1 piece of veal, 2 capons, 6 chickens, 4 flasks of wine" (*ASF*, "Carte Strozziane," series 1, 30, fol. 134v.). On Cristina's intercession to secure Galileo's brother-in-law a job in the Medici administration, see *GO*, vol. 10, no. 205, p. 227.

Medicean Stars to Cosimo, who reciprocated the gift with a gold medal and chain of relatively small value compared to the value of the Stars.[123] But Cosimo's real reward was to let Galileo into a privileged patronage relationship by nominating him his chief mathematician and philosopher.

To judge from this evidence, the beginning of a permanent patronage relationship with a great patron was marked by the patron's *acknowledging but not reciprocating the client's gift*. In this way the patron accepted some sort of debt to be paid not *una tantum* but through a range of privileges distributed over time in some regular fashion—as through a salary. In this case, the client's gift acted as an investment. In fact, it seems that the most conspicuous gifts were given when the patron did *not* enter into a closer patronage relationship. For example, the several physicians summoned to Florence to give their opinions on the illnesses of Prince Don Carlo de' Medici in the fall of 1604 and of Cosimo II in 1614 were treated like foreign aristocrats while in Florence and then loaded with gifts (the usual gold chain among them) and sent home in Medici carriages.[124] But although Galileo's friend Fabricius of Acquapendente was richly rewarded by the Medici on both occasions, he did not receive the position of *Protomedico* at the Medici court in 1607 when he asked for it through Galileo.

Generous gifts seem to have characterized discontinuous patronage relationships. In fact, all the gifts given by the Medici to Galileo (before 1610) and to the various visiting physicians were of the kind called for by court etiquette regulating the *reception of visitors*.[125] I would go so far as to suggest that the Medici artistic workshops were largely gift factories charged with the production of elegant and precious artworks to be

123. Actually, it seems that what the grand duke gave Galileo was the honor of having the Medici artists produce such a medal (carrying Cosimo's effigy on one side and the Medicean Stars on the other) but that Galileo was to provide the gold for it (see *GO*, vol. 10, no. 326, p. 368). In a sense, Cosimo's gift to Galileo was similar to a great aristocrat's allowing his/her coat of arms to be placed on the client's house (at the client's expense). See also note 101 above.

124. *ASF*, "Diari di etichetta di guardaroba 1," fol. 180: "1 September 1604. Signor Fabricius of Acquapendente, physician from Padua to cure Prince Don Carlo, is [hosted] at our expense in the palace with three at his table and two servants, served by two of our footmen. The following day he was sent to lunch to Poggio [a Caiano] and to dinner at the Villa Ferdinanda. On the fourth he left to return to Padua with one of our litters and two pack mules and four coach horses, accompanied and paid for by Alessandro Berghi, groom of Madame, up to Padua, [and was] given as a gift a rich necklace, wool cloth, black satin, and other finery of good value." In 1614, Acquapendente was invited back to Florence with other doctors to heal Cosimo II. Their reception, treatment, and gifts were comparable to those given them in 1604 (*ASF*, "Miscellanea medicea 437," fols. 34–35).

125. See Fantoni, "Feticci di prestigio"; and Giacomina Calligaris, "Viaggiatori illustri ed ambasciatori stranieri alla corte sabauda nella prima metà del Seicento: Ospitalità e regali," *Studi piemontesi* (1975): 151–63.

distributed to European princes through the Medici diplomatic and political networks or during visits of dignitaries at the Florentine court.[126] Galileo's distribution of his many telescopes through the Medici diplomatic network fell, I think, in this same category.[127] Sporadic gifts from clients and countergifts from princely patrons constituted visit-related rituals. Galileo could be seen as a nomadic client who came to visit (and brings gifts to) the Medici once a year, during the summer. Similarly, the rich gift from the Gonzaga came on the occasion of Galileo's visit to Mantua.

The place of gifts within the logic of patronage explains the role of spectacular scientific production in Galileo's career. Galileo needed to produce or discover things that could be used as gifts for his patrons. The Medicean Stars are a perfect example, but their outstanding role in Galileo's career should not make us neglect the many other gifts he produced and distributed. His brokers at the Medici court kept reminding him to take "new things to show" whenever he visited Florence,[128] and Galileo tried to keep their attention by announcing in his letters some of the novelties he would bring.[129] The ball floating between two liquids requested by Antonio de' Medici, the armed lodestone for Cosimo, the various astronomical discoveries passed around in the form of enigmas (enigmas were literally considered as challenging gifts), the military compass, the mysteriously fluorescent "Bologna stone," his various books, the telescope, and the microscope could be effectively presented as gifts. Even Galileo's letters were sometimes received and circulated as gifts not unlike today's preprints and offprints.

More personal patronage relationships were reflected by less sporadic patterns of gift-exchange. In the case of court patronage, this happened when the client stopped being an occasional visitor and became a courtier—a member of the prince's *familia*. In these cases, the princely patron reciprocated the client's gift not with valuable objects but with a contract and a stipend. As suggested by Mascardi and Ciampoli, it was the client who, by giving important gifts, tried (not always successfully) to engage the patron in a long-term patronage relationship—a process resembling the sale of titles. Under certain conditions, such an option was probably best for both parties.

It may have been more convenient to the Medici to give Galileo the

126. Paola Barocchi, "Introduzione" to the reprint of Giovanni Maggi, *Bichierografia* (Florence: SPES, 1977), 1: pp. i–xiv; Sir Robert Dallington, *Descrizione dello Stato del Granduca di Toscana nell'Anno di Nostro Signore 1596* (Florence: All'Insegna del Giglio, 1983), p. 70.

127. Westfall, "Galileo and the Jesuits," in *Essays on the Trial of Galileo* (Vatican City: Vatican Observatory, 1989), p. 35.

128. *GO*, vol. 10, no. 73, p. 84; no. 119, pp. 142–43.

129. For instance: ". . . and when I will come there this June I will carry to the Grand Duke things of infinite amazement regarding this matter" (ibid., no. 277, p. 302); and "[dur-

much-desired title of philosopher and a thousand scudi a year than to pay him back with a gift. What kind of gift could it have been? All the Europe that counted knew of Galileo's discoveries and that he had dedicated them to the Medici. And many important personages had been informed of this directly by the telescopes and copies of the *Sidereus* that Galileo had sent as gifts through the Medici diplomatic networks. In a sense, Galileo had slowly enticed the Medici in a controlled potlatch. Once he had acknowledge Galileo's dedication of the Medicean Stars, Cosimo's generosity and nobility had come under the scrutiny of all those kings, queens, dukes, and cardinals who had been given gifts of telescopes by Galileo, and it would have been difficult for him to come up with an adequate countergift.

Although this type of patronage played a crucial role in Galileo's career by securing him the position and title of court philosopher (a crucial resource for the legitimation of Copernicanism and mathematical physics), only a few clients could have access to such patronage. Because of the peculiar status and image of the absolute prince, his patronage was not only difficult to get but one also needed to play by somewhat different rules in getting it. The crucial difference was that the absolute prince did not need to acknowledge his subjects' gifts.

In his 1624 court treatise, Matteo Pellegrini discusses an intriguing feature of this peculiar form of gift-exchange, one that provides a key to the legitimation of political absolutism and to the patronage dynamics of a baroque court:

> To the eminence of Majesty is not attached any obligation toward subjects. What the great princes receive is called favor. Their affection toward their subjects is called *gratia* not because they are grateful to their subjects, but because it is something that is delivered freely and not out of duty. The powerful can ignore the sweat of thousands without fearing to be called unjust.[130]

Because of his recognized absolute power, the prince is potentially able to step out of any potlatch without staining his honor. If the prince decides to go along with the game, he can represent himself as doing so only by free choice. In a sense, the absolute prince is able to represent himself as somebody who has won all potlatches (since he is a prince, this might be historically plausible) so that it would be unthinkable that he would not win the next one also. The legitimacy of absolute power (vis-à-vis his subjects) is rooted in this subtle but radical form of induction: That he is previously undefeated is turned into his ontological undefeatability. This was

---

ing the next visit to Florence] I will have with me some improvements of the telescope and perhaps some other invention" (ibid., no. 257, p. 271).

130. Pellegrini, *Che al savio è convenevole il corteggiare libri IIII*, pp. 2–3.

the quantum leap that differentiated a normal patron from an absolute prince.[131]

It is only on this assumption that the prince can turn down challenges without fear of stain. He can either ignore them in a condescending fashion or represent them as insults. Interestingly enough, with the development of political absolutism, authors writing on duels begin to argue that challenges to a prince are no longer legitimate.[132] The prince is no longer the first among the nobles. He is no longer bound to the code of knightly honor; he is above it. To challenge him is no longer a noble gesture but becomes a crime of *lèse-majesté*.[133] Such a representation of the unique condition of the absolute prince may be the reflection at the level of the discourse of power of a specific sociohistorical change. As the challenge of the great feudal lords and other magnates to the process of political centralization had been defeated either on the battlefield or through duels of conspicuous expenditure at court, it was ruled out also at the level of court discourse. Any type of resistance to the absolute ruler was "criminalized" by the discourse of the absolutist court.

Another fundamental inversion is closely connected to this discursive turn. Although the prince's power is crucially rooted in the "gifts" he gets from subjects, the direction of the debt is turned upside down so that the prince is represented as owing nothing to the subjects while they owe everything to him.[134] Subjects are obliged to give to their absolute prince, but, as noted by Pellegrini, these gifts cannot be thought of as challenges

131. In ways that I find difficult to articulate, these and other considerations on the discourse of the absolute prince have been informed by Louis Marin's *Portrait of the King* (Minneapolis: Minnesota University Press, 1988).

132. Giancarlo Angelozzi, "Cultura dell'onore, codici di comportamento nobiliari e stato nella Bologna pontificia: Un'ipotesi di lavoro," *Annali dell'Istituto Storico Italo-Germanico in Trento* 8 (1982): 314–15.

133. Also, one of the recurrent arguments about duels in general is that they were no longer expressions of the aristocratic code of honor (one to which the prince would have been bound in a previous period), but instead became affronts to the power of the absolutist prince and state (Richard Herr, "Honor Versus Absolutism: Richelieu's Fight Against Dueling," *The Journal of Modern History* 27 [1955]: 281–85). See also Francis Bacon's views on this question in François Billacois, *The Duel* (New Haven: Yale University Press, 1990), p. 32.

134. This intriguing anomaly in the gift-exchange between the absolute prince and his subject allows for another interesting inversion: the prince maintained his power image by being in debt (of gifts) to those whom he accepted as clients, and by offering them in exchange the possibility of long-term privileges. This scenario recalls—*mutatis mutandis*—that of a modern state running on public debt yet maintaining its sovereignty. After having given enough to (or invested enough in) the prince, the client was given an annual "dividend" and/or a title. For interesting reflections on the symbolic dimensions of a similar process of delegation and empowerment, see Pierre Bourdieu, "Delegation and Political Fetishism," *Language and Symbolic Power* (Cambridge, Mass.: Harvard University Press, 1991), pp. 203–19.

(that is, something deserving responses),[135] but are to be considered only as one-way tributes:

> The great princes act as if they had everything. What other people do for them is not called *beneficio* but dutiful obligation. To acknowledge it is a sign of *gratia*, not of debt. Private citizens are generous when they give; princes are generous also when they accept. The greater the importance of the services offered to the great princes, the greater the gratia they return, if they appreciate them. . . . Oh ungrateful fate! To serve princes is to oblige oneself to them. The generosity of private individuals toward the princes obliges only those who give—although the princes may complacently accept their generosity.[136]

The absolute prince resembles a god. More precisely, a Protestant god; one who is infinitely powerful and great so that his terrestrial clients cannot expect him to react to what they are doing to please him. This may explain why Pellegrini and many other writers of treatises on the court used the term *gratia* to describe the attributes of the perfect courtier. In fact, for a courtier to receive a reward from his prince was structurally analogous to a Christian's receiving grace from God—the similarity being that, in both cases, one's actions could not guarantee the achievement of grace. The only thing a courtier could do was to behave *as if* he had grace—to show through exterior signs that he had been "chosen."[137] That is what courtly nonchalance (*sprezzatura*) is about. In any case, the courtier's "salvation" depended solely on the prince.

As discussed by Louis Marin in a different context, the use of religious metaphors within the discourse of a baroque court suggests that the legitimacy of the power of the absolute ruler was not explained away but rather turned into a "mystery"—a mystery of state.[138] To ask why the prince was absolutely powerful, although his power derived directly from

135. Pellegrini differentiates these gifts from those the prince receives from the subjects by calling them "favors."

136. Pellegrini, *Che al savio è convenevole il corteggiare libri IIII*, pp. 27–28. Pellegrini draws another important difference between normal clients and patrons, and absolute princes. The princes cannot be "honored" by their subjects. In terms of status, princes are incommensurable with clients or lower patrons. Subjects can only *serve* their absolute patrons. The prince *honors* (by giving *favors*) but does not serve (ibid., p. 32).

137. Obviously, this is a very powerful tool of social and cultural control for the prince. It allows him to foster conformity without having necessarily to reward it.

138. Marin, *Portrait of the King;* and Ernst Kantorowicz, "Mysteries of State: An Absolutist Concept and its Late Medieval Origins," *The Harvard Theological Review* 48 (1955): 65–91. Of course, this was the case with official or ceremonial representations. Diaries of courtiers or reports of political journalists presented a much less divine picture of the prince's power.

his subjects' gifts to him, was met in the same way that a religious person would handle a query about the nature of God: it is a mystery beyond our comprehension. If a challenge to the absolute prince becomes a crime of *lèse-majesté,* queries about the sources of his power are deflected by representing them as pointing to a sacred mystery. It is not the prince's problem if the subjects cannot understand why he is absolutely powerful. All subjects can do is to feel insignificant as they contemplate this awesome mystery. If they do not, then they are heretics, criminals like those who dare to challenge the prince's power.

This, I think, provides an interesting angle from which to view the many but always elusive meanings of courtly *sprezzatura* in contemporary court treatises. I would suggest that *sprezzatura,* the "essence" of courtliness, had to be presented as something ultimately defying verbal description. It had to remain an opaque notion because it epitomized the very "mystery" of the prince's power and of the courtiers' identity as shaped by it. In fact, the prince remained powerful insofar as he managed to avoid probes into the legitimacy of his power. His power was effective only insofar as its nature and legitimacy were not probed but only acted out in court behavior and culture. Thus, both the opacity of *sprezzatura* and the "mystery" of the source of the prince's power were aimed at shielding his power from being discursively probed.

As we will see, the power discourse of political absolutism had fundamental consequences for the type of tactics Galileo could develop as he legitimized his scientific discoveries and socioprofessional identity by tapping into the power discourse of the Medici. In particular, Galileo's puzzling effacement of himself as the author of the exceptional discovery of the satellites of Jupiter, in the dedication of his *Sidereus nuncius* to Cosimo II, may reflect the structure of this discourse. Galileo's representing himself not as an author but as a mere *trait d'union* between the Medicean Stars (which, according to him, had been always in the sky and always belonged to the Medici) may indicate that the only way a client could entice an absolute prince was by giving him something great while pretending to not ask anything back. As noted by Pellegrini, anything one might give to the great prince was already necessarily his, or due to him. Great princes behaved as if they had everything. Consequently, one could not say that what was being donated to the absolute prince was something the client had produced on his or her own.

This discourse of self-effacement was not peculiar to Galileo. In dedicating the *Saggi di naturali esperienze* to Grand Duke Ferdinand II in 1667, the members of the Accademia del Cimento told him that

> we should wish only to be able to offer you something that is not your own, so that we might at least flatter ourselves that we had

brought you some slight recompense and expressed some thanks of our own choice to Your Highness that would not be entirely your own or from necessity. But now we are perforce content to have in our hearts such just and proper sentiments, since the fruit of these new philosophical speculations is so strongly rooted in Your Highness' protection that not only what our Academy produces today but everything that matures in the most famous schools of Europe . . . will likewise properly be due to Your Highness as the gift of your beneficence.[139]

Had Galileo (or the Cimento) emphasized his authorship or his role as exceptional discoverer, his gift might have backfired. Paradoxically, one had to donate exceptional gifts to an absolute prince in order to be accepted as a distinguished client, and yet one could not present his exceptional gifts as really coming from him because that gesture might be read by the prince as an unacceptable "challenge," thereby jeopardizing the client's access to patronage, legitimation, and credibility. The client would appear to be an insignificant nobody whose out-of-control ego confused him to the point of forgetting that nobody could really present the prince with anything that was not already his.[140] Therefore, within court patronage, one could gain legitimation as a scientific author only by effacing one's individual authorial voice. To be a legitimate author meant to represent oneself as an "agent" (maybe a "prophet") of the prince.

It required great skill to walk the tightrope between effacing oneself as an author and, at the same time, intimating that one's work and discoveries were exceptional gifts worthy of the appreciation and acknowledgment of an absolute prince such as Cosimo. But even if Galileo displayed appreciable skill in court acrobatics, the discourse of the absolute prince was a mixed blessing for such a scientist. While providing a type of social and cognitive legitimation that no other source of authority could confer, the discourse posed important constraints on the ways in which the identity of the scientific author could be fashioned. As Tasso put it in his dialogue on the court, the courtier's honor and reputation depended on the prince as completely as a river's existence depended on the spring.[141]

139. *Saggi di naturali esperienze fatte nell'Accademia del Cimento* (Florence: Cocchini, 1667), pp. 3–4, as translated in W. E. Knowles Middleton, *The Experimenters* (Baltimore: Johns Hopkins University Press, 1971), p. 87.

140. Although Kepler dedicated his *On the Six-Cornered Snowflake* not to Emperor Rudolph II but to his advisor, John Matthew Wacker, and did not deny that what he had discovered was made possible by his patron, he also emphasized the "nothingness" (and extreme ephemeralness) of his gift. As he put it, "I can readily guess that the closer a gift comes to Nothing the more welcome and acceptable it will be to you" (Johannes Kepler, *On the Six-Cornered Snowflake* [Oxford: Clarendon Press, 1966], p. 3).

141. Tasso, *Il malpiglio*, 13.

In other words, the baroque court was no place for modern individualism and authorship.[142] If we extend Pellegrini's remarks from the sphere of power to that of authorship, it becomes apparent that there was only one ultimate author at court: the prince.

Patronage and Scientific Networks

The earliest known letter by Galileo dates from January 8, 1588. It is addressed to Christopher Clavius, the chief mathematician at the Jesuits' Collegio Romano, and was delivered by Cosimo Concini, a Florentine patron of Galileo. Concini was the son and nephew of two Medici "Primi Segretari" and the brother of Concino Concini, the favorite of Queen Maria de' Medici. In 1588 Concini was a young Church official of Pope Clement VIII. Galileo told Clavius that by giving his reply to Concini, not only could he be sure that the letter would reach him safely in Florence, but he would also increase Galileo's credit with his patron. In fact, in receiving a letter for Galileo from a high-ranking Jesuit, Concini would realise the quality of his client's connections.[143] Although Clavius did not send his answer back to Florence by Concini, he wrote to Galileo that he would make sure to tell Concini of their friendship.[144] Even this earliest of documents, a simple exchange of letters on the center of gravity of solids, involved interactions between clients and patrons, or among different patrons, with the exchange of status-signs.[145]

Such an example is by no means an isolated one. In May 1600, the same Concini—by then the Medici ambassador in Prague at the court of

142. I want to stress the historical specificity of this process. The author's predicament seemed different in the Renaissance court, where (as shown by Michelangelo's ability to maintain a close relationship with his intellectual property) a few exceptional clients did not need to efface themselves rhetorically in order to be represented as legitimate authors.

143. *GO*, vol. 10, no. 8, pp. 22–23. On Cosimo Concini, see Paolo Malanima, "Cosimo Concini," *Dizionario biografico degli italiani*, 27: 730–31.

144. *GO*, vol. 10, no. 9, pp. 24–25.

145. Why would Clavius choose to be so nice with Galileo? Here is a tentative interpretation of the "microphysics of patronage" involved in the exchange: Concini is a fairly important Church official. Therefore, if Concini is "impressed" by the fact that Galileo is a friend of Clavius's, it means that he acknowledges Clavius's high status. And Concini's very act of recognizing Clavius's high status confirms it (in Concini's eyes). This, I think, is Clavius's "gain" in the symbolic exchange. But Concini obtains his share too. In fact, his own status is confirmed (or even improved) by the fact that a high-status person like Clavius explicitly acknowledges his friendship for one of Concini's clients (i.e., one who has less status than Concini). The result of this symbolic exchange is that Galileo's status is at least confirmed in relation to Clavius, and certainly improved in relation to Concini. This "gain" by Galileo has to do with the content of his letter to Clavius. In addition to the technical content (it is a theorem on the center of gravity of bodies obtained from the rotation of conic sections), Galileo refers to Clavius as a judge to whose decision he submits himself voluntarily. This act

Rudolph II—spoke to Tycho Brahe about Galileo, a brilliant but quite unknown Italian mathematician. Although at that time Tycho was interested in establishing contacts with Italian mathematicians,[146] Concini must have been effective at singing Galileo's praises if an arrogant aristocrat like Tycho decided to write Galileo to "establish the foundation of our friendship."[147] More than just a contact among astronomers was being attempted here. Galileo did not ask his patron Concini to introduce him to Tycho; it was probably Concini who showed off to Tycho by talking about the great young mathematician who was his client.

Galileo's correspondence offers several other examples in which patrons played an active role in establishing or maintaining communication among scientists. Scientific contacts were presented as part of a more general process of status enhancement which worked for both clients and patrons. And, as in the case of Concini at Prague, Galileo was not necessarily the engine behind these strategies. Contacts were often established by the patrons allegedly in the name of the clients but actually for the sake of their own image. As noted earlier, brokers pushed their clients in order to maintain or extend their networks.

Scientific disputes were initiated and managed by patrons for similar purposes. In early modern Europe (and in other non-European societies as well) to be worthy of being challenged was a sign of nobility.[148] Challenges were forms of gifts and vice versa. Although it was better to win a duel than to lose it, the very fact of being challenged was important per se, for it implied that one's status was recognized. It was honorable to die in a duel. Consequently, duels (and scientific disputes) were part of a social

---

of voluntary submission is a gift, and Clavius reciprocates by helping Galileo's image with Concini.

146. Around this time Tycho was looking for someone to write his own celebrative biography, probably to be used to increase his standing with Rudolph II (Stillman Drake, *Galileo at Work* [Chicago: University of Chicago Press, 1978], p. 50). This may have been one of the reasons for Tycho to accept Concini's suggestion and seek Galileo's friendship.

147. *GO,* vol. 10, no. 70, p. 79.

148. Pierre Bourdieu, "The Sentiment of Honour in Kabyle Society," in J. G. Peristiany, ed., *Honour and Shame* (Chicago: University of Chicago Press, 1966), pp. 191–241; and Pierre Bourdieu, *Outline of a Theory of Practice* (Cambridge: Cambridge University Press, 1977), pp. 1–29. I think that Clifford Geertz's analysis of the social significance of the cockfight in Bali indicates that betting is a kind of duel (in which the cock, rather than a person, gets killed) that is routinely needed to maintain the status pattern of the community. What is relevant to my point is Geertz's claim that it is not really important to win (also because bets are usually almost even), but it is necessary to show publicly that one bets, that one "accepts the challenge" (i.e., that one defines himself as "challengeable") (Clifford Geertz, "Deep Play: Notes on the Balinese Cockfight," *The Interpretation of Cultures* [New York: Basic Books, 1973], pp. 412–53).

economy of honor and status.[149] And the line separating scientific contacts from challenges (i.e., disputes) could be very thin.[150]

Another example of a patron-controlled attempt to initiate scientific
dialogue comes from Sagredo—Galileo's major patrician patron during
the Paduan period. On December 20, 1602, he wrote to Galileo that the
Venetian Senate (in which the Sagredo family was represented) was sending an official to England. Sagredo was planning to send a letter to
William Gilbert, the famous natural philosopher, through the Venetian
official and asked Galileo if he had any question about Gilbert's *De magnete* he would like to include in the letter. Galileo's contribution would
have been particularly welcome, for Sagredo admitted that he was not
very familiar with the *De magnete*.[151] Therefore, in trying to put Galileo in
touch with Gilbert, Sagredo was actually trying to show off through the
skills of his client. It worked. In February 1603, Gilbert wrote to his friend
Barlow that:

> There is heere a wiselearned man, a Secretary of Venice, he came sent
> by that State, and was honourably received by her Majesty, he
> brought me a lattin letter from a Gentleman of Venice that is very
> learned, whose name is Johannes Franciscus Sagredus; he is a great
> magneticall man, and writeth that hee has conferred with divers
> learned men in Venice, and with the Readers of Padua.[152]

The role of the Medici ambassadors in distributing copies of the *Sidereus nuncius* and telescopes all over Europe is a further example of how
patronage helped in developing scientific communication. In general,
Galileo's books reached ambassadors, princes, and cardinals and then they
were passed to the mathematicians—usually with a request for their opinion. Kepler received the *Sidereus Nuncius* from the Medici ambassador in
Prague, Johannes Zugmann read the copy of his patron—the elector of
Cologne—and the astronomer Ilario Altobelli received Cardinal Conti's
copy.[153] Similar considerations apply to the other texts of Galileo. In 1619

149. In addition to the works on duels already cited, see Francesco Erspamer, *La biblioteca di Don Ferrante: Duello e onore nella cultura del Cinquecento* (Rome: Bulzoni, 1982).

150. On this issue, see Steven Shapin, "A Scholar and a Gentleman," *History of Science,* 29
(1991): 279–327; idem, *A Social History of Truth,* forthcoming; and Mario Biagioli, "Scientific
Revolution, Social Bricolage, and Etiquette," in Roy Porter and Mikulas Teich, eds., *The
Scientific Revolution in National Context* (Cambridge University Press, 1992), pp. 11–54.

151. *GO,* vol. 10, no. 89, p. 101.

152. Quoted in Antonio Favaro, "Adversaria galileiana, serie quarta: Giovanfrancesco
Sagredo e Guglielmo Gilbert," *Atti e memorie della R. Accademia di Scienze Lettere ed Arti in
Padova,* new series, 35 (1918–19): 12–15. Similar information is presented in Edgar Zilsel, "Or-
igins of Gilbert's Scientific Method," in P. P. Wiener and A. Noland, eds., *Roots of Scientific
Thought* (New York: Basic Books, 1957), p. 247, note 36.

153. See *GO,* vol. 10, no. 296, pp. 318–19 for Kepler's receipt of the *Sidereus nuncius* from
the Medici ambassador in Prague, Giuliano de'Medici. For Zugmann's receipt of a copy

the Jesuit Christopher Scheiner received a copy of Galileo and Guiducci's *Discourse on Comets* through an important patron of Galileo's, Leopold of Austria.[154] The distribution of Galileo's telescopes followed a similar pattern. As in the case of Kepler, they did not reach the astronomers directly but, in most cases, went to their aristocratic patrons first.[155]

Kepler's dedication of his *Conversation with the Sidereal Messenger* to Giuliano de' Medici (the Medici ambassador in Prague after Concini's transfer to the Spanish court) offers interesting clues about the ways in which scientific networks were often embedded in noble patronage networks. Kepler acknowledged that he obtained a copy of the *Sidereus* from Giuliano de' Medici and that, when called to the Medici palace in Prague on April 13, he was read Galileo's invitation to respond to the *Sidereus,* an invitation which was reinforced by the ambassador's "own exhortation."[156] It is important to note that Kepler did not receive the letter from Galileo but that it was *read to him* by the Medici ambassador. Therefore it was something in between a private communication between Galileo and Kepler and an official request from the Medici to the imperial mathematician. That the dialogue between Galileo and Kepler was being mediated and legitimized by the relationship between the Medici and the Hapsburg is further stressed by Kepler's reference to Galileo being "in the employ of the Medici" and by Kepler's dedication of his *Conversation* to "the ambassador of Prince Medici, Grand Duke of Tuscany, himself a Medici by birth," who had "sought this service from me."[157] Similarly, when in the spring of 1610 Galileo used Kepler's *Conversation* as a proof of the international recognition of his discoveries, to remove the doubts subtly conveyed to Galileo by the Medici through Vinta, he did not refer to Kepler by name but called him the "Mathematician to the Emperor."[158]

from his patron, the elector of Cologne, see ibid., no. 303, pp. 344–45. For Altobelli, see ibid., no. 294, p. 317.

154. *GO,* vol. 12, no. 1418, p. 489. It was through Leopold that Galileo received Remo's treatise on comets, and it was again through Leopold that he requested and obtained a copy of Kepler's *Epitome*—a book that was prohibited in Italy after the 1616 edict (ibid., no. 1403, p. 469; no. 1413, p. 481; no. 1417, pp. 484, 488).

155. On Kepler's using the telescope sent by Galileo to the elector of Cologne, see *GO,* vol. 10, no. 386, p. 427. Later on, he began to use the emperor's better telescope. This pattern of distribution had also been noted by Richard Westfall in "Galileo and the Jesuits," p. 35, and by Albert Van Helden in his translation of Galileo's *Sidereus nuncius* (Chicago: University of Chicago Press, 1989), p. 92.

156. Edward Rosen, ed., *Kepler's Conversation with Galileo's Sidereal Messenger* (New York: Johnson, 1965), pp. 3–4. It is interesting that at that time Galileo—although an official client of the Medici—was not yet in their service, as Kepler instead claims in this text. This may be a sign of the success of Galileo's attempts at officializing himself by using the Medici networks to distribute telescopes and copies of the *Sidereus nuncius.*

157. Ibid., 4.

158. *GO,* vol. 10, no. 307, p. 349. Vinta too referred to Kepler not by name but as "Mathematician of the Emperor" when he discussed with Galileo the necessity of Galileo's trip to

The formality displayed on both sides testifies that Cosimo II and Rudolph II were involved in the exchange between Galileo and Kepler. Moreover, from a letter of Martinus Hasdale to Galileo we see that Kepler was also asked by the emperor himself to express his opinion on the *Sidereus nuncius* (which Rudolph had received from the Medici ambassador).[159] A somewhat similar pattern emerged in December 1618 when the king of France asked his mathematician, Jacques Aleaume, to conduct observations of a then visible comet. Aleaume replied that his telescope was not good enough for the task and that *the king should ask the grand duke* because it was only he, through Galileo, who could answer his queries.[160]

Kepler and Galileo communicated as clients (and therefore as representatives) of the emperor and the grand duke and not only as independent scientists. Also, Galileo did not just write to Gilbert. His message, whatever it may have been, reached him through Sagredo and the Venetian Senate. In short, Galileo's and Kepler's exchanges were legitimized (and sometimes fostered) by their patrons.[161] Similarly, as we will see in the debate on the sunspots, Galileo and Scheiner did not communicate directly but rather through two high aristocrats, Welser and, to a lesser extent, Cesi. Evidently, Concini, Giuliano de' Medici, Sagredo, Cesi, and Welser were not mailmen, and Galileo did not use preexisting diplomatic or aristocratic communication networks simply because they were practically convenient.

In strongly hierarchical and status-bound societies such as those of early modern Europe we cannot draw the line between status and credibility. The use of diplomatic connections—of diplomats who partook of the status of the prince they were representing—gave Galileo credibility.[162]

Rome early in 1611 to obtain the final legitimation of his discoveries. Vinta realized that it was not so much Kepler's personal expertise but rather his title to give Galileo (and the Medicean Stars) credibility (*GO,* vol. 11, no. 464, p. 28). Similarly, Vinta saw the Jesuits' credibility as directly linked to their status as the pope's "Keplers." Consequently, he saw Galileo's trip to Rome as a way not to quench possible rumors about the heretic implications of Galileo's discoveries, but simply to legitimize their empirical status (ibid., no. 464, pp. 28–29).

159. *GO,* vol. 10, no. 291, p. 314.

160. *GO,* vol. 12, no. 1362, p. 428.

161. On the role of the prince in fostering scientific communication, see also Bruce Moran, "Science at the Court of Hesse-Kassel: Informal Communication, Collaboration, and the Role of the Prince Practitioner in the Sixteenth Century" (Ph.D. diss., University of California, Los Angeles, 1978); idem, "Wilhelm IV of Hesse-Kassel: Information Communication and the Aristocratic Context of Discovery," in Thomas Nickles, ed., *Scientific Discovery: Case Studies* (Dordrecht: Reidel, 1980), pp. 67–96; and idem, "Privilege, Communication, and Chemistry: The Hermetic-Alchemical Circle of Moritz of Hesse-Kassel," *Ambix* 32 (1985): 110–26.

162. *GO,* vol. 10, no. 277, pp. 298–99, 301; no. 284, p. 308.

And it was through patronage that Galileo had access to Concini, Sagredo, and Giuliano de' Medici. If it is a bit naive to consider scientific credibility as related only to peers' recognition, even in modern science, such a view is seriously misleading when used to interpret the construction of scientific credit and legitimation in early modern science. I think it would be useful to suspend for a moment the "natural" belief that Galileo, Kepler, and Clavius earned their titles simply because of the quality of their scientific work, and to consider, instead, that they also gained scientific credibility because of the titles and patrons they had. This link between credibility and status, or patronage connections, is supported by Tycho's stressing the credibility of Christoph Rothmann (whose observations of the changes in the latitude of stars Tycho cited as confirming his own) by referring to him as the "*landgrave's* mathematician."[163]

It would also be simplistic to treat these patronage networks as no more than "resources" to be tapped into by clever clients in pursuit of their goals. As the evidence suggests, clients were mobilized and fashioned by these networks as much as they themselves mobilized the networks. Recently, thanks especially to the work of Bruno Latour and Michel Callon, networks centered on laboratories have emerged as an important interpretive category for the dynamics of modern science.[164] Although my analysis of networks differs from Latour's in that it does not treat them simply as resources but as institutions of self-fashioning, it shares Latour's aim of moving away from group-based views of scientific activity and toward the understanding of scientific credibility as connected also to one's position within power/knowledge networks. In particular, this analysis of networks of patronage and correspondence suggests that, in an age devoid of laboratories and scientific institutions (except museums, botanical gardens, and theaters of anatomy) scientific networks were modulated on patronage and diplomatic networks. The power nodes of those networks were not centered on laboratories but on courts or aristocratic patrons, like Cesi, Welser, or Peiresc.[165]

---

163. Victor E. Thoren, *The Lord of Uraniborg* (Cambridge: Cambridge University Press, 1990), p. 293 and note 113 (emphasis in the original).

164. Bruno Latour, *Science in Action* (Cambridge, Mass.: Harvard University Press, 1987), esp. pp. 179–257.

165. For a "Latourian" analysis of networks of correspondence in seventeenth-century England, see Robert Iliffe, "'In the Warehouse': Privacy, Property and Priority in the Early Royal Society," *History of Science* 30 (1992):29–68, and idem, "Author-Mongering: the 'Editor' Between Producer and Consumer" (forthcoming). On Italian sixteenth- and early seventeenth-century letter-writing practices and volumes of letters, see Amedeo Quondam, ed., *Le "Carte Messaggiere"* (Rome: Bulzoni, 1981). See also Lisa T. Sarasohn, "Nicolas–Claude Fabri de Peiresc and the Patronage of New Science in the Seventeenth Century," *Isis* 84 (1993; in press).

Patronage and the Etiquette of Scientific Disputes

The relationship between social status, honor, and credibility described above informs the dynamics of early modern scientific disputes which, in fact, resembled duels.[166] This claim does not rest only on the evidence provided by Galileo's career. For instance, the bitter dispute between Tycho and Ursus over the authorship of the so-called Tychonic planetary model, indicates that Tycho did not perceive clear boundaries between personal challenges and scientific disputes.[167]

Tycho did not see the matter as what we would call a priority dispute but as a plain insult to be dealt with according to the code of honor to which he was bound as an aristocrat. Had Ursus been of noble rather than peasant origins, Tycho would have probably tried to settle the matter with a duel. In this case, as in the later dispute between Galileo and Capra over the invention of the military compass, or during the exchange of challenges (*cartelli*) between Tartaglia and Ferrari, "honor" rather than "scientific credibility" were presented as being at stake.[168]

Kepler—who happened to have been embarrassingly implicated in the dispute by Ursus' publication of a flattering letter Kepler had sent to him years before—was bullied by his patron, Tycho, into refuting Ursus' claims to restore Tycho's honor. At first, Kepler tried to avoid the unpleasant task by telling Tycho that his nobility would be better displayed by ignoring the matter altogether.[169] Although Kepler's argument may have reflected an opportunistic attitude, it was perfectly acceptable. In fact, ac-

166. For a later example of honor-based scientific disputes, see David Harley, "Honour and Property: The Structure of Professional Disputes in Eighteenth-Century English Medicine," in Andrew Cunningham and Roger French, eds., *The Medical Enlightenment of the Eighteenth Century* (Cambridge: Cambridge University Press, 1990), pp. 138–64.

167. A detailed analysis of the dispute has already been presented by Nicholas Jardine and Edward Rosen, and important new evidence has been recently uncovered by Owen Gingerich and Robert Westman. See Nicholas Jardine, *The Birth of History and Philosophy of Science* (Cambridge: Cambridge University Press, 1984); Edward Rosen, *Three Imperial Mathematicians* (New York: Abaris Books, 1986); and Owen Gingerich and Robert Westman, "The Wittich Connection: Conflict and Priority in Late Sixteenth-Century Cosmology," *Transactions of the American Philosophical Society* 78 (1988), part 7.

168. In Galileo's correspondence, *honore* is systematically used to designate what we may call both scientific credibility and honor, indicating that the scientist's distinct socioprofessional identity had not yet developed. See, for instance, *GO*, vol. 10, no. 23, p. 9. Reference to *honore* (sometimes used interchangeably with *fama*) became quite frequent during Galileo's dispute against Capra (ibid., no. 154, p. 172; no. 156, p. 174; no. 160, pp. 177–78; no. 162, p. 179). See also Enrico Giordani, ed., *I sei cartelli di matematica disfida di Lodovico Ferrari coi sei contro-cartelli in risposta di Nicolò Tartaglia* (Milan: Luigi Ronchi, 1876).

169. Kepler wrote Maestlin about the Tycho-Ursus dispute, saying that "it does not seem worthy of Tycho's stature to be so violently upset by this disparagement" (Jardine, *Birth of History and Philosophy of Science,* 19). Tycho did not agree with Kepler: "Nor it is true that I fell more strongly about this silly man than my status allows" (ibid., 23).

cording to contemporary codes of honor, any form of answer from Tycho's side would have been perceived as an implicit acceptance of Ursus' challenge—a recognition he did not deserve.

Beside entrusting Kepler with the restoration of his social and astronomical honor, Tycho tried to have Ursus prosecuted legally. However, he did not perceive him as guilty of a mere violation of what we would call intellectual property. Because Ursus had stained his honor, Tycho seems to have thought that Ursus should have been sentenced to death.[170] He convinced Emperor Rudolph II (whose mathematician Tycho was) to set up a committee of two jurists and two barons (rather than two mathematicians) to investigate Ursus' crime. Unfortunately, we cannot know what the sentence would have been, for Ursus died before the end of the investigation. However, his books were sequestered and publicly burned.

Tycho's example cannot be easily dismissed by attributing his behavior to his well-known hot temper and penchant for duels.[171] The exchange, in 1547–48, of the six *cartelli di matematica disfida* between the mathematicians Niccolò Tartaglia and Girolamo Cardano through Ludovico Ferrari resembles the Ursus-Tycho dispute. Because Tartaglia saw his honor stained by Cardano's not reciprocating his gift of the solution of third-degree equations, he attacked him in his 1546 *Quesiti et inventioni diverse*. Cardano, being of a much higher socioprofessional status, did not accept the challenge but passed it on to his "Kepler," that is, to Ferrari.[172] In his first *cartello*, Ferrari is explicit about his duties as a client to save his patron's honor: "I have decided to expose publicly your deceit or, rather, your malignant nature, not only to defend the truth, but also because this is my duty as his client, his Excellency being restrained by his status."[173] The homology between mathematical disputes and duels is strikingly confirmed by the name of one of Ferrari's witnesses who signed his first *cartello*: Mutio Iustinopolitano.[174] This is the nickname of Girolamo Muzio, one of the foremost Italian experts on duels and questions

170. Ibid., note 47.

171. As is well known, Tycho lost a good portion of his nose during a duel in December 1566 (Thoren, *Lord of Uraniborg*, 22–24).

172. Biagioli, "Social Status of Italian Mathematicians," 55; Ettore Bortolotti, "I cartelli di matematica disfida e la personalità psichica e morale del Cardano," *Studi e ricerche sulla storia della matematica in Italia nei secoli XVI e XVII* (Bologna: Zanichelli, 1944); and idem, "Le matematiche disfide e la importanza che esse ebbero nella storia delle scienze," *Atti della Società Italiana per il Progresso della Scienza* 15 (1927): 163–80.

173. (Giordani, *I sei cartelli di matematica*, 2). It is also quite interesting that Ferrari presents Cardano as a great mathematician, but not as a professional one. Differently from Tartaglia, he does not do mathematics for a living, but only "like a game, to derive some recreation and pleasure" (p. 1). In short, he was not a mechanical man. It is also significant that, in his *cartelli*, Tartaglia keeps trying to address himself to Cardano, and to refer explicitly to Ferrari only as his client (*creato*).

174. Ibid., 4.

of honor, and the author of *Il duello,* a treatise that went through many Italian editions, was translated into Spanish, and had three French editions.[175] A historian has recently called Muzio's book "the best-known of all" treatises on dueling.[176] Muzio's presence among the witnesses suggests that he was probably consulted by Cardano and Ferrari on how to counter Tartaglia's attack in a honorable way.

Cardano's use of Ferrari or Tycho's use of Kepler was not an unusual practice. In fact, Commandino's use of Tommaso Leonardi for his critique of Tartaglia can be also seen as a result of the gap between Commandino's aristocratic status and the much lower status of Tartaglia.[177] Pierre Bourdieu has found that in North African Kabyle society, noble families keep a "poor man" in their household for the specific purpose of passing on to him the challenges the family may receive from lower-class people.[178] Passing the task of responding to a challenge to a lower-status client was not just a way of matching the status of the opponents; it could be deliberately used to deliver an insult. This is what happened to Voltaire, who challenged the Chevalier de Rohan but ended up being thrashed by his lackeys.[179]

Galileo's career provides comparable examples. His patrons and friends wanted him to respond promptly to challenges, but they did not want their champion to engage in scientific duels with people below his status. In October 1612, Cigoli wrote Galileo that the attacks on his work on buoyancy "were things that should be answered by somebody young, or at least by somebody under that semblance."[180] Cesi shared Cigoli's view: "I have always been of the opinion that Your Lordship should not answer such adversaries, but that they should be taught a lesson by having them answered by youths. Those in charge of the answers could be partially or totally guided [by us] or they may even be made to adopt already composed replies."[181] Galileo must have taken their advice, to judge from a long, painstaking reply to some of the critiques of his *Discourse on Bodies in Water* published in 1615 under the name of his young client Castelli. Similarly, Galileo decided to enter the dispute on comets only indirectly by having Mario Guiducci present his views as an academic lecture and

175. For the identity of Girolamo Muzio and Mutio Iustinopolitano, see François Billacois, *The Duel* (New Haven: Yale University Press, 1990), p. 251; and Angelozzi, "Cultura dell'onore," 308, note 7.

176. V. G. Kiernan, *The Duel in European History* (Oxford: Oxford University Press, 1988), p. 48.

177. Biagioli, "Social Status of Italian Mathematicians," 64; and Paul L. Rose, "Letters Illustrating the Career of Federico Commandino," *Physis* 15 (1973): 401–10.

178. Bourdieu, "Sentiment of Honour in Kabyle Society," p. 206.

179. Kiernan, *Duel in European History,* p. 98.

180. *GO,* vol. 11, no. 778, p. 410.

181. Ibid., no. 777, p. 409.

then publish them under Guiducci's name in 1619 as the *Discourse on Comets*. Grassi adopted the same tactic by presenting his further reply as written by a fictive Lotario Sarsi, who presented himself as a student of Grassi.

Issues of scientific etiquette and honor became much more complicated in the later part of the debate on comets between Galileo and the Jesuit Orazio Grassi. Because Grassi had assumed the pseudonym of "Lotario Sarsi" and had obliquely referred to Galileo as a member of the Accademia dei Lincei, a complicated discussion developed between Galileo and his Lincean friends and patrons as to whose honor was at stake and under what persona and in what format Galileo should respond.[182] The Lincei were not the only ones concerned about issues of honor. Grassi's decision to assume the pseudonym of Sarsi was clearly aimed at shielding the honor of the Society of Jesus from problematic involvements in the dispute. This was common policy among Jesuits. Christopher Scheiner, who adopted the pseudonym of "Apelles" during the dispute on the sunspots, claimed, later on, that he was told to do so by his superiors, who were concerned with the discredit Scheiner's possible errors may have brought to the society.[183] To use a pseudonym or to pass a challenge to one's client was like wearing a mask. In different ways, all these devices provided a shield for one's honor (or one's patron's honor). As Castiglione remarked in the *Book of the Courtier*, a gentleman could engage in potentially status-tainting activities only by masquerading.[184]

The use of pseudonyms or masks could backfire. By hiding behind "Lotario Sarsi," Grassi thought he had disguised his real identity and corporate status. However, he was quickly found out and all his pseudonym did was to make him liable to be treated as a low-status person—someone to whom the gentlemanly code of honor and civility did not apply.[185] This may explain why, in 1621, the general of the Society of Jesus prohibited its members from publishing anonymously or pseudonymously.[186]

182. *GO*, vol. 12, no. 1429, pp. 498–99; vol. 13, no. 1433, p. 11; no. 1441, pp. 20–21; no. 1446, p. 23; no. 1448, p. 24; no. 1450, p. 25; no. 1456, pp. 30–31; no. 1466, pp. 37–38; no. 1467, pp. 38–39; no. 1474, pp. 43–44; no. 1476, pp. 46–47. However, Grassi argued that he had assumed that pseudonym only because Galileo had earlier used Guiducci as a shield in his *Discourse on Comets*.

183. William Shea, "Galileo, Scheiner, and the Interpretation of Sunspots," *Isis* 61 (1970): 498–99.

184. Baldassare Castiglione, *The Book of the Courtier* (New York: Anchor, 1959), p. 103.

185. Nevertheless, the Lincei thought that Galileo should not engage in this mask-bashing directly. At some point, they thought that he too should hide behind a mask. Eventually, they decided that he would not need a mask but should write the *Assayer* as a letter to a friend, saying that he had been encouraged by other friends to do so. On Galileo's "masks," see *GO*, vol. 13, no. 1450, p. 25; no. 1456, pp. 30–31.

186. This may explain why, after 1621, Grassi replied to Galileo as Lotario Sarsi only in a book (the *Ratio ponderum librae et simbellae*) that was not published in Rome but in Paris (*ARSI*, ROM 19, fol. 247). On Muzio Vitelleschi's prohibition, see Ugo Baldini, "Una fonte

The unusually aggressive and "carnevalesque" style that Galileo adopted in defending his honor and the Lincei's in the *Assayer* would have been quite unacceptable had Grassi not "worn a mask." As Galileo put it,

> Indeed, I think that to deal with him as an unknown person will be to gain a wider field in which to make my arguments plainer and explain my ideas more freely. I am taking into account the fact that many times those who go masked are either low persons who try in this guise to gain esteem among gentlemen and scholars and to utilize for some purpose of their own the dignity which attends nobility; or, they may be gentlemen who, thus recognized, lay aside the respectful decorum accorded to their rank and make free use (the custom in many cities of Italy) of speaking openly about anything with all and sundry, getting as much pleasure as anybody can in this raillery and contention devoid of all respect. . . . Consequently, I believe in addition that just as he, thus unknown, has allowed himself to say some things against me which to my face he would perhaps not say, so it ought not to be taken amiss if I, availing myself of the privilege accorded against masquerades, deal with him quite freely.[187]

Galileo felt the same way about Scheiner-Apelles. If the *Letters on Sunspots* happened to be less aggressive than the *Assayer*, it had only to do with Welser's role in the dispute. As Galileo told Cesi, the third letter on sunspots was taking him more time than expected because he wanted to expose the silliness of Apelles' claims while avoiding insults to Welser.[188] The patron, rather than the opponent, was the focus of Galileo's etiquette concerns.

Etiquette and status dynamics not only informed the ways scientific disputes were carried out but were part and parcel of their raison d'être. Disputes—like competitive gift-exchanges—were processes of self-fashioning. For instance, Cigoli claimed that Galileo's views on buoyancy had been attacked also because of his high status: "those ugly birds want to make a name for themselves not through their own value but through

poco utilizzata per la storia intellettuale: Le 'censurae librorum' e 'opinionum' nell'antica Compagnia di Gesù," *Annali dell'Istituto Storico Italo-Germanico in Trento* 11 (1985): 37, note 43.

187. Stillman Drake and C. D. O'Malley, trans., *The Controversy on the Comets of 1618* (Philadelphia: University of Pennsylvania Press, 1960), p. 170.

188. ". . . I hope to expose the silliness with which this matter has been treated by the Jesuit. I want to make this resentment known, but the desire to do so without insulting Signor Welser is causing me no small difficulty and is the cause of my being late" (*GO*, vol. 11, no. 792, p. 426).

the choice of adversary."[189] Cigoli's views reflect contemporary doctrine on duels. Lodovico Carbone, a specialist on duels, wrote in 1583 that "the more insolent are the young because . . . young men seek glory by violating others' honor."[190] Similarly, Domenico Mora, in *Il cavaliere* (1589) stated that "everybody seeks distinction, one mark of which is to offend fearlessly."[191] A Roman supporter of Galileo, Monsignor Agucchi, wrote to him in July 1613 that "One cannot be honored if he is not challenged, and reputation grows from opposition, especially when victories follow from duels."[192]

The structural analogy one finds in this period between scientific disputes and duels (or rather duel-*like* challenges such as courtly jousts) is confirmed by another of Galileo's masked adversaries, the Anonymous Academician, who opposed him in the dispute on buoyancy. In his response to Galileo's *Discourse on Bodies in Water,* the Anonymous Academician viewed the dispute as a pleasant tournament: "One does not turn down the pleasurable sport of having a duel with him." Similarly, he presented his critique of the logical structure of Galileo's argument as a "logical lie" (*mentita loicale*), where *mentita* (to give the lie) was the customary move to deny the adversary's claim and trigger a duel.[193]

Galileo's friends and patrons were concerned with the etiquette of his ripostes but also with their timing. A slight delay in answering a challenge was not only understandable but even exciting, for it helped titillate the audience's expectations.[194] However, whether he did so out of design or necessity, Galileo tended to stretch the reply time and his friends' patience beyond normal limits.[195] In October 1612, Cigoli, who was waiting for Galileo's response to the Jesuit Scheiner in his third letter on the sunspots,

189. Ibid., no. 573, p. 176. Commenting on the possible reasons behind Scheiner's attacks on Galileo in his 1630 *Rosa ursina,* Fulgenzio Micanzio—Paolo Sarpi's friend—expressed very similar views (*GO,* vol. 14, p. 299).

190. Lodovico Carbone, *De pacificatione et dilectione inimicorum* . . . (Florence: Sermartelli, 1583); quoted in Bryson, *Point of Honor in Sixteenth-Century Italy,* p. 29.

191. Quoted in Bryson, *Point of Honor in Sixteenth-Century Italy,* p. 28.

192. *GO,* vol. 11, p. 532.

193. *GO,* vol. 4, p. 171; On the *mentita,* see Scipione Maffei, *Della scienza chiamata cavalleresca libri tre* (Rome: Gonzaga, 1710), pp. 58–70.

194. This supports the structural analogy between gifts and challenges. As noted by Pierre Bourdieu, gifts have to be returned neither too soon nor too late for the gesture to be effective.

195. The Lincei and the other Roman supporters of Galileo pressed him to answer the various challenges he received on issues ranging from the questions on the irregularities of the lunar surface, to the discovery of the sunspots, and to the calculation of the period of the Medicean Stars. They also urged him to print his response in fear that his priority would otherwise be questioned (*GO,* vol. 11, no. 572, p. 175; no. 573, p. 176; no. 587, p. 212; no. 788, p. 419).

told him that, "If you have not answered yet, do it soon because all your friends believe that they [the letters on sunspots] should appear as soon as possible. So move on and send to the Marquis [Cesi] whatever you want so that he may give it to the distributors."[196] Another month went by without Galileo's reply. A quite nervous Cesi told him to "move on because it does not seem smart to let Apelles [the Jesuit Scheiner] take over, because I am sure that he is not sleeping now that he has seen your second letter."[197]

Ripostes were supposed to be more than timely. Patrons and brokers wanted their clients to engage in disputes and expected them to respond "heroically," that is, fearlessly and skillfully, because, in doing so, the clients would increase their (and their brokers' and patrons') honor and status. Once it is contextualized within these patronage dynamics, Galileo's well-known aggressive and sarcastic style ceases to be just a character trait.[198]

The patrons' and brokers' interest in scientific tournaments is particularly evident in the correspondence between Galileo and his Roman supporters leading to the publication of the *Assayer*.[199] In May 1620, Ciampoli, speaking also for Cesi and Cesarini, told Galileo that "To all three of us who are lovingly concerned with the reputation of Your Lordship, it seems that a riposte should be necessary and that it should be delivered as soon as possible."[200] Indisposed by a long illness, Galileo did not reply for another year. In June 1621, Cesarini pressed him further: "I want, therefore, to deliver to you my warmest invitation not to wait further to redeem your most luminous glory from the ignorant lies of the malignant. Although your silence is caused by necessity, it allows false and vain literati to triumph."[201] In November, Ciampoli wrote again to Galileo: "Signor Don Virginio and myself are waiting with infinite desire to receive the discourse on comets. Please, do us the favor to solicit the copyist so that we will not have to be tormented by this burning thirst any longer."[202] However, although the Lincei kept sending comparably

196. Ibid., no. 786, p. 418.

197. Ibid., no. 790, pp. 422–23.

198. In his "Galileo and the Jesuits," p. 39, Richard Westfall has also linked Galileo's "egotism" to the dynamics of the patronage system.

199. To trace the escalation of the "duel frenzy," read these letters in succession: *GO*, vol. 12, no. 1429, pp. 498–99; vol. 13, no. 1433, p. 11; no. 1441, pp. 20–21; no. 1446, p. 23; no. 1448, p. 24; no. 1450, p. 25; no. 1456, pp. 30–31; no. 1466, pp. 37–38; no. 1467, pp. 38–39; no. 1474, pp. 43–44; no. 1476, pp. 46–47; no. 1477, p. 47; no. 1501, pp. 68–69; no. 1512, p. 79; no. 1513, p. 79; no. 1514, p. 80; no. 1516, p. 82; no. 1518, p. 84; no. 1520, p. 86; no. 1523, p. 89; no. 1524, p. 90; no. 1536, p. 99.

200. *GO*, vol. 13, no. 1467, p. 39.

201. Ibid., no. 1501, p. 68.

202. Ibid., no. 1513, p. 79.

emphatic exhortations to Galileo, reminding him of the need to respond to save both his honor and that of the Accademia, no riposte materialized.[203] In May 1622 an increasingly anxious Cesarini wrote: "I shall dare to solicit you to publish the riposte to Sarsi, something that, in many ways, you owe the world, but in particular to take back from the ignorant their claims of victory."[204] Finally, in October, the Lincei's anxieties and "burning thirsts" were over.[205] The *Assayer* went to press soon after. However, the spectacular attack on Grassi that the Lincei had sought and obtained from Galileo did not settle the matter. The *Assayer* may have met the expectations of Galileo's patrons and friends in rescuing their honor, but it certainly triggered further duels (or perhaps vendettas) from the Jesuits' side.

In short, although patronage was a social system that allowed scientific practitioners to gain status and credibility by promoting and legitimizing their exchanges, its dynamics were not necessarily conducive to the type of dialogue found in later institutionalized science. By this I do not mean that later science was generically more polite. While scientists continued to enmesh themselves in bitter disputes, the development of scientific institutions as international forums for the discussion and legitimation of scientific knowledge led to the elaboration of protocols of interaction and communication through which the community could regulate itself. Scientists might still be at war, but they were also developing some sort of "diplomacy" for conflict resolution. I believe that the emergence of this new "diplomacy" was directly connected to the transition from an environment in which the patron was the manager of scientific interactions to one in which the community of increasingly *interdependent* practitioners became more of a self-regulated body.

As suggested by the Lincei's propelling Galileo into a confrontation with the Jesuit Grassi, patrons and especially brokers tended to initiate scientific disputes for the sake of their status and image. Sometimes, mathematicians were summoned as the champions of their respective patrons. As we have seen in the case of scientific communication, conspicuous challenges did not usually take place *directly between scientists*, but tended to be exchanges between scientists as representatives of their patrons.[206]

203. Ibid., no. 1476, pp. 46–47; no. 1501, pp. 68–69; no. 1516, p. 82; no. 1518, p. 84; no. 1520, p. 86; no. 1523, p. 89; no. 1524, p. 90.

204. Ibid., no. 1523, p. 89.

205. Ibid., no. 1536, p. 99.

206. When challenges developed directly between scientists, as in the case of Tartaglia and Ferrari, the challenger made sure that the printed *cartelli* would be sent to prestigious figures, patrons of the adversary, and well-known mathematicians so as to force the adversary to respond or to lose his honor. Also, Ferrari printed the names of the twenty-three people he had elected as "witnesses" at the end of his first *cartello* (Giordani, *I sei cartelli di matematica*

Galileo's early correspondence informs us of a number of disputes set off and administered by patrons. The debate on sunspots was initiated by Mark Welser and fostered by the Lincei, while that on floating bodies was managed by Cosimo II. Similarly, the "Letter to the Grand Duchess" represented Galileo's response to questions on the relationship between Copernican astronomy and the Scriptures that Grand Duchess Cristina had posed to him through his client Castelli.[207] Although less directly, his 1624 "Response to Ingoli," too, was the result of patronage dynamics. In fact, Galileo's text was a long answer to questions Ingoli had posed to him during a public dispute in a Roman salon in front of patrons during his visit to Rome in 1615–16.[208]

These dynamics did not apply only to Galileo's career. His correspondence has also recorded the traces of other less conspicuous or shorter scientific disputes, like the one Sagredo tried to initiate between "his friar" (Paolo Sarpi?) and a Jesuit from Ferrara, Rocco Berlinzone (another pseudonym) in the spring of 1608.[209] Similarly, Sagredo challenged Scheiner (and several other mathematicians) to solve a mathematical problem on what we would now call "time zones" by delivering the problem not to Scheiner directly but through his patron, Welser. Then, when Sagredo became irritated with what he perceived as Scheiner's combination of ignorance and lack of civility, he did not express his disappointment to the Jesuit himself but to his patron.[210]

The dispute on sunspots highlights other aspects of the patron's role in scientific exchanges. Mark Welser was the force behind the dispute. He wrote Galileo for the first time in October 1610, enclosing with his letter a critique of Galileo's description of the lunar mountains by a physician from Augsburg, Georgius Brengger.[211] Welser was a patrician, a major patron of the arts and a political figure in Augsburg, a correspondent of Kepler, a good friend of Clavius and of other Jesuits, and an important financier for Emperor Rudolph II.[212] Probably because of his political

---

disfida, 5–6). I suggest that this procedure would not have been necessary had the dispute been triggered and managed by a high-status patron.

207. For an analysis of the content and rhetorical structure of the "Letter," see Janet Dietz-Moss, "Galileo's 'Letter to Christina': Some Rhetorical Considerations," *Renaissance Quarterly* 36 (1983): 547–76.

208. "Galileo's Reply to Ingoli," in Maurice Finocchiaro, *The Galileo Affair* (Berkeley: University of California Press, 1989), pp. 154–97. Ingoli's initial tract is in *GO*, vol. 5, pp. 403–12.

209. *GO*, vol. 10, nos. 185, 186, pp. 203–4.

210. *GO*, vol. 11, no. 826, p. 459. For Sagredo's response, see *GO*, vol. 12, no. 993, pp. 45–46.

211. *GO*, vol. 10, no. 420, p. 460.

212. On Welser, see R. J. W. Evans, "Rantzau and Welser: Aspects of Later German Humanism," *History of European Ideas* 5 (1984): 257–72; Antonio Favaro, "Sulla morte di Marco

and financial status, Welser was on good terms with the Medici. Providing another case of overlap between scientific and political or diplomatic networks, Welser's first letter to Galileo reached him through Curzio Picchena, a Medici *Segretario* and a good friend of Welser.[213]

In a sense, Brengger and Galileo entered into a debate through their patrons—Welser and the Medici. Evidently, Galileo *had to* respond to Brengger's letter; he had to reciprocate this gift to his fame coming from north of the Alps. And, in fact, it was as a gift that Welser presented Galileo with Brengger's critique of the *Sidereus Nuncius*:

> I have gladly complied with the desires of a friend of mine by sending you the enclosed paper, because I thought that it would be not unpleasant to see that even here beyond the Alps your books are being read with great attention, *and that the very existence of disagreement testifies to this* . . .[214]

And Galileo acknowledged Brengger's critique as a gift from Welser, a gift that—as he said—turned him into a client of Welser's:

> I have always sought the occasion to dedicate myself to the service of your great virtue. Therefore I was most happy to receive from you the critiques from the most erudite Signor Brengger. In fact, even in case his criticisms prove unanswerable, I would still be more pleased by the errors in my work than by the truths, *since it was through my errors that I gained such a great patron.* . . .[215]

That this response fit a typical pattern is supported by a similar letter Galileo wrote to Prince Leopold de' Medici in March 1640 after the prince had asked Galileo his opinion about Fortunio Liceti's *Litheosphorus,* a book in which the philosopher challenged certain passages of Galileo's *Sidereus nuncius* on the luminosity of the moon:

> I do not perceive [Liceti's] contradictions and oppositions as dismissable and to be left unanswered, but rather as plausible and worthy to be highly appreciated and esteemed by me because they earned me such a honorable and illustrious enrichment as the ap-

Velsero e sopra alcuni particolari della vita di Galileo," *Bullettino di bibliografia e storia delle scienze matematiche e fisiche* 17 (1884): 252–70; Giuseppe Gabrieli, "Marco Welser Linceo augustano," *Rendiconti della Reale Accademia Nazionale dei Lincei,* Classe di Scienze Morali, Storiche e Filologiche, serie VI, 14 (1938): 74–99. Apparently he went bankrupt when Rudolph refused to repay him a major loan (*GO,* vol. 20, pp. 556–57).

213. *GO,* vol. 10, no. 424, p. 466.
214. Ibid., no. 420, p. 460 (emphasis mine).
215. Ibid., no. 424, p. 465 (emphasis mine).

pearance of that most humane and courteous letter Your Most Serene Highness has sent me.[216]

A patronage relationship could be established through the gift of a challenge—an offer that, as the saying goes, Galileo could not refuse. Like Prince Leopold, Welser was not a patron one could dismiss easily. Not only was his economic and political power remarkable, but he was very well connected to the Jesuits and to Clavius in particular. As we can see from Cesi's eagerness to enlist Welser among the Lincei (and Galileo's promptness in having him elected to the Accademia della Crusca), he was somebody one did not want to have as an enemy.[217]

Galileo and Welser went through a similar ritual of gift-exchange about a year later, in January 1612, when Welser wrote again to Galileo to inform him of the observations of sunspots made by a client of his, the Jesuit mathematician Scheiner. Under the pseudonym of Apelles, Scheiner had communicated his discoveries in the form of three letters addressed to Welser and then published by that patron.[218] An exchange of letters between Galileo and Apelles via Welser followed. Galileo's three interventions were then published by the Accademia dei Lincei.[219]

Apelles' observations were presented to Galileo as challenges to his priority in the discovery of the sunspots as well as to his ability to interpret

216. *GO,* vol. 18, no. 3982, p. 166. For Leopold's letter, see ibid., no. 3981, p. 165. The book in question was Fortunio Liceti, *Litheosphorous, sive De lapide Bononiensi, lucem in se conceptam ab ambiente claro mox in tenebris mire conservante* (Udine: Schiratti, 1640). On this dispute, see also the previously unknown letters between Jacopo Soldani and Prince Leopold published in Mario Biagioli, "New Documents on Galileo," *Nuncius* 6 (1991): 157–69.

217. On Cesi's eagerness to enlist Welser in the Lincei and then use him to develop a section of the Lincei in Germany, see *Il Carteggio Linceo,* (hereafter *CL*), pt. 2, sect. 1, published as *Memorie della Reale Accademia Nazionale dei lincei,* Classe di Scienze Morali, Storiche & Filologiche, series 6, 7(1939), no. 132, p. 242; no. 136, p. 245; no. 140, p. 250; no. 147, p. 258; no. 238, p. 353; no. 257, pp. 372–73; no. 259, p. 375. Welser was elected to the Accademia della Crusca on 4 September 1613 (Severina Parodi, *Catalogo degli Accademici dalla Fondazione* [Florence: Sansoni, 1983],p. 56). On Galileo's and Salviati's roles in Welser's election, see *CL,* pt. 2, sect. 1, no. 275, p. 390; no. 284, p. 396; no. 291, p. 402.

218. Christopher Scheiner, *Tres epistolae de maculis solaribus scriptae ad Marcum Velserum* (Augustae Vindelicorum, 1612); reprinted in *GO,* vol. 5, pp. 23–32. Later in the same year, Apelle published a longer version, the *De maculis solaribus et stellis circa Iovem errantibus, accuratior disquisitio ad Marcum Velserum,* also reprinted ibid, pp. 35–70.

219. *GO,* vol. 11, no. 667, p. 289; no. 672, p. 293; no. 683, pp. 303–4; no. 741, p. 374; no. 771, pp. 402–3; no. 776, pp. 407–8; no. 794, pp. 427–28; no. 799, pp. 433–34; no. 806, p. 440; no. 817, p. 452; no. 832, pp. 464–65; no. 851, p. 486; no. 884, pp. 516–17; no. 938, pp. 587–88; no. 959, pp. 609–10. On the exchange between Scheiner and Galileo, see Antonio Favaro, "Oppositori di Galileo III: Cristoforo Scheiner," *Atti del Reale Istituto Veneto di Scienze, Lettere ed Arti* 78 (1918–19): 1–107; Bellino Carrara, S.J., "L'Unicuique Suum' nella scoperta delle macchie solari," *Memorie della Pontificia Accademia Romana dei Nuovi Lincei* 23 (1905): 191–287; 24 (1906): 47–127; and William R. Shea, "Galileo, Scheiner, and the Interpretation of Sunspots," *Isis* 61 (1970): 498–519.

them. Nevertheless, they were not presented as wild attacks but rather as gifts or tributes to Galileo's fame. Welser claimed that Galileo had broken the ice with astronomical discoveries and that it would have been cowardly for German mathematicians not to take up the challenge.[220] What Welser suggested was that Galileo's discoveries had been perceived both as gifts and challenges which—by having been met—reinforced Galileo's honor. Similarly, Welser saw Galileo's critical replies to Apelles as tributes as well. He thought that one of Galileo's answers was "written with reasons so good, so solid, and explained so properly, that Apelles—although you contradict most of his opinions—*should feel much honored.*"[221]

A crucial role of the patron in creating and managing scientific disputes emerges in these ritualistic exchanges between Galileo and Welser. Galileo did not perceive Brengger's and Scheiner's critiques as gifts coming from them, *but from Welser.* Similarly, Brengger and Scheiner were not receiving Galileo's "gifts" directly from him but rather from Welser. Welser was more than a *trait d'union.* His power—and the fact that both contenders were clients of his—warranted the legitimacy of the exchange. What in a different context may have appeared as intemperate attacks were given the legitimate status of duel-like exchanges. In fact, in an environment in which communication and credibility were established through patronage networks, the status of being a legitimate insider or dismissible outsider (someone who delivered "wild attacks") depended on one's location within those networks.

Galileo did not answer delle Colombe, Horky, and Sizi's attacks on the *Sidereus nuncius;* that is, he ignored all those works that were perceived as dismissable by the community (as in Sizi's case) or that were not dedicated and protected by patrons Galileo could not ignore.[222] It is also worth noticing that Delle Colombe, after having seen his "Contro il moto della terra" dismissed by Galileo, tried to force Galileo into a dispute on the irregularities of the moon's surface by having his views endorsed by Clavius and by circulating them among high Church officials such as Cardinal de Joyeuse—patrons Galileo could not dismiss.[223] Analogous strat-

220. *GO,* vol. II, no. 637, p. 257.

221. Ibid., no. 683, pp. 303–4 (emphasis mine).

222. Sizi's work was quickly dismissed by Kepler, the Roman Jesuits, and della Porta (*GO,* vol. II, no. 517, pp. 90–91; no. 559, p. 157). On Galileo's benign reaction to Sizi (a Florentine gentleman and a well-placed courtier at Paris), see Stillman Drake, "A Kind Word for Sizzi," *Isis* 49 (1958): 155–65.

223. *GO,* vol. II, no. 534, p. 118. according to delle Colombe, Clavius seemed to share some of his own views. Delle Colombe's strategy apparently had some chance to succeed, for only six months later it became clear that Clavius would not have entered a dispute on delle Colombe's side (ibid., no. 602, pp. 228–29). But delle Colombe was more successful in his strategy with Cardinal de Joyeuse and his Maggiordomo Gallanzoni. In fact, the cardinal—after

egies were adopted by the Roman philosopher Lagalla, who tried to have Galileo answer his critique of the *Sidereus nuncius*—the *On the Phenomena in the Orb of the Moon*—through his patron Cesi. As Cesi put it: "Lagalla wants a reply and has pressured me to write you on this matter . . . so that he could have full satisfaction."[224] Similarly, the mathematician Johann Remus Quietanus reached Galileo through the intercession of two common patrons—Leopold of Austria and Cesi.[225] Leibniz's triggering of the so-called Leibniz-Clarke correspondence by writing to Caroline, the princess of Wales, in an attempt to get at Newton, may reflect similar patronage-laden tactics.[226]

The patron did more than simply legitimize scientific disputes; in doing so, he also bound the contenders to engage in them. If they did not, they would put themselves on the same level of those people who—although having honor—did not reciprocate gifts or did not accept challenges to duels. In other words, they would step outside of the patronage

---

having read a copy of the letter sent by delle Colombe to Clavius—had Gallanzoni write Galileo asking for an answer (ibid., no. 546, pp. 131–32). A long answer came soon (ibid., no. 555, pp. 141–55), in the form of a private letter to Gallanzoni and Cardinal de Joyeuse (*mio Padrone*). One could speculate that if delle Colombe would have sent his critique of Galileo to the cardinal rather than to Clavius, de Joyeuse may have felt compelled to print the exchange. But probably delle Colombe could not have written directly to him, for he was not known to the cardinal. Again, delle Colombe was unsuccessful because of his poor patronage connections.

However, delle Colombe's letter (also supported by Brengger's critique of Galileo's views on the lunar mountains) had some impact—at least as a catalyst. In fact, if we may judge from Galileo's correspondence, there seems to have been a great deal of discussion in Rome in the second half of 1611 on the irregularities of the moon's surface. The pattern of the debate is quite confused because it was carried out mostly through an intricate exchange of letters and relatively uninformed comments of bystanders. Also, the discussion overlapped with the reverberations of the dispute on buoyancy at Florence and the beginning of that on the sunspots (ibid., no. 534, p. 118; no. 541, pp. 126–27; no. 545, pp. 130–31; no. 546, pp. 131–32; no. 550, p. 137; no. 555, pp. 141–55; no. 560, p. 158; no. 568, p. 169; no. 572, pp. 174–75; no. 573, p. 176; no. 576, pp. 178–208; no. 584, pp. 210–11; no. 585, p. 211; no. 587, p. 212; no. 588, pp. 213–14; no. 597, p. 223; no. 599, p. 226; no. 602, pp. 228–29; no. 612, p. 237; no. 625, p. 248; no. 632, p. 253; no. 651, pp. 268–69; no. 654, pp. 272–74; no. 665, p. 285. It seems that this debate was eventually absorbed and superseded by the more spectacular one on sunspots, which monopolized the attention of the Roman environment in 1612. To conclude, delle Colombe may have played a major role in triggering this debate, but his name was not attached to it because of his weak patronage links to the Roman environment.

224. Ibid., no. 665, p. 285.

225. *GO,* vol. 12, no. 1368, p. 433; no. 1374, p. 439; no. 1406, p. 471; no. 1417, p. 484.

226. H. G. Alexander, ed., *The Leibniz-Clarke Correspondence* (Manchester: Manchester University Press, 1956); and Steven Shapin, "Of Gods and Kings: Natural Philosophy and Politics in the Leibniz-Clarke Dispute," *Isis* 72 (1981): 187–215. Another example of the role of the patron in managing disputes is that of Sagredo challenging Apelles through Welser (*GO,* vol. 11, no. 826, p. 459).

networks that were granting them status and credibility. They would lose not just a patron but face as well—they would "disappear."

In a sense, the high status of the patron was transferred to the mathematicians, who in the social structure of the time may not have had status and honor enough as private individuals to make them eligible for challenges. It was this transfer of honor from the patron to the client that bound the latter to answer challenges, an ethic that they would not necessarily have had to adhere to as private individuals.[227] For instance, according to Sagredo, the "duel" he was trying to set up between "his friar" and Rocco Berlinzone failed to materialize because Berlinzone could excuse himself (without losing his honor) by claiming that the friar was a heretic and, consequently, honor-less and not worth challenging.[228]

Therefore, it should not be surprising that, in a patronage environment, texts generated within a given dispute tended to be dedicated to the same patron. Either it was the patron who had initiated the dispute or it was the practitioners who dedicated their texts to him, hoping to have their claims taken seriously and possibly responded to by their adversaries. All the contenders in the dispute on buoyancy at the Medici court in 1611–13 were clients of the Medici. Galileo dedicated his *Discourse on Bodies in Water* to Cosimo II, while his opponents, trying to force Galileo to confront their criticism, dedicated their replies to various other members of the Medici family. In 1659–60, Cosimo II's son, Prince Leopold, found himself at the center of the dispute on the appearance of Saturn and received the dedication of the texts of the participants in the dispute: Christiaan Huygens, Eustachio Divini, and Honoré Fabri.[229]

Noncommittal Patrons and Unending Disputes

Networks of aristocratic and princely patronage played a crucial role in early modern scientific life. They allowed for communications among scientists, framed their socioprofessional ethos, provided parameters to distinguish legitimate from illegitimate practitioners, gave them access to

227. However, while binding his client to the obligations of an honorable confrontation, the patron kept himself outside of the "duel."

228. "I have had a brief response from Messer Rocco Berlinzone in Ferrara, who does not want to dispute with my friar, and he excuses himself by saying that the friar shows himself to be a heretic rather than a religious person": Sagredo to Galileo, 22 April 1608 (*GO*, vol. 10, no. 185, p. 203). According to Bryson's *Point of Honor in Sixteenth-Century Italy* (pp. 25–26), to be a heretic meant to be without honor. Also, "to call one a heretic was among the strongest insults that could be conveyed by words" (p. 37).

229. Albert Van Helden, "Eustachio Divini Versus Christiaan Huygens: A Reappraisal," *Physis* 12 (1970): 36–50; and idem, "The Accademia del Cimento and Saturn's Ring," *Physis* 15 (1973): 237–59.

social status and credibility, and fostered, publicized, and legitimized their debates.[230] However, if the patronage system provided scientific practitioners with a social system in which they could operate legitimately, it also constrained their tactics and discourse in ways that, judged by present standards, may be seen as limiting. Because of the patrons' usual noncommittal attitude about the claims being debated, patronage was not a social system in which scientific disputes could necessarily reach closure. As a social system based on honor and status, patronage could hardly allow the patron (especially a princely one) to side with one disputant and to put his honor at stake by doing so.

Let me first present a few examples of the patrons' noncommittal attitudes toward their clients' claims and then propose an interpretation of the social roots of the phenomenon.

The Medici's patronage of Galileo did not automatically imply their endorsement of his opinions or discoveries. To a great patron, a client's victory was certainly more welcome than a defeat, but to have one's client challenged was already honorable enough. It was a gift, a recognition of the patron's status through the visibility of the client. It almost seems as if patrons, especially important ones, viewed the outcome of disputes in statistical terms. They seem to have expected that if their client did not win one game he would win the next.[231] As noted before, one of the reasons patrons were interested in clients in general was to keep their patronage networks alive. Quite similarly, I would say, patrons needed challenging (if not victorious) clients at least to "keep the action going" around themselves. Sometimes, this was literally the case because disputes took place in the patrons' courts, salons, and dining rooms.

Consequently, what interested most patrons was the "good sportmanship" displayed during the "duel" rather than the bloody ending.[232]

230. Obviously, I do not endorse the demarcation between legitimate and illegitimate practitioners as introduced by the patronage system, but I simply indicate that the patronage system *did* produce the demarcation—one that helped manage scientific practices.

231. Geertz's analysis of the social significance of the Balinese cockfight is relevant here. He detects two structures of betting around the fight. One, much less legitimate, deals with small amounts of money bet at fairly high odds. People bet this way trying to make money. But the much more conspicuous and legitimate betting is not done for money. High-ranking members of the community bet this way as if to perform a civic ritual that confirms their status. Because the odds in this kind of betting are basically even, it does not really matter whether one wins or loses since, in the long run, wins and losses will even out.

232. The transition from bloody duels and spectacles (typical of ancient and medieval-feudal aesthetics) to controlled games in which the process (the "sport") is more important than the outcome is a crucial aspect of the development of court culture and modernity. Historians and sociologists who study the development of modern manners and etiquette as part of the "civilizing process" have focused on it. Norbert Elias's work is an example of this orientation. In particular, his study of the sociogenesis of fox hunting seems particularly fitting here. The gentlemen involved in fox hunting take pleasure in the sport of chasing and

The pope's behavior during the dispute on comets between Galileo and Grassi exemplifies this point very well. Although Urban VIII had the *Assayer* read to him during meals, he was excited not by its technical points but by the display of Galileo's witty literary style in the section that became known as the "Fable of Sound."[233] In particular, he appreciated Galileo's poking fun at dogmatic philosophers who ruined the pleasure one ought to derive from investigating nature (the "philosophical sport") by narrow-mindedly pursuing dogmatic claims. The patron's aesthetic of the "good sport" is also reflected in the theatrical quality of some court scientific disputes. For instance, Cardinals Barberini and Gonzaga participated personally in the dispute on buoyancy at the Medici court in 1611. Barberini sided with Galileo, Gonzaga with the philosopher Papazzoni.[234] Reports of the dispute do not seem to focus on the truth-value of the claims being debated but rather on the style, wit, and elegance of Galileo's and Papazzoni's performances.

In this respect, scientific debates were no different from those which went on in literary academies or noble salons. Giovanni Aurelio Augurelli, a humanist from Rimini, described a salon discussion on the meaning of an emblematic painting: "Everyone has a different opinion and nobody agrees with anybody else. *This is more delightful than the pictures themselves.*"[235] The patron's concern was with the form rather than the content of a debate; Cosimo II did not discuss the technical details of the dispute on buoyancy with Galileo but simply criticized him for his failure to employ proper etiquette during the first part of the dispute.[236]

The patrons' privileging the aesthetics of process over the epistemological status of the claims may help us to understand why a dispute like that on buoyancy at the Medici court in 1611–13 did not end with a winner

---

finding the fox, not in its killing, which, in fact, is the hounds' task. Similarly, I think, a baroque patron may have felt that it was improper to kill the fox, i.e., to declare one defeated. I want to stress the adjective *baroque*. In earlier periods, the patron may have found pleasure in seeing the actual "intellectual killing" of one of the contenders. Tycho's aggressiveness probably reflects an older, feudal ethic and aesthetic (Norbert Elias, "An Essay on Sport and Violence," in Norbert Elias and Eric Dunning, *Quest for Excitement* [Oxford: Blackwell, 1986], pp. 150–174).

233. *GO,* vol. 13, no. 1593, p. 145; no. 1594, p. 146. Robert Westman has analyzed the implications of this public reading of the *Assayer* in a paper presented at the Conference on the Scientific Revolution at Keeble College, Oxford, July 1990.

234. *GO,* vol. 11, no. 684, p. 304; no. 699, p. 326.

235. Salvatore Settis, *Giorgione's Tempest* (Chicago: University of Chicago Press, 1990), p. 128 (emphasis mine). As indicated by this quote, emblems were a form of "intellectual challenge" that was interesting *per se.*

236. Galileo Galilei, *Discourse on Bodies in Water,* trans. Thomas Salisbury (London: Leybourn, 1663); reprint, Stillman Drake, ed., (Urbana, Ill.: University of Illinois Press, 1960), pp. 2–3.

and a loser. Patrons did not have any specific interest in ending a dispute if it kept offering a good spectacle. On the other hand, disputes could simply die off because they became uninteresting to the patron, without any of the contestants having been crowned the winner. In short, if patrons were at the center of disputes (and a dispute could be legitimate only if that were the case), they did not necessarily act as the needle on the scale. Patrons may have welcomed the dedication of disputing or disputed texts, but they did not feel obliged to uphold any of the claims put forward in them. When a patron was put in that position—as in the case of Prince Leopold de' Medici, who was elected by Huygens as judge of his hypothesis about Saturn's rings—he would try to get out of it as elegantly as he could.[237] People who bet on fighting cocks do not seek revenge in case their bird is killed. Similarly, although Cardinals Barberini and Gonzaga took sides in the dispute on floating bodies at the table of Cosimo II, they did not pursue Galileo or his opponent, the philosopher Papazzoni, once the game was over.

Welser's ambiguous stand about the truth-value of Galileo's claims fits this pattern. Even before he entered into correspondence with Galileo, Welser followed Galileo's astronomical discoveries with a mix of excitement and skepticism and frequently asked Clavius for his opinion. In his correspondence with Clavius about Galileo's discoveries (as well as about a new system for squaring the circle and a new star observed in 1604) Welser's caution does not seem that of somebody who is carefully evaluating the evidence in order to make a decision. Rather, he acts as an intellectual broker who does not want to take a stance but enjoys starting discussions with his friends by relaying the opinions he receives from Clavius.[238] Welser maintained the same cautious and noncommittal stance not only during the debate on sunspots but also in connection with the fluorescent Bologna stone and Galileo's early replies to Brengger about lunar mountains.[239] He did not praise Galileo's replies as neces-

237. Van Helden, "Accademia del Cimento and Saturn's Ring," pp. 242–44.

238. Welser to Clavius, 25 October 1602; Welser to Clavius, 31 October 1603; Welser to Clavius, 10 October 1608; Welser to Clavius, 5 December 1608; Welser to Clavius, 14 August 1609; Welser to Clavius, 12 March 1610; Welser to Clavius, 7 January 1611; Welser to Clavius, 11 February 1611. These letters (preserved in the archive of the Pontificia Università Gregoriana, Rome) are being published in the correspondence of Clavius, edited by Ugo Baldini and Pier Daniele Napolitani. I wish to thank Ugo Baldini for having provided me with transcriptions of these letters.

239. On Galileo's replies to Brengger, see GO, vol. 11, no. 452, p. 14; no. 453, p. 14. On the Bologna stone, see ibid., no. 549, p. 136; no. 554, p. 140. On the sunspots, see ibid., no. 637, p. 257; no. 638, pp. 257–58; no. 662, pp. 281–82; no. 771, p. 402. Also, Welser was usually the receiver and distributor of the works of Galileo's extreme adversaries, like Sizi (ibid., no. 503, p. 77).

sarily true, but rather as well argued, very convincing, clearly explained, and very enjoyable.[240]

The aesthetics of "good sport" is especially noticeable in Welser's remarks, but it can be found elsewhere in Galileo's correspondence. In June 1612, Cardinal Barberini thanked Galileo for having sent him a long letter about his recent observations and interpretation of sunspots. Although he praised Galileo's ingenuity, he did not commit himself to his cause. Rather, he concluded by asking Galileo to keep him informed so that, at social events, he would be able to speak intelligently on the current debate.[241] When, a few days later, Galileo sent him more material, the cardinal thanked him again, telling him, "I see that you touched soundly on new and curious things and that, with your rare ingenuity, you reached the best understanding to be had in such a short time and with so few observations."[242] However, Barberini's praise was not only explicitly conditional in relation to the current state of the debate and observations, but he also reminded Galileo that "in any case, the judgment on these matters should not come from me but from persons more competent in these matters."[243]

A more extreme version of this noncommittal attitude is found in the reports written by a Roman courtier, Antonio Querengo, to Alessandro d'Este describing Galileo's visit to Rome in 1616. Querengo began by telling of Galileo's brilliant performances (*discorsi stupendi*)[244] in the Roman salons:

> Your Most Illustrious Highness would take great pleasure from listening to Galileo arguing, as he often does, in the midst of fifteen or twenty adversaries as they attack him cruelly. . . . He stays there as if he were in a fortress and laughs at them, *and although the novelty of his doctrine is not convincing,* he demonstrates the vanity of most of the arguments with which his opponents try to defeat him. On Monday . . . he performed wonderfully; and what I liked most of all was that—before answering the adversaries' arguments—he amplified and reinforced them with apparently very powerful evidence which then made his adversaries look more ridiculous when he eventually destroyed their positions. . . .[245]

But Querengo's excitement at Galileo's argumentative skills, rather than at the truth of his doctrine, did not last long. Two months later Coper-

240. Ibid., no. 683, pp. 303–4; no. 775, p. 407; no. 776, p. 408. Similarly, even Galileo's friend Ciampoli, who sided with him during the dispute on buoyancy, described the court debate between Galileo and Papazzoni as "quelle gratiose dispute" (ibid., no. 820, p. 453).

241. Ibid., no. 690, p. 318.

242. Ibid., no. 697, p. 325.

243. Ibid.

244. *GO,* vol. 12, no. 1156, p. 212.

245. Ibid., no. 1170, pp. 226–27 (emphasis mine).

nicus was put on the Index, and the pious Querengo wrote to Alessandro d'Este that

> The disputes of Signor Galileo have vanished like alchemic vapors since the Holy Office has declared that those who uphold that opinion [Copernicus's] go manifestly against the infallible dogmas of the Church. We are reinforced in our belief that, by avoiding spinning tops in our brains, we can stay still and in our place rather than flying with the earth like many ants on top of a balloon in the air. . . .[246]

As we have seen with Welser and Cardinal Barberini, Querengo did not endorse Galileo's Copernican cosmology but appreciated the good sport of the dispute so long as he was allowed to offer it. However, we should not see Querengo's attitude as merely opportunistic or superficial. It could be found, in a more sedate version, even among Galileo's supporters who, in fact, were more appreciative of Galileo's argumentative *skills* and philosophical *style* than of his specific *claims*.

Ciampoli, whom we have seen vigorously pressing Galileo to answer Grassi on the comets, was enthusiastic about what he saw as the philosophical elegance of the *Assayer* but tended not to commit himself about the truth-value of Galileo's empirical claims. He praised the simplicity with which Galileo was able to account for the various cometary phenomena, but concluded his praises with: "However, I, who understand little, can *admire* it better than *argue* about it."[247] What Ciampoli was most appreciative of was Galileo's philosophical nonchalance and creativity, his "new things," "paradoxical propositions," and "precious gems."[248]

He expressed similar views about Galileo's performance in the "gracious disputes about water" with Papazzoni at the Medici court in 1612. While acknowledging that he had been much more impressed by Galileo's "acute experiences" than by the Aristotelian's responses based on "dry distinctions" between "*per accidens* or *secundum potentiam* or *secundum quid*," he was explicit in casting his views as nonjudgmental opinions (*in via di discorso*) based on personal taste rather than on the assessment of the specific empirical dimensions of the debate. As he put it, he liked Galileo's performance better "for whatever reason" (*qual se ne sia la cagione*) and stressed that his appreciation for Galileo should not be seen as a criticism (*depressione*) of Papazzoni.[249] However, Papazzoni heard of Ciampoli's remarks, took them personally, and confronted him about them.

Ciampoli's defense is interesting, for it casts light on the viewers' attitudes about philosophical disputes. He countered Papazzoni's queries by asking him "what new customs [about disputes] he wanted to introduce among Italy's ingenious minds so as to prevent people who were observ-

---

246. Ibid., no. 1186, p. 243.
247. Ibid., no. 1399, p. 466 (emphasis mine).
248. Ibid., no. 1429, p. 499.
249. *GO*, vol. II, no. 820, pp. 453–55.

ing a disputation between two doctors from saying: 'I like that one better; the answers of the other do not satisfy my taste, etc.,'"—a practice he saw as the standard one.[250] Papazzoni replied by saying that he would have not minded such a remark had it come from a university student. Instead, he was hurt by hearing it pronounced by somebody of Ciampoli's status— somebody "who is in high esteem with great cardinals and princes" and whose conversation they esteem as a "unique delicacy."[251] This shows how careful one had to be not to say that a person was "right" or "wrong" in a courtly context. Commenting on the "spectacle" of Galileo's thought processes was more rewarding than evaluating his empirical claims. The former led to pleasant courtly conversations (events in which the talkers could display their courtly skills while commenting on those of the disputants).[252] The latter led to unpleasant disputes in which nobody would have pleasure and somebody's honor might end up threatened or tainted.

Ciampoli's attitudes about Galileo's work were not unusual. Like Ciampoli, Cesi was a staunch supporter of Galileo's way of doing philosophy but not necessarily of his empirical claims. Cesi's allegiance to Galileo was rooted in their common commitment to "freethinking." In Cesi's case, this commitment had social as well as epistemological roots. In a letter of July 1611 to Galileo, Cesi criticised the Roman Aristotelian philosopher Lagalla for being unable to get out of his Scholastic modes of thinking. Interestingly enough, he saw Lagalla's "philosophical prison" to be cognitively damaging as well as socially distasteful. As he put it: "dignified intellects are bound to freedom."[253] Noble minds are free by definition, while being bound to a philosophical system displays one's subordinate intellectual position. Nobles and gentlemen ought to be freethinkers if they want to think in a way fitting to their social status. *Mens libera in corpore libero.* Cesi must have perceived Galileo's work as the epitome of this "noble thinking"—a philosophical *orientation* (but not a *system*) he wanted to embody in his Accademia dei Lincei.

Cesi's commitment to freethinking is best exemplified by his response to Paolo Antonio Foscarini's *Lettera sopra l'opinione de' Pittagorici e del*

250. Ibid., p. 454.   251. Ibid., p. 455.

252. The relationship between courtly science and the art of conversation has been introduced and explored by Jay Tribby in his studies of seventeenth-century Italian, and French courtly and gentlemanly science. On the subject, see Jay Tribby, "Of Conversational Dispositions and the *Saggi*'s Proem," in Elizabeth Cropper, ed., *Documentary Culture: Florence and Rome from Grand Duke Ferdinand I to Pope Alexander VII* (Florence: Olschki, forthcoming); and idem, "Stalking Civility: Conversing and Collecting in Early Modern Europe," *Rhetorica,* forthcoming; and idem, "Cooking (with) Clio and Cleo: Eloquence and Experiment in Seventeenth-Century Florence," *Journal of the History of Ideas* 52 (1991): 417– 39. See also Denise Aricò, "Retorica barocca come comportamento: Buona creanza e civil conversazione," *Intersezioni,* 1 (1981): esp. pp. 338–39, 342; and Giorgio Patrizi, ed., *Stefano Guazzo e la civil conversazione* (Rome: Bulzoni, 1990).

253. *GO,* vol. 11, no. 560, p. 158.

Copernico, a piece in which the Carmelite friar argued for the compatibility of Copernican astronomy with the Scriptures.[254] Writing to Galileo in March 1615, Cesi protested that "The author assumes that all our companions [Lincei] are Copernicans, despite the fact that that's not true. All that we are committed to as a group is freedom in natural philosophy."[255] With this, Cesi was not saying that Galileo should not be a Copernican. Freethinking did not mean that one could not have strong beliefs. What Cesi meant was that the Lincei as a group should not be perceived as Copernicans. Each one of them could freely develop whatever *personal* beliefs he wished, but those beliefs should not be turned into an enslaving dogma. When that happens (or when one uses a very rigid, self-righteous attitude in expressing one's views) no leeway is left to negotiate claims which have to be accepted or rejected.[256] As shown by Simon Schaffer and Steven Shapin in their study of the debate between Boyle and Hobbes, dogmatism was seen as a potential threat to the dialogue within the republic of letters and to social intercourse in general.[257] One may add that dogmatism was unacceptable to patrons, for it put their honor at stake with the client's claims.

I suggest that the source of the patrons' clear or nuanced noncommittal attitudes may be found in status dynamics. Honor was the patrons' greatest asset (especially if they were important princes). Honor was an opaque category; it had to be so in order to be socially effective. One could not pinpoint the "essence" of honor. As shown by many court treatises, honor (or *sprezzatura*) was precisely that whose essence could not be reduced to words. However, people knew how honor could be achieved and lost. They agreed that aristocrats received it through their family's indefinitely long history and could lose it in an instant. In fact, to have a great honor meant to have a very sensitive one; one that could be upset by a breeze. Great patrons were as vulnerable as they were powerful.[258] A

254. Paolo Antonio Foscarini, *Lettera del R.P.M. Paolo Antonio Foscarini Carmelitano sopra l'opinione de' Pittagorici e del Copernico della mobilità della terra e stabilità del sole e del nuovo Pittagorico sistema del mondo* (Naples: Scoriggio, 1615). On Foscarini, see Stefano Caroti, "Un sostenitore napoletano della mobilità della terra: Il padre Paolo Antonio Foscarini," in Fabrizio Lomonaco and Maurizio Torrini, eds., *Galileo e Napoli* (Naples: Guida, 1987), pp. 81–121; and Bruno Basile, "Galileo e il teologo 'Copernicano' Paolo Antonio Foscarini," *Rivista di letteratura italiana* 1 (1983): 63–96.

255. *GO,* vol. 12, no. 1089, p. 151.

256. For instance, Cesi criticized Galileo for having assumed an *ex professo* voice in his *Discourse on Bodies in Water,* and invited him to adopt a softer style for the future (probably a reference to the *Letters on Sunspots*) (*GO,* vol. 11, no. 737, p. 370). It is interesting that Cesi's caution did not seem to be triggered by concerns with possible theological implications of Galileo's claims, for the treatise on buoyancy was far from having scriptural implications.

257. Schaffer and Shapin, *Leviathan and the Air Pump,* pp. 23–109.

258. Paradoxically, power and fragility were inherently related. To be powerful meant being fragile, and to be fragile meant being powerful. Or, to put it differently, to be powerful meant having a lot to lose.

slightly impolite statement or veiled criticism might be read by them (or by those around them) as an insult that needed a riposte. Therefore, to take side in a dispute was a prelude to accepting a challenge and, given how sensitive their honor was, it was very easy for great patrons to be drawn into "duels."

In short, great patrons needed to shield themselves from challenges precisely because they were so sensitive to them. In the case of a patron's "participation" in scientific disputes, this necessity took the form of non-committal attitudes. Great patrons were conventionalists, not logical empiricists. These distancing devices were used in all spheres of a great patron's or prince's life. The intricate etiquette of baroque courts resulted precisely from the need to manage the intense interaction between absolute princes, aristocrats, and lower-status people while maintaining the appropriate distances and avoiding "pollution".259 As we know, etiquette blunders were usually experienced as personal insults.260 Just as the honor of the patron could legitimate scientific disputes only by being present but *not* challenged, the power (or honor) of the prince was effective only if it was *not* "touched." To touch it would have been to short circuit the entire system.261 That would have stained not just the patron's or prince's honor, it would have undermined the foundations of the entire legitimating process.262

259. Similarly, in his studies of deference and demeanor in contemporary society, Goffman has noticed that "the higher the class the more extensive and elaborate are the taboos against contact" (Erving Goffman, "The Nature of Deference and Demeanor," *American Anthropologist* 58 [1956]: 481).

260. However, there were important exceptions to this economy of status and honor. High-status people were not always in danger of seeing their honor stained by their own or others' words or actions. For instance, during carnival an aristocrat could mask him/herself and behave in ways that would have been otherwise unacceptable. Also, we find court jesters, who—precisely because of their liminal status—can say what other people cannot (and are rewarded for that). Similarly, we sometimes find aristocrats and princes who violate the rules of etiquette, speaking their minds or behaving impolitely. These exceptions do not violate the pattern but rather confirm it. In fact, the people who act this way either are recognizably liminal (the jester, the "old duke," the "eccentric queen mother," etc.) or are explicitly trying to violate rules to demonstrate their power. These are not "wild" exceptions but rather behaviors that (when acted by the right people or at the right time) play a role in the maintenance and evolution of the rules of etiquette. Of course, the use of pseudonyms during disputes, or the use of fictional literary genres to convey one's views (like Galileo's *Dialogue*), reflects similar strategies.

261. Quite literally, it would have exposed the vacuity of the prince's power. In a sense, the prince's power was a "bluff" that was effective (for all those who played that game) only insofar as it was not called.

262. The fact that the princely patron's noncommittal attitude was not accidental and limited to natural philosophy is indicated by comparable attitudes that patrons commonly displayed in dealing with requests they routinely received from their clients. As noted by Saint-Simon, it was rare for Louis XIV to say more than "I shall see" when anyone petitioned him (quoted in Norbert Elias, *Court Society* [New York: Pantheon, 1983], p. 131). Similarly,

Lower-ranking people were not in this predicament. To the young men mentioned by Carbone or the liminal philosophers mentioned by Cigoli, a challenge was a reasonable investment. They had little to lose and much to gain by making challenges. Their main problem was to be taken seriously. These dynamics, however, did not lead to a situation in which the powerful were sitting ducks routinely attacked by aggressive, honor-seeking social climbers. As we have seen, absolute rulers managed to shield themselves from challenges by "civilizing" their subjects. Court society, etiquette, and a discourse that "criminalized" challenges to the prince were all aspects of what Norbert Elias has termed the "civilizing process"—one that paralleled the establishment of political absolutism.

I suggest that these dynamics of honor, power, challenges, and duels (and their avoidance) may help us to contextualize the repeated invitations of Galileo's patrons (from Cosimo II to Cesi and Urban VIII) to present his arguments as hypotheses, to write dialogues rather than treatises, to argue *ex suppositione*. These dynamics may also explain why his patrons often treated Galileo's live debates as theatrical events. The princely patrons' attempts to frame their clients' claims as hypothetical or fictional reflect a discourse that was structurally homologous to the one that preserved the honor and power of the absolute prince from being undermined through challenges of various sorts. Galileo's having to write about Copernican astronomy as a hypothesis and having to use the "soft" genre of the dialogue should be seen as one instance of this much broader "discourse of effacement" aimed, in this case, also at preserving the power image of the "Papal Prince."[263]

---

the Lincean Johannes Faber, the pope's specialist in simple medicaments, could not persuade Urban to take an explicit stand about a lectureship he was seeking at the Sapienza. The pope never said "yes" or "no" but gave Faber only benign, vague reassurances and let him receive promising statements (and see them vanish) from his brokers (*CL,* pt. 2, sect. 2, no. 696, p. 828; no. 714, pp. 842–43; no. 721, p. 850). As with epistemological claims, great patrons did not want to take stances on anything unless it was important toward maintaining their position and status.

263. The link between *raison d'état* and a scientific discourse that would not threaten the power of the prince is described in Federico Cesi, "Del natural desiderio di sapere et Istitutione de' Lincei per adempimento di esso," an essay in which he sketched out the program of the Accademia dei Lincei, reproduced in: Gilberto Govi, "Intorno alla data di un discorso inedito pronunciato da Federico Cesi fondatore dell'Accademia de' Lincei e da esso intitolato: Del natural desiderio di sapere et Istitutione de' Lincei per adempimento di esso," *Memorie della Reale Accademia dei Lincei,* Classe di Scienze morali, storiche e filologiche, series 3, 5 (1879–80): 244–61; for the reference to *raison d'état,* see p. 257. Very similar points are made in the *Praescriptiones Lynceae Academiae Curante Joanne Fabro Lynceo Bambergensi* (Terni: Guerrero, 1624), p. 7. Concerns with raison d'état must have been common in the academies of the period if we find them explicitly mentioned in the statutes of the Umoristi—Rome's most important literary academy (Piera Russo, "L'Accademia degli Umoristi, fondazione, strutture e leggi: Il primo decennio di attività," *Esperienze letterarie* 4 [1979]: 59).

However, an upwardly mobile client such as Galileo did not necessarily share his princely patrons' "discourse of effacement." On the contrary, he often stressed (unless pressed to do otherwise) the truth-value of
his claims. Like Carbone's duel-prone youths, Galileo had to "attack" and
cast his arguments not as hypotheses or fictions but as true claims, if he
wanted to go up in the social scale, become a philosopher, gain status and
credibility, and legitimate the new worldview. Galileo's climbing put his
honor (and his opponent's) at stake at each step. Instead, the princely patrons (the Medici and the pope) whose power and honor he was tapping
into to achieve legitimacy did not usually need to *increase* their power
through their clients' challenges but wanted to *maintain* it by preventing
it from being put to the test. This irresolvable tension structured Galileo's
later career.[264] His trial of 1633 may have been a predictable result of these
dynamics.

To conclude, I want to suggest that although the patrons' lack of commitment to the contenders' claims was essentially a result of their *power*, it
was often represented as a sign of their *objectivity*.[265] For instance, the gap
in social status between Welser and his competing clients—a gap that had
him avoid commitment to either of them—was represented as disinterestedness. A *distance* in *social status* (and therefore "honor") was represented as the *distancing* that allows for *objectivity*. In a sense, patrons were

264. I would say that there were no rigid protocols according to which one could assess
the "reality" or "fictionality" of a given claim. As shown by Urban VIII's reaction to Galileo's
Dialogue, the patron's perception of the boundaries between realist and fictional or hypothetical claims was highly contextual. It depended also on the patron's perception of the stability of his/her own power at that time. Actually, I do not think that the patron needed to
believe that a client's claim was hypothetical in order to entertain it. All that he/she needed
was that the claim could be represented as such in that given context. As shown by Torquato
Accetto's 1641 treatise, dissimulation was no crime in a baroque court. (Torquato Accetto,
Della dissimulazione onesta, in S. Caramella and B. Croce, eds., *Politici e moralisti del Seicento*
[Bari: Laterza, 1930]). Courtly dissimulation had already been discussed and approved in
Stefano Guazzo, *La civil conversazione* (Brescia: Bozzola, 1574). For instance, in the *Assayer,*
Galileo excused his quite aggressive style by claiming to be attacking not a real person but a mask
whose true identity was unknown. However, the book's audience knew perfectly well that
Lotario Sarsi was Orazio Grassi and that, therefore, Galileo's text was an *ad hominem* attack
on the Jesuit. In short, everybody knew that the *Assayer* was part of a personal duel in which
the participants' honor was really at stake, but they could pretend that it was a nice, polite
tournament. Although the book was dedicated to the pope, his honor was not involved.
The dissimulated "fictionality" of a client's claim produced a double meaning that, to
some extent, worked well for both the patron and the client. The client could attack in ear-
nest while pretending to be just gaming. This allowed the client to display skills and arguments to those willing to appreciate them while, at the same time, leaving the patron's honor
out of the duel's "closed field." Similarly, the patron could participate in a "risky" (and therefore more entertaining and "distinctive") spectacle in relative safety. In fact, if things got a bit
out of control, the patron could still—as shown by Urban VIII—blame his client for having
transgressed the proper discursive boundaries.
265. *GO,* vol. II, no. 554, p. 140; no. 771, p. 402; no. 776, p. 408.

bound to be "objective" by a peculiar kind of *noblesse oblige*. Even when patrons like Ciampoli or Cardinal Barberini excused themselves by citing their lack of specific competence, those statements did not reflect humility but social boundaries. Being competent meant being *professional*, that is, *technical*, but patrons could not be technicians—they were gentlemen or aristocrats. Therefore, what was presented as the patrons' *impartiality* was actually the manifestation of a social *boundary*—almost a taboo.

Great Patrons and Claims of Objectivity

Although patrons (especially princely ones) were extremely careful in putting their honor on the line for their clients' claims, there were rare but informative exceptions to this rule. The Medici eventually endorsed Galileo's discovery of the satellites of Jupiter largely because he managed to present them as remarkable dynastic emblems of the house of Medici. Because Galileo's gift enhanced their image, the Medici were willing to take some risk with it. However, a patron's endorsement of a client's claims was not the only way in which a client could gain cognitive legitimacy. Clients could enhance their status and credibility just by entering into patronage relationships with important patrons. In this section, we will consider the specific mechanisms of this process by looking at a few examples of interactions between clients and great patrons.

Pierre Bourdieu and Jean-Claude Passeron have approached the French university system not just as an educational institution but as one which reproduces social hierarchies, and, within this framework, they have studied the sociogenesis of the myth of the autonomy of academic knowledge. According to them, the "illusion of the absolute autonomy of the educational system" is strongest when the professors are no longer paid piecemeal by their students but become a structured professional body dependent only upon the state.[266] This is the setting in which the "ideology of disinterestedness" develops best. Such ideology is functional for both the professors and the state. By claiming the total autonomy of their cultural productions, the professors present their knowledge as "pure" and "disinterested," and therefore nonarbitrary and legitimate. The more the professors stress their autonomy (also from the state), the more they serve the interests of the state, which uses the university and the culture it reproduces to, in turn, reproduce social structures and hierarchies. But such reproduction is effaced by the very fact that it is presented as education into "pure" culture. Briefly, "purity" develops out of the replacement of the particular students by the state as the professors' patron.

266. Pierre Bourdieu and Jean-Claude Passeron, *La reproduction: Éléments pour une théorie du système d'enseignement* (Paris: Minuit, 1970), p. 82.

Alain Viala's study of the emergence of the literary author as a legitimate social role in seventeenth-century France has introduced a distinction between ranks of patronage relationships: *clientélisme* and *mécénat*.[267] Usually, *clientélisme* developed around small patrons. Their clients (tutors or historians) performed fairly routine and low-visibility tasks and were usually paid piecemeal. These clients tended to choose low-risk, low-speed career strategies. The client-patron bond was generally weak, and clients of such patrons tended to enter simultaneously into a number of patronage relationships. Texts produced within such patronage relationships were usually mainstream. The client was not expected to remain faithful to such a patron.

Mécénat was something very different from *clientélisme*. It was much rarer, too. Great patrons (the king or a *grand* like Condé) were interested in brilliant, controversial, highly visible, and *galant* authors. These writers were not rewarded through salaries and were not asked to work in the sense of taking care of specific tasks. They received "gratifications." Clients of such patrons were expected to be loyal to them.[268] They were their patrons' champions. Their careers were of the high-risk, fast-track type. They were perceived as having the status of nobility also because their cultural "aggressiveness" was perceived to fit the aristocratic ethics and aesthetics: "In fact, we can properly speak of *literary heroism:* his glory as writer granted him nobility in the same way military exploits had formerly turned a free man into a knight."[269]

Again, as with Bourdieu and Passeron's university professors, Viala's writers were perceived as being noble or disinterested by serving one great patron instead of many petty ones and by adopting the cultural mannerisms that went with it. Great patrons needed to represent these clients as nobles because, being aristocrats themselves, the patrons could not have their status confirmed by means of clients who were perceived to be paid by them to do so. Della Casa's *Galateo*—a classic etiquette handbook—teaches clients to adopt an aura of liberality while serving their patrons: "what is done for duty is perceived as such by the patron who, consequently, feels little gratitude toward that client. Instead, those clients who go well beyond their duties are perceived as giving something of their own and are therefore loved and perceived as munificent."[270] Both great patrons and their clients wished to present the relationship as a voluntary rather than a utilitarian one. As with the university professors and the

267. Alain Viala, *Naissance de l'écrivain* (Paris: Minuit, 1985), pp. 51–84.

268. In "Gift-Giving and Patronage in Early Modern France," a study of French early modern aristocratic society, Sharon Kettering has noticed the same pattern described by Viala about loyalty (p. 137) and about the giftlike nature of "gratifications" (pp. 140–41).

269. Viala, *Naissance de l'écrivain,* p. 222 (translation mine).

270. Giovanni della Casa, *Galateo* (Venice, 1558; reprint Turin: Einaudi, 1975), p. 34 (translation mine).

state, the mutual interest of Viala's patrons and authors was better served when the economic dimensions of the patronage relationship were effaced or denied.

In a statement that reminds us of Galileo's letter to Vinta, Michelangelo was proud to claim that never in his life did he have to open a workshop (*aprire bottega*), but that he always worked at court, where he had only one princely patron rather than many less important ones.[271] Similarly, he claimed that he was *not* a painter or sculptor in the sense those terms had come to have in association with urban guilds. To stress his unique independence and socioprofessional status, Michelangelo told a Bolognese painter that he was obliged to his patron (Pope Julius II) as much as the painter was to his suppliers of colors.[272] Like Viala's *écrivain galant* who did not want to be identified with clerks, or Galileo who asked for the title of "Philosopher," Michelangelo wanted to stress his uniqueness. Michelangelo was Michelangelo.

And it is not accidental that this Renaissance artist who achieved the greatest status of all—the one who was called "divine"—had developed a quite dialectical patronage relationship with the most important Maecenas of the time—Pope Julius II—and repeatedly stressed his artistic independence while denying the utilitarian motivations of his work.[273] When, under Paul III, Michelangelo was put in charge of the *fabrica* of St. Peter, he asked the pope to include in the contract that he would work "for love of God and not for any economic reward."[274] The myth of the artist that developed around Michelangelo's "divinity" (like the other myths developed around academic "autonomy" and writers' "nobility") was one which benefited both artists and patrons by elevating art and making it a more prestigious status symbol for patrons. Therefore, what attracted Galileo to Florence was not only nostalgia, a great salary, and no teaching load, as commonsense interpretations may suggest; it was probably the unique chance to become a "philosopher"—the Michelangelo of mathe-

271. "I was never a painter or a sculptor like those who set up a shop for that purpose. I always refrained from doing so out of respect for my father and brothers" (Michelangelo, letter of 2 May 1548; quoted in Peter Burke, *Culture and Society in Renaissance Italy 1420–1540* [New York: Scribner's, 1972], p. 69). The mechanical connotation of "aver bottega" is confirmed by Giorgio Vasari, who referred to a minor painter as "one of those who keep an open shop and stand there in public, working at all sorts of mechanical tasks" (ibid., 69).

272. "I have the same obligation to Pope Julius, who gave me [the bronze for the statue,] as you have to the apothecaries who give you colors with which you paint" (quoted in Giorgio Vasari, *Vita di Michelangelo,* ed. Paola Barocchi [Milan-Naples: Ricciardi, 1962], 1: 34).

273. On the tense relationship between Michelangelo and Julius II, and the various popes that followed Julius, see ibid., pp. 28, 31–34, 38–41, 52–53, 55, 71–72, 74, 92–93.

274. Ibid., 84. According to Vasari, although the pope "sent him money for that job on several occasions, he was never willing to accept it" (ibid.). See also pp. 97–98 for a similar instance.

matics. But the most interesting aspects of this process are perhaps those related to the status of Galileo's discipline.

By finding a patron whose status was great enough to oblige him to repress the economico-utilitarian link between himself and a high-visibility client, Galileo was able to present himself, his method, and his discipline as "disinterested" and, therefore, "objective."[275] By doing so, he could try to shed the low social connotation that mathematics received from its link to mechanics and other practical disciplines. It was by re-pressing the utilitarian dimensions of patronage (something to which Co-simo was bound by his honor) that Galileo could present himself as disinterested and objective and, consequently, could represent his find-ings as truthful. Galileo's representation of his method and claims as ob-jective was the result of the same patronage dynamics that produced Michelangelo's "divinity." Moreover, such examples display another rele-vant common denominator: they all exemplify the patronage dynamics that allowed for the legitimation of a *new socioprofessional identity*.

The analogy between Galileo and Michelangelo (also in relation to the academic use of their images) is particularly telling. Michelangelo's "divin-ity" was institutionally appropriated by the Florentine artists who trans-formed him into a sort of patron saint of their profession. His extraordin-ary status as an artist was instrumental in allowing them to present the entire artistic profession as having high social status—something it was previously missing. It was the newly founded Accademia del Disegno (the first official fine arts academy) that organized and choreographed the Flo-rentine funeral of Michelangelo in 1564.[276] Michelangelo could not have died at a better time. His death was a "mirabil congiuntura," a crucial re-source for the legitimation of the newly institutionalized visual arts. As Vasari put it, "It is true that it was great luck that Michelangelo did not die before our Academy was created."[277]

As a hypothesis, I would suggest that the funerary monument that Galileo's supporters tried (unsuccessfully) to construct in Santa Croce im-mediately after Galileo's death may have been aimed at something simi-lar.[278] Galileo was to become the "patron saint" of the new breed of mathematicians—the "philosophical ones." The extraordinary status Galileo had obtained as a mathematician was probably supposed to help

275. As mentioned before, Galileo's large salary was public knowledge. However, it was presented as a sign of the grand duke's disinterested generosity and power rather than the payment for a fitting dedication.

276. For a description of the funeral, see Vasari, *Vita di Michelangelo*, 1: 132–91, esp. pp. 135–49, 174–79, 185.

277. Ibid., 185.

278. *GO*, vol. 18, no. 4194, p. 378; no. 4196, p. 379; no. 4197, pp. 379–80; no. 4202, p. 382; and Giovanni Battista Nelli, *Vita e commercio letterario di Galileo Galilei* (Lausanne, 1793), 2: 874–85.

improve the mathematicians' status in the same way Michelangelo's "divinity" had helped the former painters, architects, and sculptors to become "artists" (in the modern sense of the word). Moreover, like Michelangelo, who excelled at all the three arts of *disegno* (painting, sculpture, and architecture), Galileo had mastered most of the branches of mathematics (astronomy, mechanics, optics, fortification, and hydraulics), knew well traditional philosophy, and had begun to introduce a new natural philosophy. In short, Michelangelo and Galileo were the perfect patron saints for their respective disciplines. It is interesting that Galileo and Michelangelo's tombs are now in the same church, facing each other, and that both mausoleums are monuments to the professions of the two men. Michelangelo is surrounded by muses symbolizing architecture, painting, and sculpture (the three arts of *disegno*), while Galileo is surrounded by more muses symbolizing the mathematical sciences which he was able to bring up to the status of philosophy.

This option of social legitimation was provided only from the patronage relationship between a highly visible client and a great patron. To engage in this peculiar form of patronage, the client must offer something very unusual, new, and controversial (so as to qualify him as a "challenger," a *free*-thinker, somebody who is not *dependent* on a philosophical tradition). Moreover, to be perceived by a great patron as "noble" enough to qualify for a "voluntary" relationship (almost a kinship) with him, the client must be challenged by many "worthy" people. Galileo certainly qualified on these grounds. Also, he was challenged on an issue related to the very "honor" of the Medici dynasty: the Medicean Stars. In a sense, by accepting Galileo's *gift* of the dedication, Cosimo turned Galileo into a Medici defender.[279]

Galileo was bound to defend his prince's honor, and he did so in the best possible way, that is, conspicuously and successfully. Hence, Cosimo's socioprofessional "ennoblement" of Galileo as his "scientific knight," that is, *Filosofo e Matematico Primario del Granduca di Toscana*.[280]

279. It is significant that Vinta called Galileo's dedication of the satellites to the Medici a "generous and *heroic*" gift (emphasis mine) (*GO*, vol. 10, no. 266, p. 284). Galileo agreed with Vinta about the "heroicness" of his discovery and dedication (ibid., no. 277, p. 298). Also, Galileo's "heroicness" fit well with the courtly appreciation for nonservile minds (as we have discussed in connection to Cesi's ethos).

280. We find five months between Galileo's dedication to the Medici (*GO*, vol. 10, no. 265 [13 February 1610], pp. 282–84) and the grand duke's official deliberation (ibid., no. 359 [10 July 1610], pp. 400–401.). Some indications of the possibility of a post at the Medici court were passed on by Vinta to Galileo during his visit to the court in Pisa during the Easter vacation. During these months, Galileo informed Vinta of the various attacks he had addressed regarding the reality of the Medicean Stars (ibid., no. 307, pp. 348–53). However, while accepting with pleasure the dedication, the Medici did not move eagerly to have Galileo at court. With the uncommittal attitude typical of patrons, they seemed to observe

In this sense, Galileo's nomination was already implied (unless he should "die" dueling) from the time of Cosimo's acceptance of the dedication. And Cosimo did not accept such a dedication just because of the spectacular nature of the discoveries or because of their Copernican significance, but because of the fundamental role that Jupiter played in the recently developed mythology of the Medici family. As we will see in the next chapter, Jupiter was an emblem for Cosimo I, the founder of the dynasty, and the Medicean Stars were represented as emblems of Cosimo I's dynastic progeny. In short, they were a confirmation of the naturalness of the Medici rule.

In this game of mutual legitimation Galileo could not bind any other patron more easily than the Medici. Although princely patrons like the Medici tended to avoid commitment, this time they had an important incentive in stressing Galileo's "nobility" and disinterestedness: by helping legitimize his discoveries they would have legitimized and naturalized his important contribution to the Medici self-legitimizing dynastic imagery. The analogy between this case and the symbiotic relationship (analyzed by Bourdieu and Passeron) between the state's concern with the reproduction of social structures and hierarchies, and the professors' desire to have their culture be represented as "autonomous" is, I think, telling.

And Galileo could not obtain such an "ennoblement" from any patron lesser than Cosimo II. That is what he told Vinta when he described to him his dissatisfaction with the employment conditions he experienced with the Venetian Republic. As he saw it, the problem was not with Venice but with republics in general. A number of visual artists before Galileo had come to the same conclusions.[281] The employment conditions Galileo requested from the Medici did not describe the life style of a proto-scientist but rather that of a *noble*. The reason he did not want to teach was not simply related to his desire for free time to dedicate to his research. He pointed also to an issue of status: he did not want to have a position that would mark him as a mechanical person, that is, as one who was *obliged* to work. Also, he did not want to teach many *generic* students, but *only* his *Signori,* when they desired to benefit from his teachings. He was interested in *mécénat,* not *cliéntelisme.*

Actually Galileo succeeded in obtaining the status of a court gentleman because—in addition to receiving the title of philosopher—he was included in the category of *familiari senza provisione* (people of patrician status who had full access to the court but were not paid as court workers)

---

how the dispute was developing. But Galileo thought that their distance was a bit excessive, and that a stronger or quicker endorsement from their side would help resolve the dispute (ibid., no. 307, p. 349; no. 339, p. 379).

281. Burke, *Culture and Society in Renaissance Italy 1420–1540,* p. 230.

rather than in the category of *Artisti, architettori et altri manifattori* (in which we find artists, craftmen, engineers, architects, teachers of mathematics, and geographers).[282]

This link between great patrons, high-visibility, controversial clients, and the disinterestedness (or objectivity) consequently attributed to the client's cultural production may explain Galileo's concern (which he shared with Michelangelo) in later trying to develop a patronage relationship with the greatest patron of his time—the pope. As indicated by the trial of 1633, his hope for an even higher legitimation went unrealized.

Patronage and Commitment to a Theory

After this analysis of the patronage system, I would like to go back to some of the points raised at the beginning of this chapter and argue that patronage does not need to be treated as something external to Galileo's scientific concerns, commitments, and choices. Although this issue will be analyzed in more detail later, I want at least to indicate how an understanding of patronage dynamics may provide a more complex perspective on Galileo's defense of his astronomical discoveries of 1609–10 as well as on his becoming a full-fledged Copernican.

Galileo the Copernican and Galileo the courtier were not two distinct personae. Similarly, it would be unrewarding to think of Galileo's commitment to Copernican astronomy as the driving force of his scientific life and to treat his patronage concerns as an obstacle. While it is true that patronage dynamics and court discourse posed serious constraints on a client who was making *hard* claims in general (and not just about the heliostatic cosmos), these constraints should not lead us to claim that patronage had only a censoring effect on science.

Quite to the contrary, patronage was a *productive* system. It propelled clients, fostered and structured their communication and debates, rewarded novelties, legitimized knowledge claims that would have been unacceptable elsewhere, and gave clients the resources to legitimize unconventional socioprofessional identities. Obviously, such a system

---

282. *ASF,* "Depositeria generale 389," fol. 89r., and *ASF,* "Guardaroba medicea 309," fol. 38v. Galileo's salary was not paid by the Medici treasury—the Depositeria generale—but by the University of Pisa. This can be interpreted in two ways. Perhaps the Medici wanted to avoid charging such a large salary on the court budget; or—given that Galileo did not actually work or have any specific role at court—they really meant to have him at court as a "gentiluomo non provvisionato" while giving him the salary from Pisa as a sort of sinecure. This second interpretation would find support in what Viala has observed in France. There, the "grands" would reward their top clients not through salaries drawn from payroll funds, but with payments coming from different, less salary-connoted sources. The reward of a great client was called "gratification" and came from "une rubrique budgétaire spéciale," while the salaries of lesser clients were simply called "émoluments ordinaires" (Viala, *Naissance de l'écrivain,* pp. 56–57).

had its rules and its parameters about what a legitimate discourse should look like—parameters that were different from the ones that regulate today's science. Therefore, although this framework could accommodate only certain types of discourses, the tensions between Galileo's "scientific" and "courtly" concerns should not be viewed as a clash of two irreconcilable "worlds" but rather as a fundamental tension between two aspects of the same system.

In this case, I suggest that an analysis of Galileo's defense of his astronomical discoveries of 1609–10 indicates that his pre-1609 *belief* in Copernicanism may not have been necessarily the primary driving force behind his passionate defense of his findings. Rather, the increasing *commitment* to Copernican astronomy that Galileo developed in those years may have resulted also from the patronage dynamics that pushed him to defend his discoveries and produce even more of them.[283] Much historiography dealing with the question of Galileo's Copernicanism has limited itself to considering Galileo's statements about Copernicus in his books, manu-

---

283. For a different critique of Galileo's Copernicanism seen as an explanatory category of his entire career, see Maurice Finocchiaro, "Galileo's Copernicanism and the Acceptability of Guiding Assumptions," in Arthur Donovan, Larry Laudan, and Rachel Laudan, eds., *Scrutinizing Science* (Dordrecht: Kluwer, 1988), pp. 49–67; and Stillman Drake, "Galileo's Steps to Full Copernicanism and Back," *Studies in History and Philosophy of Science* 18 (1987): 93–105. Drake and Finocchiaro approach the question of Galileo's Copernicanism by following very different paths. Finocchiaro traces Galileo's increasing commitment to Copernicanism by looking mostly at his astronomical thought, while Drake takes Galileo's mechanics (his notion of indifferent motion) and his theory of tides as the key to his attitudes about Copernican astronomy. My sociological account of Galileo's increasing commitment to Copernicanism is largely congruous with Finocchiaro's internal analysis but is at odds with Drake's narrative, which has problems. First, by relying on negative evidence and introducing *ad hoc* hypotheses, Drake makes his thesis almost unfalsifiable. Second, an apologetic agenda seems to inform his narrative. Basically, Drake seems to be more concerned with refuting what he perceives as the historians' unfair critique of Galileo because he tried to present heliocentric astronomy as a reality in 1632 when he had only probable arguments for it. Drake wishes to show that in fact by 1632 Galileo was not a full Copernican anymore but that, as the Church had ordered him in 1616, he held Copernican astronomy only as a probable hypothesis. He does so by arguing that from 1590 to 1595 Galileo had indeed moved from an adoption of Copernican astronomy as a hypothesis to its endorsement as the real description of the cosmos. However, according to Drake, Galileo knew very well that he had no final proof for his belief and therefore, when he received the admonition of 1616, he gladly receded to the hypothetical position of the 1590s—a position that he held until his death. In short, when in 1616 he adopted the Church's order to treat Copernicanism as a hypothesis, Galileo was not dissimulating his real beliefs but was acting as a good scientist who realized he did not have the proof his realist argument required. What happened in 1633 was therefore a scam fabricated against him. Galileo was really treating Copernicanism as a hypothesis, but the Inquisition "overruled the intelligent, logical, and legal acts of their own officers" (p. 105.). What Drake sees as even worse is that historians and philosophers of science have continued to misinterpret the scam of 1633, unjustly criticizing Galileo for having presented as true a thesis that was only probable. By establishing how Galileo became a full Copernican by 1595 but then retreated from that position in 1616, Drake wants to show that "his modern judges will appear hardly more wise than those who convicted him of heresy" (p. 105).

scripts, and letters, and has consequently developed an unnecessarily narrow perspective on the issue. Galileo is presented as having been a
Copernican since the 1590s or, more tenably, as having been a potential
Copernican for a long time, with the transition from potentiality to actuality resulting simply from the increasing availability of pro-Copernican
evidence. In short, Galileo's Copernicanism is presented as a "natural"
outcome—sometimes a teleological one. This perspective ignores the fact
that Galileo's drive to produce more discoveries, some of which happened
to support the Copernican hypothesis, reflected also his patronage concerns. Similarly, this focus on Galileo pays only little attention to the ways
in which the *interaction* (also framed by patronage dynamics) between
Galileo and his critics helped shape his intellectual commitments.

Although it is true that by 1609 Galileo had already expressed Copernican sympathies in private letters and that later he defended the physical
reality of his discoveries very energetically, the Copernican significance of
his discoveries did not play a major role in the *early* phases of the debate on
the *Sidereus nuncius* (a book in which Galileo did not commit himself explicitly to Copernicanism).[284]

In fact, Galileo's discoveries contradicted the beliefs of the Aristotelian philosophers but did not need to be *necessarily* perceived as evidence
for the Copernican hypothesis. Writing to Galileo as late as May 1611,
Paolo Gualdo drew a clear distinction between the acceptance of Galileo's
discoveries as observations and as evidence in support of the Copernican
hypothesis: "It seems to me that you have already acquired glory through
the observation of the moon, of the four planets, and similar things, without having to defend something so contrary to human intelligence and
mental disposition. . . ."[285]

The perception of the discoveries varied with the audience's socioprofessional identity. While Aristotelians saw the discoveries as philosophically threatening, technical astronomers were less disturbed by
them. I think that what bothered astronomers like Magini (who initially
attacked Galileo) was not the Medicean Stars but Galileo's sudden fame.
As they were not supposed to engage in philosophical issues, technical
astronomers were not (at least in principle) to be bothered by the way
Galileo's discoveries were undermining the philosophers' beliefs. Not

284. The passages explicitly or implicitly connected to beliefs of heliocentrism are on
pp. 31, 36, and 84 of Galileo's *Sidereus nuncius,* trans. Albert Van Helden (Chicago: University of Chicago Press, 1989). For interpretations that question a straightforward Copernican
agenda of these passages, see Finocchiaro, "Galileo's Copernicanism and the Acceptability
of Guiding Assumptions," 57–58; Stillman Drake, *Telescope, Tides, and Tactics* (Chicago: University of Chicago Press, 1983), p. 223, note 5; and Wade L. Robinson, "Galileo on the Moons
of Jupiter," *Annals of Science* 31 (1974): 165–69.

285. *GO,* vol. II, no. 526, pp. 100–101.

only could they fit several of Galileo's novelties into their Ptolemaic *mathematical* framework, they could also account for the phases of Venus by adopting Tycho's system—one that left the earth at the center of the cosmos. For instance Simon Marius (a German astronomer who claimed to have discovered Jupiter's satellites simultaneously with Galileo) presented his discoveries as fitting a Tychonic framework.[286] Baliani, writing to Galileo in January 1614, expressed similar feelings: "I seem to see that Your Lordship approves of Copernicus's opinions. Nevertheless I would think that the observations done with the telescope about Venus, the Medicean Stars, and the sunspots proved the fluidity of celestial matter, and that, therefore, they make Tycho's opinion more probable."[287]

Similarly, without being Copernicans, the Jesuits eventually gave a very strong endorsement of Galileo's telescopic observations.[288] When, in the spring of 1611, they were asked by Cardinal Bellarmine for an official evaluation of Galileo's various discoveries, not only did they confirm them with only two minor qualifications, but they stressed the accuracy of Galileo's observation of the phases of Venus—the discovery that was most deleterious to Aristotelian philosophy: "It is *most* true that Venus waxes and wanes like the Moon."[289]

Probably the Jesuits did not perceive Galileo's discoveries as threatening because, as they did a few years later, they may have thought of framing them within the Tychonic system.[290] Although the Jesuits tried to domesticate several of Galileo's own *interpretations* of the discoveries, they probably saw them as important resources for the legitimation of mathematics in relation to philosophy—a battle Clavius had been waging for

286. Simon Marius, *Mundus iovialis* (Nuremberg: Laur, 1614); translated by A. O. Prickard in "The 'Mundus Jovialis' of Simon Marius," *The Observatory* 39 (1916): 367–81, 403–12, 443–52, 498–503. The passages in which Marius presents himself as a Tychonic are on pp. 372, 376, 379, and esp. p. 447.

287. *GO,* vol. 11, no. 973, p. 21.

288. Ibid., no. 437, pp. 484–85. I tend to interpret their previous reluctance to endorse Galileo's claims not as a sign of their philosophical and cosmological conservatism, but rather as a sign of competitiveness. Galileo's discoveries had caused such excitement that several mathematicians tried either to deny them or to claim that they had achieved them before Galileo.

289. *GO,* vol. 11, no. 520, p. 93 (emphasis mine). The two qualifications regarded the interpretation (but not the observation) of the irregular appearance of the moon and the shape of Saturn, which they saw as oblong rather than three-bodied.

290. Actually, the Jesuits' Tychonic leanings began to emerge soon after Galileo's discoveries. Referring to the 1611 edition of Clavius's classic *Commentary on Sacrobosco's Sphere,* Scheiner wrote in 1612 that "moved by these phenomena recently discovered (though ancient in themselves), Clavius advised astronomers to start thinking of some other cosmic system" (Christopher Scheiner, *De maculis solaribus et stellis circa Iovem errantibus, accuratior disquisitio ad Marcum Velserum* [*GO,* vol. 5, 68–69], as translated in Shea, "Galileo, Scheiner, and the Interpretation of Sunspots," p. 502.

decades within the Society.[291] If the Jesuit *mathematicians* were quite zealous at confirming Galileo's discoveries and eager to give him a public triumph in Rome in the spring of 1611 (in the presence of several cardinals) it was not because of the Copernican implications of his discoveries but because the discoveries reinforced the cognitive claims of mathematicians over those of the philosophers—a battle for disciplinary legitimation the Jesuit mathematicians shared with Galileo.[292]

The connection between Galileo's discoveries and Copernicus was not an automatic one. Depending on one's beliefs, socioprofessional iden-

291. Peter Dear, "Jesuit Mathematical Science and the Reconstruction of Experience in the Early Seventeenth Century," *Studies in History and Philosophy of Science* 18 (1987): 133–75; Ugo Baldini, "La nova del 1604 e i matematici e filosofi del Collegio Romano," *Annali dell'Istituto e Museo di Storia della Scienza di Firenze* 6 (1981): 63–98; idem, "Additamenta Galilaeana I: Galileo, la nuova astronomia e la critica dell'aristotelismo nel dialogo epistolare tra Giuseppe Biancani e i revisori romani della Compagnia di Gesù", *Annali dell'Istituto e Museo di Storia della Scienza di Firenze* 9 (1984): 13–43; William A. Wallace, *Galileo and His Sources* (Princeton: Princeton University Press, 1984), pp. 126–48; Alistair C. Crombie, "Mathematics and Platonism in the Sixteenth-Century Italian Universities and in Jesuit Educational Policy," in Y. Maeyama and W. G. Saltzer, eds., *Prismata* (Wiesbaden: Steiner Verlag, 1977), pp. 63–94; and Adriano Carugo and A. C. Crombie, "The Jesuits and Galileo's Idea of Science and Nature," *Annali dell'Istituto e Museo di Storia della Scienza di Firenze* 8 (1983): 3–68. I think that the partial overlap between the Jesuits' and Galileo's agendas may explain some of the tensions that later emerged between them. While the Jesuits endorsed the new discoveries in order to legitimize their position within the society, they could not—because of institutional constraints—totally endorse Copernicanism as Galileo wanted to do. Not only could they not support Galileo in his Copernican developments, but it may be that their endorsement of Tycho made things worse and propelled Galileo toward more radical positions. In fact, by becoming Tychonic, the Jesuits could make sense of the new discoveries in a way that would establish their independence from the Scholastic philosophers without upsetting them too much. Thus the Jesuits' Tychonic domestication of Galileo's discoveries was more insidious to him than the outright opposition to them by traditional philosophers. While Galileo's discoveries were devastating to Ptolemy, they were not so to Tycho. That explains, I think, Galileo's dismissive attitude in the *Assayer* about Tycho's model as a system. His exclusion of the model from the "chief world systems" is more evident in the *Dialogue*. In short, Galileo's increasing hostility toward the Jesuits may reflect not only his disagreement with their positions, but the fact that their agenda was ruining his plans by providing a more traditional interpretation of his discoveries. Also, Galileo was facing a problem of "product differentiation." He could not be Tychonic because that would have deprived him of his originality. He wished instead to be recognized as an *astronomo filosofo*.

292. The *Avviso di Roma,* which reported the event, is quoted in full in J. A. F. Orbaan, *Documenti sul barocco in Roma* (Rome: Società Romana di Storia Patria, 1920), 2: 284. The Jesuits' celebratory attitude about Galileo's discoveries reflects the same disciplinary tensions that had several astronomers cheering at the discovery of the new star of 1604—one that challenged Aristotelian cosmology rather than Ptolemaic astronomy. For instance, a friend of Galileo, the astrologer Altobelli, saw the nova as shattering the beliefs of the "half-philosophers": "It pleases me that Your Lordship has noticed this new monster in the sky, [something] to drive crazy the Peripatetics who until now have believed so many lies about that new and miraculous star, devoid of motion and parallax" (*GO,* vol. 10, no. 106, p. 117). Altobelli continued his attack on the "half-philosophers" in a subsequent letter (ibid., no. 107, pp. 118–20).

tity, and patronage outlook, the Copernican dimensions of the discoveries could be legitimately emphasized or effaced. Moreover, these various interpretations were not produced in the calm of some philosophical recess but in the midst of hot disputes. The participants' opinions changed as the debate evolved and their positions in it were modified accordingly. This is reflected in the different ways in which Galileo's discoveries were challenged. The meaning of the observations was *modified* in the process of their being attacked and defended by different people with different resources and agendas.

In fact, the early challenges and Galileo's related responses focused on the very existence of the things he claimed to have observed rather than on their Copernican significance. The reliability of the telescope, rather than the philosophical or theological plausibility of the Copernican system, was the main target of Galileo's adversaries and the subject of his replies in the period immediately following the publication of the *Sidereus nuncius*.[293] Judging from Galileo's correspondence and the printed attacks on the *Sidereus nuncius*, it seems that Copernicanism had begun to emerge as an important issue in the public debate between Galileo and his opponents only at the beginning of 1611.[294] Martinus Horky's *Brevissima peregrinatio contra nuncium sidereum*—the first (and quite virulent) tract to

293. A reading of Galileo's 450-letter correspondence from January to December 1610 supports this point. Although most of this correspondence deals with or mentions some aspect of his astronomical discoveries, Galileo expresses his belief in Copernicus to Giuliano de' Medici, Castelli, and Clavius only after his discovery of the phases of Venus in December. Of the very many letters Galileo received in that year, only six mention Copernicus (and four of them are from either Kepler or Hasdale). Among the letters not addressed to Galileo but dealing with him, we find only two (by Kepler and Maestlin) that mention Copernicus in connection to Galileo's discoveries. In short, it seems that Galileo was very cautious about voicing his opinions about Copernicus, even in his private letters, until he discovered the phases of Venus almost a year after his first observations. Moreover, most of the people who linked Galileo's discoveries to Copernicanism were Copernicans to begin with. This suggests that the vast majority of Galileo's audience did not see, or care about, the Copernican significance of his discoveries. For a survey of the responses to the *Nuncius*, see Albert Van Helden's "Conclusion" in his translation of Galileo's *Sidereus nuncius*, pp. 87–113.

294. The first printed theology-based attack on Galileo to contain theological arguments was Francesco Sizi's *Dianoia*, which did not, however, challenge the Copernican implications of Galileo's discoveries. Rather, Sizi argued against the existence of the satellites of Jupiter based on the Bible's claim about the number of existing planets. The first theological attack came in Ludovico delle Colombe's "Contro il moto della terra." However, delle Colombe's piece was not printed but circulated in manuscript form, mostly in Florence. Thus the perception of the Copernican dimensions of Galileo's discoveries, and theological attacks on them, were far from common, nor did they come from distinguished philosophers or theologians. Other debates on the *Sidereus nuncius* focused on Galileo's interpretation of the irregular surface of the moon and its anti-Aristotelian implication. However, even in this case, the philosophers seemed generally busy defending the Aristotelian system from the devastating effect of Galileo's discoveries rather than attacking the Copernican astronomy that may have been supported by them.

be printed against Galileo's discoveries in the summer of 1610—did not bring up the issue of Copernicanism but attacked Galileo's empirical claims as untenable.[295]

This delay in the emergence of the Copernican dimensions of the discoveries should not surprise us. It makes perfect sense that Galileo began to be attacked for the Copernican implications of his discoveries only *after* the reliability of his telescope began to be accepted. In a sense, attacks on his Copernicanism were a sure sign of his enemies' taking his telescope and discoveries seriously. However, the legitimation of Galileo's instrument and discoveries that propelled the debate to the next stage (one characterized by the deployment of philosophical and then theological arguments against Copernicanism) was framed by patronage dynamics.[296] As we will see in the next chapter, Galileo did not win over his audience by explaining the optical behavior of the telescope, to prove that it did not deceive the viewers, or by introducing a full-fledged alternative epistemology to Aristotle's in order to legitimize the use of instrument-mediated sense data.[297] Instead, he tried to bring to closure the debate over the reliability of the telescope (to "blackbox" it, as the sociologists would say) by tying the Medici's image to his discoveries and by mobilizing the resources available through their diplomatic networks.[298] We should remember that Kepler's opinion on Galileo's discoveries was requested from him by the emperor and by the Medici ambassador. Sim-

295. Horky's pamphlet is reprinted in *GO*, vol. 3, pp. 129–45.

296. However, it is important to notice that mathematicians and philosophers each reacted differently to Galileo's discoveries. The philosophers (and theologians) tended to attack Galileo on Copernicanism, probably because they saw him invading the domain of natural philosophy. Instead, the mathematicians tended first to question the reliability of the telescope, but, once convinced, they usually desisted from their attacks. Magini provides an interesting case. He initially opposed Galileo's discoveries very energetically, but then turned around and even tried to adopt Galileo's patronage strategies. In fact, the patronage success of Galileo's telescope made him realize that optical instruments could become court marvels and enhance the status of their inventor. Interestingly enough, Magini's postconversion letters to Galileo often referred to a large concave mirror that he had sent to the emperor, and to the negotiations about its purchase. Magini's emphasis on the details of the transaction seems to suggest that he wanted to show Galileo that he too could produce such marvels and be rewarded for them (*GO*, vol. 10, no. 400, pp. 437–38; no. 404, pp. 442–43; no. 408, p. 446; no. 444, p. 496). In sum, mathematicians and philosophers tended to share an initial reaction to Galileo's discoveries, but their attitudes parted soon afterward. These responses may be accounted for by considering the disciplinary differences between mathematics and philosophy.

297. On these issues, see Paul Feyerabend, *Against Method* (London: Verso, 1975), pp. 99–143.

298. The notion that Galileo tried to develop an alternative audience for his discoveries and natural philosophy, rather than convince the traditional philosophers, had already been stressed by Paul Feyerabend (see ibid., pp. 141–43). On the "blackboxing" of instruments in connection with the closure of scientific disputes, see Harry M. Collins, *Changing Order* (London: Sage, 1985); and Latour, *Science in Action*.

ilarly, all the telescopes and copies of the *Sidereus nuncius* Galileo had distributed to princes and cardinals through the Medici diplomatic networks made sure that plenty of "people who counted" had seen the satellites of Jupiter. Interestingly enough, some of these telescopes were not requested but were voluntarily sent by Galileo (with the Medici's approval) in a clever attempt to enlist those European princes in his camp.[299] Once the European courtly *monde* had observed the Medicean Stars, Galileo hoped he did not need to worry too much about the various tracts of Horky, Sizi, and delle Colombe against his discoveries.[300]

For instance, in July 1610 Martin Hasdale wrote to Galileo from the imperial court at Prague informing him that the increasingly frequent challenges to his discoveries (fueled by reports of Galileo's failure to convince Magini and others during his visit to Bologna) had almost overwhelmed "poor Kepler." Fortunately, Galileo's opponents were much meeker now because the Emperor had begun to endorse Galileo's claims.[301] Hasdale's letter indicates that it was Rudolph's endorsement rather than either Magini's skepticism or Kepler's support that settled the dispute on Galileo's claims (though Rudolph may have not been able to see the satellites without Kepler's help).[302] Galileo's blackboxing of the telescope and his observational claims resulted also from his ability to connect himself to

---

299. *GO*, vol. 10, no. 277, p. 301.

300. Galileo's correspondence of that period indicates that a primary goal of his defense was to prevent Cosimo II from developing doubts about the reality of the planets Galileo had named after his family (*GO,* vol. 10, no. 307, p. 349; no. 339, pp. 379–82). With all the copies of the *Nuncius* and the many telescopes distributed by the Medici to the major European courts, an international scandal would have broken out if the planets had turned out to be spurious. For Galileo that would have been the end of his career. And it would have been very unlikely that in those circumstances he could have gone back to Padua.

301. *GO*, vol. 10, no. 360, p. 401. However, because of his status, the emperor would not put his endorsement in writing. Therefore, the impact of his support of Galileo's discoveries was largely local. I would suggest that Rudolph's willingness to support Galileo's observations (though only within the local context of his court) was also connected to the fact that it was not his honor (but the Medici's) that was at stake with Galileo's claims. Rudolph's quite unconventional philosophical taste may have also contributed to his willingness to recognize Galileo's revolutionary discoveries once he looked through the telescope's eyepiece.

302. I am suggesting that Kepler may have had a crucial role in Rudolph's "seeing" the satellites of Jupiter by providing him with what Wittgenstein and Kuhn have called an "ostension." In fact, because of the problems encountered by early users of telescopes (narrow field of view, various distortions, and the unconventionality of instrument-mediated evidence) as well as the revolutionary nature of the claims themselves, the recognition of Galileo's discoveries may have entailed the extension or renegotiation of one's conceptual and perceptual repertoire—a different sorting out of "signal" from "noise." Such a process may have required an ostension. Galileo himself knew that ostensions helped. When he sent the first telescope to Florence, he made sure that Enea Piccolomini (a friend of his and a supporter of his claims) saw it. Then, during his visit to Pisa at Easter in 1610, Galileo himself showed the satellites to Cosimo. Similarly, writing to Matteo Carosio in May 1610, he claimed he did not have problems convincing people of the truth of his discovery provided he

Galileo's Self-fashioning

97

the socially powerful through the Medici diplomatic networks and in this way to bypass or overpower the professional communities that may have opposed those claims (the philosophers and some of the mathematicians). The type of evidence produced by Galileo's telescope was also crucial to the success of his prince-oriented strategies. While the spectacularity of his claims and the "exoticness" of the telescope made princes eager to look through it, the nontechnical nature of the visual evidence provided by that instrument may have been digestible to them. One could even argue that the princes' lack of an extensive knowledge of philosophy and optics may have facilitated their perception of Galileo's claims as trustworthy.[303]

However, Galileo's new patronage-produced visibility and status were not only important resources to rebuff attacks on the reality of his discoveries: *they were crucial also in eliciting them.* I would suggest that usually Galileo was not attacked because he was a Copernican but because of his (and his discoveries') extreme visibility and his success in becoming the mathematician and philosopher of the grand duke. People seeking status attacked those who were perceived as having high status. Similarly, in discussing Galileo's involvement in several challenges about the quality of telescopes, Albert Van Helden has pointed to the "political" rather than simply scientific agenda of those debates. In particular, he has argued that Galileo's behavior could be understood also in terms of his attempts to maintain and reinforce his status as the authority on all matters telescopic, a role that was a direct result of his conspicuous position at the Medici court.[304]

Actually, Galileo and his adversaries followed analogous tactics that

---

could show the satellites of Jupiter to them in person (*GO*, vol. 10, no. 313, p. 357). It seems that those who replicated Galileo's observations tended to share some of his scientific agenda (Kepler and the Jesuits), or had been provided with ostensions by somebody who had already endorsed Galileo's discoveries (like Kepler), or accorded him credibility because of friendship, status, or the endorsements he had received from practitioners they considered authoritative. Of course, this hypothesis resonates well with Kuhn's emphasis on the role of "tacit knowledge" in allowing scientists to connect perceptions to concepts, as well as with the recent literature on replication of experiments through the transfer of "skill," as in Harry Collins's *Changing Order*. On these issues, see Albert Van Helden's discussion of Galileo's "didactical approach" in his "The Telescope and Authority from Galileo to Cassini," *Osiris* 9 (1993; in press), and Mario Biagioli and Albert Van Helden's forthcoming volume on the dispute on the sunspots.

303. I am not claiming that the evidence provided by the telescope was "self-evident"—it became so only after its "blackboxing." By saying that such evidence was "nontechnical," I mean that those who—like most princes—had a limited "philosophical baggage" could more easily undergo the readjustment of their perceptual and conceptual categories necessary to "see" the objects Galileo claimed to have discovered. On Galileo's use of visual evidence in astronomy, see Albert Van Helden and Mary Winkler, "Representing the Heavens: Galileo and Visual Astronomy," *Isis* 83 (1992): 195–217.

304. Albert Van Helden, "Galileo and the Telescope," in Paolo Galluzzi, ed., *Novità celesti e crisi del sapere* (Florence: Giunti Barbéra, 1984), pp. 156–57.

led them to bind each other in a feedback loop of challenges and counter-challenges. Galileo tried to achieve visibility and the status of philosopher by presenting his discoveries as challenging the "honor" of the dominant philosophy. In turn, his newly achieved status (something he had to defend without hesitation as the philosopher and mathematician of the grand duke) elicited challenges from competitive fellow mathematicians as well as from the philosophers whose worldview and socioprofessional role he had challenged.

The high status he had achieved through patronage tactics made him more vulnerable but, at the same time, provided him with resources to counter those attacks. In short, he was becoming more powerful and more vulnerable at the same time. This feedback loop increased Galileo's visibility as well as the ranks of his opponents and the insidiousness of their attacks. The more Galileo increased his status and legitimized his instrument by meeting challenges through mobilizing (and increasing) the resources provided by his patronage connections, the more his competitors tried to attack him on grounds that he could not control through his resources.[305] While claims about the unreliability of the telescope could be countered by listing the several European princes who had pleasurably used it, the claim that Galileo was trying to use his discoveries to refute the teaching of the Scriptures could not be countered simply by the Medici or Rudolph II saying that Galileo was a good Catholic. Attacks on Galileo based on the Copernican (and later theological) implications of his discoveries reflected his success at countering attacks on other grounds. In a sense, Galileo's Copernicanism was brought up by his successful handling of his adversaries—something that he had to (and could) do because of his strategic patronage situation.

To put it differently, by operating under the pressure put on him by his new socioprofessional position, Galileo produced *more* discoveries, such as the phases of Venus (a crucial piece of evidence against Ptolemy and Aristotle) and the "three-bodied" Saturn, and eventually entered the dispute on sunspots.[306] Not only did these new discoveries and disputes

305. Even Galileo's Roman "triumph" of 1611 should be understood in terms of his connection to the Medici. Galileo went to Rome as an official envoy of the Medici who not only paid for his expenses but gave him several helpful letters of introduction to the Roman high society. Thus I suggest that the warm and sometimes spectacular receptions he was granted should also be seen as tributes to the Medici, whom he represented as their "scientific ambassador." Also, his trip to Rome was not a vacation. Vinta was very clear about the relation between Galileo's visit to Rome and the legitimation of his discoveries (*GO*, vol. 11, no. 464, pp. 28–29).

306. For instance, Stillman Drake interprets Galileo's discovery of the phases of Venus as having been driven not necessarily by his desire to prove Copernicus right, but more probably by his fear that Santini, Clavius, or Kepler would get there first because the quality of their instruments was rapidly improving (Stillman Drake, "Galileo, Kepler, and Phases of Venus," *Journal of the History of Astronomy* 15 [1984]: 198–208).

CHAPTER ONE

confirm the reliability of the telescope and further upset his opponents, pushing them to counterattack Galileo on his Copernican agenda, they also gave him more resources to meet the attacks. In the process, his initial conceptual sympathies for Copernican astronomy were strengthened by his further patronage-driven discoveries. Eventually, Copernicanism became incorporated into his professional identity to become increasingly important in directing his further moves. By defending his newly acquired honor, Galileo became a full-fledged Copernican.

Also, as indicated by the previous analysis of the honor dynamics typical of the patronage system, by 1610 dropping Copernicanism was not a real option for Galileo. Not only was he sympathetic to Copernican astronomy, but, because of his newly obtained court position, he was expected to maintain a high profile by producing new philosophical claims and engaging in controversial debates. By giving up Copernicanism, Galileo would have been undaring—he would have become "normal." More important, a *realist* reading of Copernicus allowed Galileo to live up to the title he desired so much: that of philosopher.[307]

As I hope I made clear, I do *not* wish to argue that Galileo was not a Copernican when he made his discoveries. Rather, I suggest we problematize categories such as "Copernican" and "Copernicanism."[308] The letters Galileo exchanged with Kepler and Jacopo Mazzoni in 1597 show that, at the time, Galileo was a Copernican sympathizer but not yet a committed defender of the Copernican hypothesis.[309] Some historians have tended to focus on a letter by Galileo in which he told Kepler that he had believed in the Copernican theory for some time, but they have paid less attention to the fact that a subsequent letter in which Kepler tried to involve Galileo in a sort of Copernican crusade went unanswered.[310] It is one thing to be conceptually attracted to the Copernican system and another to be a committed supporter of the Copernican cause. Although I believe that by the time he wrote the "Letter to the Grand Duchess" in 1615

307. The Tychonic system did not offer the same philosophical self-fashioning opportunities as did the Copernican system. Since the former was a largely mathematical model (its physical interpretation would have been quite nightmarish), its endorsement would not have provided a powerful resource toward presenting oneself as a "philosopher."

308. Westfall, "Science and Patronage," esp. pp. 26–29. For a critique of Westfall's views of Galileo's Copernicanism, see Biagioli, "Galileo's System of Patronage," 42–45. For a different problemization of Copernicanism, see Robert Westman, "The Melanchthon Circle, Rheticus, and the Wittenberg Interpretation of the Copernican Theory," *Isis* 66 (1975): 165–93.

309. Galileo's letter to Mazzoni is in *GO*, vol. 2, pp. 197–202, esp. p. 198, where he says he "considered the opinion of the Pythagoreans and of the Copernicans about the motion and location of the earth . . . as much more probable than the other one of Aristotle and Ptolemy."

Galileo was a full Copernican, we should not lose sight of the patronage-laden process that resulted in that commitment.[311] Copernicanism, patronage, and philosophical self-fashioning went hand in hand. At least at this juncture.

310. *GO*, vol. 10, no. 57, pp. 67–68; no. 59, pp. 69–71.

311. Prior to the "Letter to the Grand Duchess," Galileo's strongest endorsement of Copernicus is in a March 1614 letter to Giovanni Battista Baliani (*GO*, vol. 12, pp. 34–35).

TWO

Discoveries and Etiquette

THE LONGEST AND MOST BOMBASTIC ENTRY in the philosophy section of the catalog of the spring 1610 Frankfurt book fair reads:

Sidereal messenger, unfolding great and very wonderful sights and displaying to the gaze of everyone, but especially philosophers and astronomers, the things that were observed by Galileo Galilei, Florentine patrician and public mathematician of the University of Padua, with the help of a spyglass lately devised by him, about the face of the Moon, countless fixed stars, the Milky Way, nebulous stars, especially about four planets flying around the star of Jupiter at unequal intervals and periods with wonderful swiftness; which, unknown by anyone until this day, the author was recently first to detect.[1]

The short book behind this not-so-subtle advertisement had a very brief but remarkable history, and its publication radically changed Galileo's life and scientific career.

When, in the summer of 1609, Galileo succeeded in constructing a telescope that was remarkably better than those previously built in northern Europe, he was a professor of mathematics at Padua. With this new instrument he made a number of astronomical discoveries that contradicted the dominant Aristotelian cosmology and could be used to support the claims of the Copernicans. In the spring of 1610 he presented his exceptional discoveries in the *Sidereus nuncius,* which he dedicated to

1. *Catalogus universalis pro nundinis Francofurtensibus vernalibus de anno MDCX* (Frankfurt: Latomi), C3v. ("C" does not mean "carta" but designates section "C" of the catalogue. There are no page numbers). The entry is a transcription of most of the frontispiece of Galileo's book. Except for a very minor change, this translation is from Galileo Galilei, *Sidereus nuncius,* trans. Albert Van Helden (Chicago: University of Chicago Press, 1989), p. 26.

Cosimo II de'Medici, the grand duke of Tuscany. He announced that the surface of the moon was far from smooth, as the philosophers had claimed, and that the number of stars was much greater than had previously been believed. He also made the explosive claim that there were four more planets—which he called Medicean Stars—than the dominant cosmology recognized, and that these circled Jupiter, not the Earth. The *Sidereus nuncius* brought him international visibility and opened for him the doors of Medici patronage. By September 1610 Galileo was back in Florence, with no teaching duties and with the remarkable stipend of a thousand scudi a year.

The award of a thousand-scudi stipend was very unusual in comparison to the stipends of other important artists and officials of the Medici court. Although it is difficult to produce absolute comparisons of courtiers' incomes for they usually exceeded their salaries, Galileo's stipend appears to have been at least three times that of any highly paid artist or engineer and one and a half times that of a *Primo Segretario* such as Belisario Vinta or Curzio Picchena.[2] Galileo's stipend was comparable to that of the *Maggiordomo Maggiore*—the highest court official. Even the sculptor Gianbologna—the most famous among the Medici artists at the beginning of the century and one who was repeatedly courted by two emperors—made in 1606 less than half the salary Galileo would receive a few years later.[3] As far as I can tell, Galileo's salary was among the ten highest of the grand duchy of Tuscany at that time.[4]

2. This is because certain courtiers had bonuses such as meals, wood, candles, and horses in addition to their salaries (*ASF,* "Despositeria generale 389," pp. 5, 11). For a comparison between Galileo's salary and that of other state officials, and for an analysis of the various sources of officials' income, see R. Burr Litchfield, *Emergence of a Bureaucracy: The Florentine Patricians, 1530–1790* (Princeton: Princeton University Press, 1986), pp. 190–200.

3. Hugh Trevor-Roper, *Princes and Artists* (London: Thames and Hudson, 1976), pp. 109–12, 130. Giambologna made 300 scudi per year in 1602 (*ASF,* "Miscellanea medicea 474," fol. 3) as well as in 1606 (*ASF,* "Guardaroba medicea 279," fol. 13). He appears as the highest-paid artist in both those two *ruoli.*

4. The highest court salary in 1588 was that of Orazio Rucellai—the *Maggiordomo Maggiore*—who made 1,000 scudi per year (*ASF,* "Depositeria generale 389," p. 1); Belisario Vinta, a *Segretario,* made 480 scudi per year (ibid., p. 5); and Ostilio Ricci, court mathematician, made 144 scudi (ibid., p. 9). Rucellai's was still the highest salary in 1599 (*ASF,* "Guardaroba medicea 255," fol. 2r). In 1609, the second highest salary was that of the *Maggiordomo* Iacopo de' Medici, who made 600 scudi per year (*ASF,* "Guardaroba Medicea 301," fol. 1r). In 1624, the highest salary at court was that of the new *Maggiordomo Maggiore,* Piero Guicciardini, who made 1000 scudi per year (*ASF,* "Depositeria generale 396," fol. 36). Matteo Neroni, the court cosmographer, made 120 scudi (ibid., fol. 115). The salaries of the chief commanders of the Tuscan infantry, artillery, and cavalry ranged between 1000 and 2500 scudi per year (see "Relazione delli Clarissimi Signori Giovanni Michiel et Antonio Tiepolo Cavalieri ritornati Ambasciatori dal Granduca di Toscana alli 9 novembre 1579," in Arnaldo Segarizzi, ed., *Relazioni degli ambasciatori veneti al Senato* (Bari: Laterza, 1916), 3: 256–59, 269.

Having been socialized in a culture that takes for granted the scientific importance of Galileo's astronomical discoveries of 1609–10, we may think it natural that the Medici rewarded him so lavishly. But Galileo did not become philosopher and mathematician to the grand duke because of his contributions to the proof of the Copernican hypothesis. The Medici court was not the Nobel Prize committee *avant la lettre* and Cosimo II was no Copernican. Westfall has argued, quite correctly, that the Medici rewarded Galileo's discoveries not because of their technological usefulness or scientific importance, but because they prized them as spectacles, as exotic marvels.[5] And the Medici must have perceived the satellites of Jupiter as truly exceptional marvels since Galileo's attempts to move to the Medici court, repeatedly frustrated before 1610, were quickly and generously welcomed after these discoveries. The explanation for this exceptional reward is hardly to be found in the appreciation of the scientific significance of Galileo's discoveries by contemporary mathematicians and philosophers. Instead, it is by looking at a quite different audience, that of the Medici court, and at Galileo's representation of his discoveries as fitting court discourse that we can understand why Cosimo II called Galileo back to Florence.

Although courtiers were generally incompetent in astronomy and mathematics, Galileo must have considered the court an important space for his work if, after 1604, he repeatedly tried to leave the university and move there.[6] And it was more than the good salary and freedom from teaching that attracted him. By moving to court, he also hoped to avoid the constraints of the disciplinary hierarchy characteristic of the university, a hierarchy in which mathematicians were subordinated to philosophers in terms of both salary and professional status.[7] Philosophy, it was held, dealt with the real causes of natural phenomena, while mathematics could only deal with their "accidents," that is, with their quantitative as-

---

5. Richard Westfall, "Scientific Patronage: Galileo and the Telescope," *Isis* 76 (1985): 11–30; and idem, "Galileo and the Accademia dei Lincei," in Paolo Galluzzi, ed., *Novità celesti e crisi del sapere* (Florence: Giunti Barbéra, 1984), p. 199.

6. *GO*, vol. 10, no. 97, pp. 106–7; no. 99, p. 109; no. 131, pp. 154–55; no. 190, pp. 210–13; no. 209, pp. 231–34; no. 211, p. 235. See also Richard Westfall, "Scientific Patronage," pp. 13–17.

7. This pattern has been identified by Robert S. Westman in "The Astronomer's Role in the Sixteenth Century: A Preliminary Study," *History of Science* 18 (1980): 105–47, and elaborated in his "The Copernicans and the Churches," in David C. Lindberg and Ronald L. Numbers, eds., *God and Nature* (Berkeley: University of California Press, 1986), pp. 73–113. For the Italian scenario, see Mario Biagioli, "The Social Status of Italian Mathematicians, 1450–1600," *History of Science* 27 (1989): 41–95. The connection between Galileo's concern with the title of philosopher and the legitimation of mathematical physics as philosophy was also discussed by Eugenio Garin, "Galileo the Philosopher," *Science and Civic Life in the Italian Renaissance* (New York: Anchor, 1969), pp. 123–25, and Michael Segre, "Galileo as a Politician," *Sudhoffs Archiv* 72 (1988): 75.

pects. Consequently, mathematicians were not entitled to produce legitimate *physical* interpretations of natural phenomena.[8]

But if a mathematician as mathematician could not become a philosopher in the university, he could do so at court, where one's social and cognitive status was determined less by one's discipline than by the prince's favor. As a Roman court proverb put it, the courtier is like paper money whose exchange value is set by the prince only.[9] Consequently, the court was a social institution in which Galileo could obtain the title of philosopher which, in turn, would give him the standing to argue legitimately for the philosophical significance of the Copernican theory and for the mathematical analysis of natural phenomena.

Stars in Context

Some reasons for the Medici's interest in the moons of Jupiter are easy to grasp. As Galileo asserted in the dedication of the *Sidereus nuncius,* these new planets were monuments to the Medici dynasty.[10] Moreover, they were monuments of exceptional durability and worldwide visibility—at least for audiences equipped with good telescopes. But there were other reasons behind the Medici's enthusiasm for Galileo's discoveries, reasons fully apparent only to a Florentine audience familiar with the mythology the Medici had been articulating since Cosimo I established the dynasty in the middle of the sixteenth century. In this mythology, a correspondence was drawn between the cosmos and Cosimo, and Jupiter was regularly associated with Cosimo I, the founder of the dynasty and the first of the "Medicean gods."[11] Consequently, while Galileo could have dedicated the newly discovered planets to any patron, the Medici were in the position to fully appreciate (and reward) the mythological significance of Galileo's discoveries.

Although the Medici had been the de-facto rulers of an allegedly republican Florence since the fifteenth century, the dukedom itself was of much more recent origin. In fact, Cosimo I became duke of Florence in 1537 and was made grand duke of Tuscany only in 1569. During the 1540s he

8. Peter Dear, "Jesuit Mathematical Science and the Reconstitution of Experience in the Early Seventeenth Century," *Studies in History and Philosophy of Science* 18 (1987): 133–75; Nicholas Jardine, *The Birth of History and Philosophy of Science* (Cambridge: Cambridge University Press, 1984), pp. 225–57; and Robert S. Westman, "Kepler's Theory of Hypothesis and the 'realist dilemma,'" *Studies in History and Philosophy of Science* 3 (1972): 233–64.

9. Francesco Liberati, *Il perfetto Maestro di Casa* (Rome: Bernabò, 1658), p. 9.

10. Galilei, *Sidereus nuncius,* pp. 29–33.

11. Giorgio Vasari, *Ragionamenti di Giorgio Vasari sopra le invenzioni da lui dipinte in Firenze nel Palazzo di loro Altezze Serenissime con lo Illustrissimo ed Eccellentissimo Don Francesco de' Medici* [published posthumously by Vasari's nephew in 1588], in Gaetano Milanesi, ed., *Le opere di Giorgio Vasari* (Florence: Sansoni, 1882), 8: 85.

had to create the political and administrative structure of the new state, along with a new political mythology that would stabilize the Medici rule and present it as a dynastic one.[12] After becoming duke of Florence, Cosimo needed to establish a court out of almost nothing. The powerful Florentine families were to be transformed from former political leaders into a docile court aristocracy, and the new mythology that represented the ducal rule as natural and necessary was to indicate the role the powerful Florentine families had to assume within it.[13]

Cosimo's strategy was to represent the Medici rule as Florence's manifest destiny. The city's horoscope, so commonly cast since the Middle Ages, was normalized to suggest the astrological necessity of Medici rule by linking that rule to the history and fate of the city.[14] New Medici-oriented histories and Medici-sensitive reinterpretations of ancient myths were commissioned, while Medici-related imagery was introduced in Florentine art.[15] Most important, Medici-controlled academies—among them the Accademia Fiorentina and the Accademia del Disegno—were established to manage this cultural program.[16]

12. Standard works on the period are Riguccio Galluzzi, *Istoria del granducato di Toscana sotto il governo della Casa Medici* (Florence: Cambiagi, 1781); Furio Diaz, *Il Granducato di Toscana: I Medici* (Turin: UTET, 1976); and Giorgio Spini, ed., *Architettura e politica da Cosimo I a Ferdinando I* (Florence: Olschki, 1976).

13. R. Burr Litchfield, *Emergence of a Bureaucracy: The Florentine Patricians 1530–1790* (Princeton: Princeton University Press, 1986). Glimpses of Florentine courtly life can be found in P. F. Covoni, *Don Antonio de' Medici al Casino di San Marco* (Florence: Tipografia Cooperativa, 1892); Gaetano Pieraccini, *La stirpe dei Medici di Cafaggiolo* (Florence: Nardini, 1986); Graziella Silli, *Una corte alla fine del Cinquecento* (Florence: Alinari, 1927); Gaetano Imbert, *La vita fiorentina nel Seicento* (Florence: Bemporad, 1906); and Angelo Solerti, *Musica, ballo e drammatica alla corte medicea dal 1600 al 1637* (Florence: Bemporad, 1905). Solerti's book reproduces large sections of the official manuscript court diary.

14. The relationship between the city's horoscope and Medici fate up to Cosimo I is elaborated in Janet Cox-Rearick, *Dynasty and Destiny in Medici Art* (Princeton: Princeton University Press, 1984). On the early-Renaissance city horoscopes in Florence, see Richard Trexler, *Public Life in Renaissance Florence* (New York: Academic Press, 1980), pp. 73–84.

15. Probably the best example is Benedetto Varchi, *Storia fiorentina*, ed. Gaetano Milanesi (Florence: Le Monnier, 1857–58), 3 vols. On Medici imagery in art, see Cox-Rearick, *Dynasty and Destiny*, p. 231.

16. The Accademia Fiorentina, established in 1540, was the first academy sponsored and controlled by the Medici. It coordinated Cosimo I's cultural politics based on the normalization of Florentine culture around the axis of linguistic identity. See Sergio Bertelli, "Egemonia linguistica come egemonia culturale e politica nella Firenze Cosimiana," *Bibliothèque d'Humanisme et Renaissance* 38 (1976): 249–83; and Cosimo di Filippo Bareggi, "In nota alla politica culturale di Cosimo I: L'Accademia Fiorentina," *Quaderni storici* 23 (1973): 527–74. The Accademia del Disegno, established in 1564 and run by a "lieutenant" appointed by Cosimo, was also a component of his cultural system. Its main functions were to coordinate the work of visual artists working for the Medici and to make sure that the codes of the Medici cultural politics were respected. In fact, the artists of the Accademia del Disegno managed large political spectacles ranging from weddings to funerals to visits of foreign dig-

Although Cosimo did not go so far as to commission a family history in the form of a Greek-style theogony, he had classical theogonies allegorically reinterpreted to resemble the history of the house of Medici. This mythological program was best articulated in Vasari's frescoes decorating the Apartment of the Elements and the Apartment of Leo X in the Palazzo della Signoria—the first Medici court palace, later known as Palazzo Vecchio.[17]

The project's basic schema is clear enough. The Apartment of the Elements was a kind of Olympus divided into several rooms, each dedicated to a specific god (Hercules, Jupiter, Ops, Ceres, Saturn) or to a predivine entity such as the primordial "elements" (fig. 1). Right below the Olympus of the Apartment of the Elements we find the Medici pantheon—the Apartment of Leo X. Each room of the Apartment of Leo X is dedicated to a member of the Medici family who was instrumental in establishing the dynasty (fig. 2).

Each room dedicated to a Medici in the Apartment of Leo X was put, as Vasari says, in a plumb-line relation with a god-dedicated room in the Apartment of the Elements directly above it. The frescoes of each room downstairs present a mythologized history of the member of the Medici family the room honors. Each history was made to mirror as closely as possible the classical theogony of the corresponding god. The Room of the Elements, the primordial entities that allowed for the formation of all things, corresponded to the Room of Leo X, the Medici pope who made the emergence of the Medici dynasty possible. As Vasari put it, "There is nothing painted upstairs that does not correspond to something painted downstairs."[18] The heavenly order legitimized and naturalized the earthly one. Appropriately elegant stairs ensured communication between the two floors.

Vasari describes in detail the intricacies of the entire Medici mythology as presented in these frescoes.[19] What we need to consider here is the specific correspondence established in them between Jupiter (the greatest of the gods) and Cosimo I (the founder of the grand duchy of Tuscany), for that mythological relation played a crucial role in Galileo's patronage tactics.

---

nitaries. The Accademia del Disegno was a kind of "department of public relations" of the Medici court. For bibliographical references, see note 22 below.

17. Ettore Allegri and Alessandro Cecchi, *Palazzo Vecchio e i Medici* (Florence: SPES, 1980), pp. 55–182. The letters between Vasari and Cosimo's humanistic advisors on the iconography and emblematics of the Apartments are in Karl Frey, ed., *Il carteggio di Giorgio Vasari* (Munich: Muller, 1923), vol. 1, no. 220, pp. 409–12; no. 221, pp. 412–14; no. 232, pp. 436–37; no. 234, pp. 438–41; no. 236, pp. 446–50. The official nature of the mythological narrative of the two Apartments is confirmed by its being designed by Vincenzo Borghini—the first "lieutenant" of the Accademia del Disegno.

18. Vasari, *Ragionamenti*, 85.   19. Ibid.

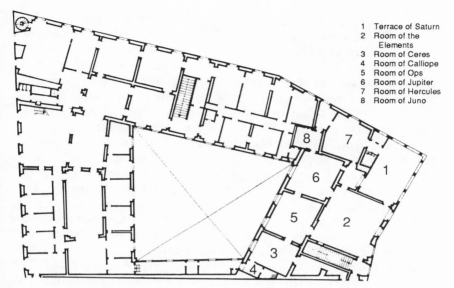

1 Terrace of Saturn
2 Room of the
  Elements
3 Room of Ceres
4 Room of Calliope
5 Room of Ops
6 Room of Jupiter
7 Room of Hercules
8 Room of Juno

Fig. 1. Apartment of the Elements, adapted from Ettore Allegri and Alessandro Cecchi, *Palazzo Vecchio e i Medici* (Florence: SPES, 1980), p. xxv.

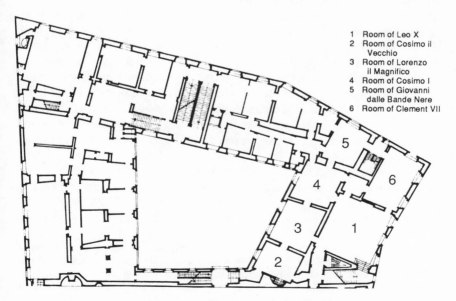

1 Room of Leo X
2 Room of Cosimo il
  Vecchio
3 Room of Lorenzo
  il Magnifico
4 Room of Cosimo I
5 Room of Giovanni
  dalle Bande Nere
6 Room of Clement VII

Fig. 2. Apartment of Leo X, adapted from Ettore Allegri and Alessandro Cecchi, *Palazzo Vecchio e i Medici* (Florence: SPES, 1980), p. xxi.

The correspondence between the room of Jupiter and that of Cosimo I is the pivot for the mythological narratives developed throughout the paintings of the two apartments. The paintings in the Room of Jupiter, which present his childhood, are in fact tied to Cosimo as well. Born of Ops and Saturn, the child Jupiter was saved from the father's cruelty (Saturn tended to eat his offspring) by his mother, who hid him in a cave in Crete. There, the infant Jupiter was reared by two nymphs. One of them, Amalthea, was represented as a goat and was allegorically associated with divine providence, while Melissa, the other nymph, was an allegory of divine knowledge. The message was that Cosimo absorbed, literally, those virtues in the cradle. In memory of Amalthea, Jupiter added the sign of Capricorn to the zodiac. The seven stars of Capricorn became emblems of the seven virtues, three theological and four moral. Quite conveniently, Capricorn happened to be Cosimo's sign, thereby confirming the destiny uniting the first grand duke and Jupiter. Thus Cosimo was endowed with divine providence and knowledge by Jupiter and received the seven virtues from Capricorn.

In the dedication of the *Sidereus nuncius* to Cosimo II, Galileo himself introduced the analogy between the Medicean Stars and Cosimo I's virtues, some moral, others "Augustan." He claimed that young Cosimo obtained those same virtues (which, according to Galileo, he displayed all the time) directly from Jupiter, which was just above the horizon at the moment of his birth. Those virtues were "emanating" from the four stars which—like innate virtues—always revolved very closely around Jupiter and never abandoned him. Therefore, given the link between Jupiter and Cosimo I, Galileo was suggesting that Cosimo I passed on his (and Jupiter's) virtues to his successor through the Medicean Stars, and that Galileo himself, by revealing these stars, was somehow midwife to this astrologico-dynastic encounter. The correspondence between the Medicean Stars and the four moral virtues was accepted by the Medici's humanistic advisors: even in the thirty years following Galileo's condemnation, the four moral virtues were used as painterly allegorical representations of the four stars.

These mythologies were more than a sign of the Medici's imaginative pretentions. They constituted the "master narrative" that informed the imagery used in public political ceremonies and festivals as well as the subject matter of court poetry, theater, painting, and opera.[20] They offered a

20. Genealogies of the gods were a common genre to celebrate ruling families. On the use of this genre in theatre, see Cesare Molinari, *Le nozze degli dei* (Rome: Bulzoni, 1968). On the use of mythological imagery and emblems in Florentine civic pageantries see Annamaria Petrioli Tofani and Giovanna Gaeta Bertelà, *Feste e apparati medicei da Cosimo I a Cosimo II* (Florence: Olschki, 1969); and Arthur R. Blumenthal, *Theater Art of the Medici* (Hanover: University Press of New England, 1980). See also David Moore Bergeron, *English Civic Pageantry 1558–1642* (London: Arnold, 1971); Roy Strong, *Art and Power: Renaissance Festivals*

Discoveries and Etiquette

framework for court culture. When needed, this mythological imagery could be expanded by means of emblematic translations, as conveniently listed in sixteenth-century catalogues or dictionaries of emblems like those of Cesare Ripa, Paolo Giovio, and Andrea Alciati.[21] The entire cultural framework was maintained and articulated by Medici-controlled institutions such as the Accademia Fiorentina and the Accademia del Disegno.[22]

Court culture itself was permeated by these mythologies from the time of Cosimo I. Familiarity with them allowed the courtiers and the Florentine upper classes to engage in the game of interpreting the emblematic narratives displayed in Medici ceremonies and other political semiologies.[23] As indicated by Baldassarre Castiglione's *Book of the Courtier,* Stefano Guazzo's *Dialoghi piacevoli,* or Torquato Tasso's *De l'imprese,* skill in emblematics was a tool required of those who wanted to engage in courtly life.[24] As Castiglione put it, "Sometimes other discussions would turn on a variety of subjects, or there would be a sharp exchange of quick retorts; often "emblems" as we nowadays call them, were devised; in

1450–1650 (Berkeley: University of California Press, 1984); and Randolph Starn and Loren Partridge, *Arts of Power* (Berkeley: University of California Press, 1992).

21. Paolo Giovio, *Dialogo dell'imprese militari e amorose* (Rome, 1551); Andrea Alciati, *Emblematum liber* (Augsburg: Steyner, 1531); Cesare Ripa, *Iconologia* (Rome: Gigliotti, 1593; Lepiolo Facis, 1603 [this is the first illustrated edition]). A standard secondary source is Mario Praz, *Studies in Seventeenth-Century Imagery* (Rome: Edizioni di Storia e Letteratura, 1964). See also Peter M. Daly, *Literature in the Light of the Emblem* (Toronto: University of Toronto Press, 1979).

22. On the Accademia del Disegno, see Zygmunt Wazbinski, *L'Accademia medicea del Disegno a Firenze nel Cinquecento* (Florence: Olschki, 1987), 2 vols.; Karen-edis Barzman, "Liberal Academicians and the New Social Elite in Grand Ducal Florence," in Irving Lavin, ed., *World of Art: Themes of Unity and Diversity* (University Park: Pennsylvania State University Press, 1989), 2: 459–63; Mary Ann Jack, "The Accademia del Disegno in Late Renaissance Florence," *Sixteenth Century Journal* 7 (1976): 3–20; For bibliographical references on the Accademia Fiorentina, see note 16 above.

23. Annamaria Petrioli Tofani, "Contributi allo studio degli apparati e delle feste medicee," *Firenze e la Toscana nell'Europa del '500* (Florence: Olschki, 1983), 2: 645–61; Annamaria Petrioli Tofani and Giovanna Gaeta Bertelà, *Feste e apparati medicei da Cosimo I a Cosimo II* (Florence: Olschki, 1969); Benedetto Betti, *Ordine dell'apparato fatto da' Giovani della Compagnia di San Gio. Evangelista* (Florence: Giunti, 1574); Alois Maria Nagler, *Theatre Festivals of the Medici 1539–1637* (New Haven, Conn.: Yale University Press, 1964); and Strong, *Art and Power,* pp. 3–74, 126–52. See also David Cannadine and Simon Price, eds., *Rituals of Royalty* (Cambridge: Cambridge University Press, 1987); and Sean Wilentz, ed., *Rites of Power* (Philadelphia: University of Pennsylvania Press, 1985).

24. Stefano Guazzo, *Dialoghi piacevoli* (Venice: Bertano, 1585); Torquato Tasso, *Il conte, o vero de l'imprese* (1594), reprinted in Cesare Guasti, ed., *I dialoghi di Torquato Tasso* (Florence: Le Monnier, 1901), 3: 361–444. Skills in emblematics became so entrenched in upper culture that the Jesuits began to teach emblematics as part of the rhetoric courses (Jennifer Montagu, "The Painted Enigma and French Seventeenth-Century Art," *Journal of the Warburg and Courtauld Institutes* 31 [1986]: 307, 312).

which discussions a marvelous pleasure was had."[25] Emblematics provided the courtiers with more than an engaging parlor game; it also was a powerful tool for self-fashioning.[26] Court society affirmed its own social identity by differentiating itself from the lower classes, which—although participating as spectators of some of those public ceremonies—could not fathom their full meaning.[27] Emblematics was to court spectacles what etiquette was to court behavior: it differentiated social groups and reinforced social hierarchies by controlling access to meaning.[28]

This mythologico-emblematic framework of Medici court society and culture constituted the background for Galileo's representation of his astronomical discoveries as emblems of the Medici dynasty. If he wanted to become a courtier by differentiating himself from the other practitioners of a low-status discipline like mathematics, Galileo had to use the same codes court society had adopted to differentiate itself successfully from the noncourtly masses.[29]

Courtiers, Pedants, and Mathematicians

Galileo's understanding of the courtly cultural context did indeed differentiate him from most of the other Italian mathematicians of the time.

25. Baldassarre Castiglione, *Book of the Courtier*, trans. Charles Singleton (Garden City, N.Y.: Anchor Books, 1959), p. 17.

26. On the use of emblematics as parlour games, see Thomas Frederick Crane, "Parlor Games in Italy in the Sixteenth Century," *Italian Social Customs of the Sixteenth Century* (New Haven, Conn.: Yale University Press, 1920), pp. 263–322.

27. Various authors have noticed this process of semiological control. Roy Strong (*Art and Power*, p. 27) mentions that spectators at Cosimo I's marriage in 1566 complained about the intricacy of the imagery. After 1630, once Florentine court society became both socially and spatially enclosed, less obscure metaphors began to be utilized in court spectacles (ibid., pp. 31–32). In Vasari's *Ragionamenti* we find that even Don Francesco de' Medici mentions the obscurity of the meaning of the imagery presented by Vasari: "*Prince:* Giorgio, today you make me hear things that I never thought were under these colors and images" (p. 22). Given that the dialogue is written by Vasari, such a statement means that Vasari thought of the perceived obscurity of his imagery as a praise of his skills in managing the codes of dynastic imagery.

28. On the development of etiquette, see Norbert Elias, *The History of Manners* (New York: Pantheon, 1982); *Power and Civility* (New York: Pantheon, 1982); and idem, *The Court Society* (New York: Pantheon, 1983).

29. As shown by Maria Luisa Altieri Biagi (*Galileo e la terminologia tecnico-scientifica* [Florence: Olschki, 1965]), Galileo maintained a complex relationship with the mechanical connotations of his discipline. On the one hand, he tried to downplay or even reject them. One example is Galileo's very negative reaction to Elsevier changing the title of his 1638 *Two New Sciences* in ways that he saw as ruining the "noble" image of the book and reducing it to a "vulgar" text (ibid., pp. 22–23). On the other hand, he used engineering or common terms to poke fun at the Aristotelian and Jesuit use of philosophical language, which he considered

His exceptional career and the trajectory of socioepistemological legit-imation he pursued are also related to his cultural background (a quite unusual one for a mathematician) and to the perceptions of the patronage system associated with it.

Obtaining a court position required much more than professional competence. Not only were appropriate patronage connections necessary, but one also needed to be courtly, that is, to have adequate social skills.[30] The issue of the client's competence in courtly etiquette emerges often in Galileo's correspondence and in other texts from that time. When Galileo was asked by the Medici secretary Curzio Picchena about the physician Minadoi (a professor at Padua who was being considered for a court ap-pointment) he replied that Minadoi was "of pleasant and honest manners and customs, and, in my view, able to do well at court no less than in the classroom."[31] Similarly, in writing to Vinta on behalf of the philosopher Papazzoni, who was interested in a chair of philosophy at Pisa, Galileo remarked that he was "pleasant and of gracious conversation," intimating that he would be at ease at court, an environment which—as a major pro-fessor at Pisa—Papazzoni would have to frequent.[32]

If competence in court life and culture could not be assumed for a physician or a philosopher, it was quite exceptional when found in a math-ematician. For instance, in his description of the Roman court in 1611, Gir-olamo Lunadoro took a sudden two-page detour to sing the praises of Giovanni Battista Raimondi, a mathematician and polymath who trans-lated and commented on Euclid and Archimedes. Raimondi is described as an old sage whose profound technical knowledge of mathematics and philosophy has not turned him asocial. He is "clean and proper in the way he dresses—something that is not usual among philosophers" and, more important, he is very pleasant in conversation. He does not show off his knowledge and does not try to put down others, and when he talks about

---

stuffy (ibid., p. 34). As we will see, this ambivalence is emblematic of Galileo's new socio-professional role.

30. The adoption of the lifestyle and culture of the upper classes was also a prerequisite for artists who wanted access to social legitimation and status; see Francis Haskell, *Patrons and Painters* (New Haven, Conn.: Yale University Press, 1980), pp. 18–19.

31. *GO*, vol. 10, no. 150, p. 168 (emphasis mine).

32. *GO*, vol. 11, no. 461, p. 27 (emphasis mine). In writing a letter of recommendation for his client Niccolò Aggiunti, who was up for a chair at Padua, Galileo wrote that Aggiunti would have pleased the Venetians "not only in the teaching of mathematics, but also for the exquisite intelligence he has for the study of the humanities, that is known to be highly es-teemed particularly by the Venetian nobility. As far as I can judge, I do not think [Aggiunti] is second to anybody in this subject, being a gifted writer of both prose and verse, and en-dowed with eloquence and quickness of wit to honor any eminent chair." This letter is not included in *GO*, but was discovered and published by Maria Francesca Tiepolo in "Una let-tera inedita di Galileo," *La cultura* 17 (1979): 60.

mathematics or philosophy or theology, he does so with clarity, elegance, and propriety.[33]

Lunadoro's insistence on Raimondi's exceptionality exposes cultural assumptions about the noncourtliness of mathematicians and, sometimes, philosophers. These assumptions reflected class and disciplinary biases, but they also had some empirical evidence behind them.[34] One's ability to cross the social boundaries between the university and the court, to gain social and epistemological legitimation, could not be taken for granted. Galileo himself was certainly helped in his own crossing by being the son of a musician who was well known at court and being a member of an old family who had some degree of nobility and political visibility earlier in the Renaissance. It may not be a coincidence that Galileo and Giovanni Battista Benedetti—the only two Italian mathematicians who received (or used) the title of philosopher at court—were people who could claim some degree of nobility.[35]

Galileo was not wealthy but knew how to present himself as a *gentiluomo*. He knew Giovanni della Casa's classic etiquette textbook, the *Galateo*, and owned a number of texts on rhetoric and literary composition.[36] On the frontispieces of his books he styled himself as a "Florentine Patrician" even before becoming the "Philosopher and Mathematician of the Grand Duke." His Latin was elegant and the style of his Florentine language remarkable. He knew how to write for a courtly audience. As he wrote to Prince Leopold in 1640, he disagreed with those who

> would like to see philosophical doctrines compressed into the most limited space, and would like people always to use that stiff and con-

33. Girolamo Lunadoro, *Relatione della corte di Roma* (Rome: Frambotto, 1635), pp. 63–65. Lunadoro's text was completed in January 1611. That Raimondi was perceived not only as a mathematician but also as a gentleman is confirmed by his status within the Medici payroll (he was a Medici employee in Rome). In the *ruolo* of July 1610 he is included in the category of court gentlemen, the same one in which Galileo would later be included (*ASF*, "Depositeria generale 389," fol. 82r).

34. Biagioli, "Social Status of Italian Mathematicians"; Harcourt Brown, *Scientific Organizations in Seventeenth-Century France* (Baltimore: Johns Hopkins University Press, 1934), p. 87; Steven Shapin, "Who Was Robert Hooke?," in Michael Hunter and Simon Schaffer, eds., *Robert Hooke: New Studies* (Woodbridge, Suffolk: Boydell, 1989), pp. 253–85; and Robert Iliffe, "'In the Warehouse': Privacy, Property and Priority in the Early Royal Society," *History of Science* 30 (1992): 29–68.

35. On this issue, see Biagioli, "Social Status of Italian Mathematicians," 49–50; and Paul Lawrence, *The Italian Renaissance of Mathematics* (Geneva: Droz, 1975), p. 155.

36. Galileo was familiar with della Casa's writings. He cited him in his "Considerazioni al Tasso" (*GO*, vol. 9, p. 133). His library contained a good number of texts of rhetoric and literary composition, as well as "how to" books for the courtier, such as *Idea di varie lettere usate nella Segreteria d'ogni Principe* (Antonio Favaro, "La libreria di Galileo Galilei," *Bullettino di bibliografia e storia delle scienze matematiche e fisiche* 19 [1886]: 219–93, esp. 273–75). Galileo

cise manner, that manner bare of any grace or adornment typical of pure geometricians who would not use even one word that was not absolutely necessary. Quite to the contrary, not only do I not see it as a problem if a treatise dedicated to a specific topic includes other various materials (unless they are completely extraneous and attached to the main argument without any real coherence), but, in fact, I appreciate it. The nobility, the greatness, and the magnificence that makes our actions marvelous and excellent does not consist of necessary things (although their absence would be the greatest defect), but of those that are not necessary. . . .[37]

If he often wrote in a Rabelaisian or Ruzantian literary style marked by sarcasm and jokes that blurred into insults it was not because he came from a lower-class background. Galileo was not the smart "man of the street" who achieved success at court. Like Ruzante before him, Galileo knew how to play "popular culture," how to display spontaneity and unaffected wit to attract a high-culture audience eager to take a break from an increasingly rigid court etiquette.[38] Galileo's somewhat abrasive style was not addressed to the village marketplace but to an upper-class audience. It was an antidote to an overworked courtly *sprezzatura* which edged over into pedantry. For example, the *Dialogo de Cecco di Ronchitti,* which Stillman Drake has, I think correctly, attributed to Galileo, was written in vulgar Paduan dialect but it must have been addressed to an upper-class audience if it was dedicated to Antonio Querengo, one of Padua's most important patrons of the arts.[39]

It is to the court's rhetorical contempt for pedantry that we can trace the style of Galileo's abrasive attacks on the Peripatetics.[40] The Simplicio of Galileo's dialogues (or the philosopher of Cecco's *Dialogo*) was not only

---

learned rhetoric quite well, as shown by Maurice Finocchiaro, *Galileo and the Art of Reasoning* (Dordrecht: Reidel, 1980); and Janet Dietz-Moss, "The Rhetoric of Proof in Galileo's Writings on the Copernican System," in William A. Wallace, ed., *Reinterpreting Galileo* (Washington, D.C.: Catholic University of America Press, 1986), pp. 179–204.

37. "Lettera al Serenissimo Principe Leopoldo di Toscana," in *GO,* vol. 8, p. 491.

38. Ruzante (Angelo Beolco) was by no means a member of the lower classes. His use of vernacular and of aggressive, obscene language was a product for the upper classes, not for the village marketplace (Ludovico Zorzi, "Introduzione," in Ruzante, *L' anconitana* [Turin: Einaudi, 1965], pp. v–xi). On the possible connection between Galileo's "Ruzantian" style and his use of common language, see note 29.

39. *Dialogo de Cecco di Ronchitti da Bruzene in perpuosito de la stella nova* (Padua: Tozzi, 1605), translated in Stillman Drake, *Galileo Against the Philosophers* (Los Angeles: Zeitlin and Ver-Brugge, 1976), pp. 33–53.

40. On the literary culture of the court and the place of rhetoric in it, see Marc Fumaroli, *L'âge de l'éloquence: Rhétorique et res literaria de la Renaissance au seuil de l'époque classique* (Geneva: Droz, 1980).

Galileo's philosophical straw man but also a representative of what court culture perceived itself to be rejecting.[41] Starting with Annibal Caro and continuing with Galileo's friend Jacopo Soldani, university philosophers had been a target of the satires of court writers, humanists, and academicians.[42] In his *Contro i peripatetici* (a poem he published in 1623 to celebrate Galileo's *Assayer*), Soldani attacked the "bum from Stagira" and his "herd of sheep" for practicing "a philosophy that closes and ties down knowledge with ropes."[43] In supporting Galileo's cause, he tried to undermine the philosophers not by attacking them on scientific or methodological grounds but by playing the trope of the philosophers' stuffy, servile, and myopic knowledge—a form of culture that was unacceptable to the courtly audience he was addressing.

This trope emerges in several other courtly texts. In his 1624 treatise on the court, Pellegrini dedicated a chapter to "The Qualities of the Scholar Which Are Inconvenient to the Courtier" in which he described philosophers who are unlikely to attract princely favor because of their "rigid manners and uncivil appearance." Also, "unaccustomed to the appreciation of pleasure, they shun it and in so doing they annoy those [the princes] who, instead, seek it. Thus, their conversation is rough and unpleasant."[44] Elsewhere in the book, Pellegrini invited the philosopher seeking a career at court as advisor to the prince not to annoy him with "sophistical and tedious questions" about state matters.[45] In poking fun at boring court literati, Agostino Mascardi—the leader of Cardinal Savoia's trend-setting Accademia dei Desiosi—mentioned the term "essential" and wrote immediately after it in parentheses "Forgive me for using scholastic terminology."[46] In a move that reminds us of Lunadoro stressing that, unlike most philosophers, Raimondi was clean and well

41. The contempt for pedantry on the part of the more culturally sophisticated was by no means an exclusively Italian phenomenon. For the French scenario, see Londa Schiebinger, "Battles over Scholarly Style," *The Mind Has No Sex?* (Cambridge, Mass.: Harvard University Press, 1989), pp. 119–59, esp. p. 156. For England, see Steven Shapin, "A Scholar and a Gentleman," *History of Science* 29 (1991): 279–327.

42. Annibal Caro, in his 1543 *Comedia degli straccioni,* written for his patron Pier Luigi Farnese (nephew of Paul III), has Pilucca (a member of the Roman demimonde) end a discussion concerning a planned fraud by praising his accomplice Marabeo for his "doctrine" (i.e., his scheme): "PILUCCA: Very well, I like this doctrine. Whose is it? Of the Peripottetici [pun on Peripatetics and "potta" (cunt)] or of the Stronzici [pun on Stoics and "stronzi" (shits)]?" (Annibal Caro, *Comedia degli straccioni* [Turin: Einaudi, 1967], p. 24).

43. Quoted in Alberto Asor Rosa ed., *I poeti giocosi dell'età barocca* (Bari: Laterza, 1975), p. 167. On Galileo's literary style and its audience, see also Robert S. Westman, "The Reception of Galileo's Dialogue," in Galluzzi, *Novità celesti e crisi del sapere,* pp. 331–35.

44. Matteo Pellegrini, *Che al savio è convenevole il corteggiare libri IIII* (Bologna: Tebaldini, 1624), p. 109.

45. Ibid., p. 292.

46. Agostino Mascardi, "Discorso ottavo," *Prose vulgari* (Venice: Baba, 1653), p. 148.

dressed, the poet Giambattista Marino, in his famous poem *Adone,* presented philosophy as dirty, badly dressed, and with unkempt hair.[47] To a courtly audience, an old-fashioned Aristotelian philosopher was as "technical" (and therefore as unfashionable) as a mathematician.

That Galileo had access to court as a teenager must have helped him avoid these pitfalls.[48] He inherited from his father, Vincenzio, some of his early connections with the Florentine court as well as the knowledge of courtly etiquette.[49] Vincenzio was a well-known musician and music theorist, and a member of the Camerata de' Bardi, an institution that could be considered Florence's first music academy. That a career at court was not an unusual thought for a Galilei is shown by the life of Galileo's brother Michelangelo, who—a musician like his father—worked at various European courts.

Galileo's early literary productions were all embedded in Florentine academic and courtly culture of the period. His oration on the geometry of Dante's *Inferno,* presented in 1588 at the Accademia Fiorentina (whose consul he would become in 1620), dealt with what was probably the canonical text of that institution.[50] His critique of Tasso and praise of Ariosto were equally the product of the culture of Florentine academies.[51] Quite unoriginally, Galileo represented the official position of the Florentine Accademia della Crusca—an academy to which Galileo was elected in

47. Giambattista Marino, *L'Adone* (Paris, 1623; reprint, Turin: Paravia, 1922), p. 157 (tenth canto, no. 130). This is the same canto in which Marino praises Galileo, his telescope, and his discoveries.

48. We know that he had early access to court because it was there that he met his future mathematics teacher, Ostilio Ricci.

49. Also, participation in patrician salons like that of Morosini in Venice or of Pinelli in Padua (as well as the many visits to the Medici court over the summer as Cosimo's mathematics tutor) helped Galileo develop some familiarity with the court's proper style of argumentation and behavior. Galileo himself admitted having undergone such a process of socialization in a letter to Michelangelo Buonarroti, Jr., in December 1609: ". . . knowing now what are the very honorable manners and habits of the Florentine nobility . . ." (*GO,* vol. 10, no. 257, p. 271).

50. Galileo Galilei, "Due lezioni all'Accademia Fiorentina . . ."; in *GO,* vol. 9, pp. 29–57. Dante's work was one of the institutional foci of the Accademia Fiorentina because of its relation to the Florentine vernacular. The specific issue of the geometry of Dante's *Inferno* also received attention, notably by the architect Manetti (Antonio Manetti, "Circa il sito, forma e misura dell'Inferno di Dante Alighieri, poeta eccellentissimo," in Ottavio Gigli, ed., *Studi sulla Divina Comedia di Galileo Galilei, Vincenzo Borghini ed altri* [Florence: Le Monnier, 1855], pp. 35–114). Galileo's lectures must have received some attention if they were still remembered in 1594 (see *GO,* vol. 10, no. 54, p. 66). On Galileo's consulate at the Accademia Fiorentina, see *GO,* vol. 19, pp. 444–45.

51. Galileo Galilei, "Considerazioni al Tasso," in *GO,* vol. 9, pp. 59–148; and idem, "Postille all'Ariosto," in ibid., pp. 149–94. The dates of these two works are uncertain. Favaro seems to think that the "Considerazioni" were probably written in the 1590s (ibid., pp. 12–14).

1605—which sided with Ariosto against Tasso.[52] Debates on such issues must have been common in academies since, about two decades later, Mascardi remarked on "the silliness of those who nauseate the literati by getting into fights over the precedence between Ariosto and Tasso, and who get so deeply into the circles of Dante's [Inferno] that they cannot get out anymore."[53] Similarly, Galileo's letter to Cigoli on the relative status of sculpture and painting dealt with a topic that was routinely discussed in the Florentine Accademia del Disegno and in other artistic academic circles.[54]

Galileo's involvement with these literary activities does not mean that he contemplated a career as a writer, but rather that he needed to prove his competence with courtly and academic culture. It was an almost necessary rite of passage for young men looking for patronage and ambitious careers.[55] Moreover, being able to represent oneself as a literato was particularly important for somebody like Galileo who, as a mathematician, would have been otherwise assigned a much lower status. In fact, in Renaissance Italy writers were consistently given a higher social status than visual artists and mathematicians, and had much better chances for a career at court, a space where the notion of a "nobility of letters" was beginning to emerge.[56]

52. On Galileo's election to, and relationship with, the Accademia della Crusca, see *GO*, vol. 19, p. 221; and Paola Manni, "Galileo Accademico della Crusca," *La Crusca nella tradizione letteraria e linguistica italiana* (Florence: Accademia della Crusca, 1985), pp. 119–136. Galileo's perspectives on Ariosto and Tasso are also discussed in Erwin Panofsky, *Galileo as a Critic of the Arts* (The Hague: Martinus Nijhoff, 1954). Tasso was excluded from the *Vocabolario degli accademici della Crusca* first published in Florence in 1612 (Salvatore Nigro, "Dalla lingua al dialetto: La letteratura popolaresca," in Rosa, *I poeti giocosi dell'età barocca*, p. 66).

53. Mascardi, "Discorso secondo," *Prose vulgari*, p. 34.

54. *GO*, vol. 11, no. 713 (26 June 1612), pp. 340–43. Favaro is skeptical about the authenticity of this letter, mostly on stylistic grounds. His position has been convincingly refuted by Margherita Margani in "Sull'autenticità di una lettera attribuita a G. Galilei," *Atti della Reale Accademia delle Scienze di Torino* 57 (1921–22): 556–68. The debate on the primacy of sculpture over painting is a very frequent theme of sixteenth-century academic writings on the arts. The *Lezione di Benedetto Varchi nella quale si disputa della maggioranza delle arti*, which he read to the Accademia Fiorentina in 1547 (partially reproduced in Paola Barocchi, *Scritti d'arte del Cinquecento* [Turin: Einaudi, 1977], I: 99–105, 133–51), is an example of this genre.

55. Galileo's literary efforts must have been quite successful, if his academic friends in Florence—whom he met again during the summers spent in Florence—kept writing to him in Padua asking for his comments on sonnets or books they had written (*GO*, vol. 10, no. 52, pp. 63–64; no. 72, pp. 82–83; no. 76, pp. 86–87). References to Galileo's literary and poetic production and expertise are found in *GO*, vol. 10, no. 54, p. 66; no. 409, p. 447; vol. 11, no. 492, p. 68; no. 563, p. 164; no. 647, p. 265.

56. On the relative status of writers and artists, see Peter Burke, "Artists and Writers," *The Italian Renaissance* (Princeton: Princeton University Press, 1986), pp. 43–87. On the continuity between the social world of artists and mathematicians, see Biagioli, "Social Sta-

During these early phases of his career, Galileo was introduced not only to Florentine court and academic culture but into patronage networks as well. It is to this period of his life, to the culture he absorbed and the patrons and friends he met (and whom he kept meeting later during his regular summer visits to Florence from Padua), that we can trace most of the patronage strategies he developed later in his life.

The social groups Galileo frequented in Venice and Padua after 1592 were similar to those he was familiar with in Florence, but because Venice had no court, Paduan and Venetian culture was quite different from the Florentine, and patronage was of the patrician rather than the princely type. If Giovanfrancesco Sagredo was a patrician patron in Venice comparable to Filippo Salviati in Florence, we still cannot find the Cosimo II for Galileo's Paduan period. Salons, *casini,* and private academies rather than the court or official academies were the loci of such a patronage.[57] Moreover, although Venice was concerned with maintaining its own state myths (especially in its period of decadence at the turn of the century), these were not centered on a specific family dynasty but on the idea of the republic.[58] Consequently, Galileo's discoveries could not be made to fit those state myths in any relevant or particularly rewarding way. In fact, Galileo dedicated the telescope to the Venetian Senate as a instrument of navigation and warfare rather than as a means of viewing dynastic monuments.

The initiation into Florentine court and academic culture provided Galileo with the competence necessary to see *naturalia* as potential Medici dynastic emblems. Galileo understood that he needed an absolute

---

tus of Italian Mathematicians"; and Thomas B. Settle, "Egnazio Danti and Mathematical Education in Late Sixteenth-Century Florence," in John Henry and Sarah Hutton, eds., *New Perspectives on Renaissance Thought* (London: Duckworth, 1990), pp. 24–37. On the nobility of letters, see Alain Viala, *Naissance de l'écrivain* (Paris: Minuit, 1985); and Lunadoro, *Relazione della corte di Roma,* 5, where he discusses the position of the pope's *Cameriere d'honore* usually reserved to "people of quality" because of their birth or because they were *illustri per lettere.*

57. Krzysztof Pomian, *Collectionneurs, amateurs et curieux* (Paris: Gallimard, 1987); pp. 81–158, 213–87; Gino Benzoni,*Gli affanni della cultura* (Milan: Feltrinelli, 1978), esp. pp. 7–77; idem, "Le accademie," in G. Arnaldi and M. Pastore Stocchi, eds., *Storia della cultura veneta* (Vicenza: Neri Pozza, 1984), vol. 4, pt. 1, 131–62; Gaetano Cozzi, *Paolo Sarpi tra Venezia e l'Europa* (Turin: Einaudi, 1979), pp. 135–234; Antonio Favaro, *Amici e corrispondenti di Galileo* (Florence: Salimbeni, 1983), 1: 65–91, 191–322; 2: 703–36; idem, "Un ridotto scientifico in Venezia al tempo di Galileo Galilei," *Nuovo archivio veneto,* series 2, vol. 5 (1893): 199–209; and idem, *Galileo Galilei e lo Studio di Padova* (Padua: Antenore, 1966), 2: 69–102.

58. On Venice's decline, see Alberto Tenenti, *Piracy and the Decline of Venice 1580–1615* (Berkeley: University of California Press, 1967); James C. Davis, *The Decline of the Venetian Nobility as a Ruling Class* (Baltimore: Johns Hopkins University Press, 1962); and Richard T. Rapp, *Industry and Economic Decline in Seventeenth-Century Venice* (Cambridge, Mass.: Harvard University Press, 1976). On Venice's political rituals, see Edward Muir, *Civic Ritual in Renaissance Venice* (Princeton: Princeton University Press, 1981).

CHAPTER TWO

prince as a patron—and not just because, as he told Vinta, only a prince could have offered him the salary and leisure he was seeking. He needed an absolute prince because his marvels could best gain value and grant him social legitimation if they were made to fit the dynastic discourse of such a ruler.[59] When he discovered Jupiter's satellites at the end of 1609, he correctly realized that Venice was not the best marketplace for his marvels.

However, the understanding of patronage dynamics and of the codes of academic culture that Galileo had developed during his Florentine youth was not wasted in Padua and Venice. He managed to develop patronage relationships with powerful Venetian patricians such as Sagredo, he had access to the most respected salons, and he took an active part in Padua's academic life.[60] In 1599, Galileo was among the founding members of the Paduan Accademia dei Ricovrati, taking the name *Abbattuto* (the depressed or defeated one). Together with other colleagues, he was in charge of designing the academic impresas for that body.[61] The impresa Galileo proposed for Cosimo's wedding to Maria Maddelena of Austria in 1608 showed his mastery in emblematics and in the culture of the Medici court.

From Lodestones to Satellites

Knowing that gold and silver medals were usually struck to commemorate major dynastic events, in September 1608 Galileo wrote to Cosimo's mother, Grand Duchess Cristina, proposing an emblem for a medal.[62] The letter is a concise summary of Medici dynastic ideology and presents a subtle "scientific" metaphor for the "naturalness" of the Medici rule. Referring to the lodestone he had bought for Prince Cosimo from Sagredo a few months earlier, Galileo compared the power of a future absolute ruler like Prince Cosimo to that of the lodestone. Using the terminology of the emblematist Giovio, Galileo proposed that the "body" (i.e., the image) of the impresa be a globe-shaped lodestone that would hold a number of small pieces of iron around it.[63] The "soul" of the impresa (i.e., the motto) was *Vim Facit Amor* ("Love produces strength").

59. *GO*, vol. 10, no. 307, pp. 348–53.
60. Favaro, *Galileo Galilei e lo Studio di Padova*, 1: 36–77; 2: 1–7, 18–32.
61. Benzoni, *Gli affani della cultura* p. 176; and *GO*, vol. 19, pp. 207–8.
62. *GO*, vol. 10, no. 199, pp. 221–23. Devising emblems for medals was also a popular courtly and academic game. In Girolamo Bargagli's *Dialogo de' giuochi* (Siena: Bonetti, 1572)—a book dedicated to Donna Isabella Medici—the "game of reverses" is discussed. It is played as follows: Medals are imagined to be struck in honor of the ladies present at the gathering, and each gentleman must create a reverse worthy of the medal of one of the ladies (Crane, *Italian Social Customs of the Sixteenth Century*, p. 280).
63. Paolo Giovio, *Dialogo delle imprese militari e amorose*, ed. Maria Luisa Doglio (Rome: Bulzoni, 1978), p. 37. On the political symbolism of cosmologies during (and before) the

Galileo was well aware of the tensions underlying the representations of the Medici's absolute rule. On the one hand, the Medici wanted to stress the "naturalness" of their rule and the assent given to it by their subjects; on the other hand, they wished to emphasize the power of their rule and its lack of tolerance for deviant behavior. Galileo solved this riddle of political imagery by identifying in the sympathetic attraction between the lodestone and the small pieces of iron a fine metaphor for the Medici political agenda. According to Galileo's image, the pieces of iron (the subjects) seemed to be voluntarily driven up (elevated) toward the lodestone (the Medici power), for its force was not felt by other materials. Those pieces of iron (the subjects) wanted to be attracted. But, at the same time, such an uplifting attraction was powerful and ultimately inevitable. It was based on *love* but manifested itself as *power*. The motto *Vim Facit Amor* captures the meaning of the image. According to Galileo, the allegorical meaning of the motto was that

> as fragments of iron are lifted up and held by the lodestone (but with a sort of loving violence, for they seek the stone avidly, as if they were rushing voluntarily to it), so that it is difficult to tell whether such a tenacious bond is the result of the strength of the magnet, the natural tendency of the iron, or the loving dialectic of power and obedience, the pious and courteous affection of the Prince— represented by the lodestone—does not oppress but rather lifts up his subjects, and makes them—represented by the fragments of iron—love and obey him.[64]

Galileo then explained to Cristina that the globe-shaped lodestone was itself an allegory of Cosimo as the cosmos and of the Medici coat of arms, which contains six globes. Those analogies had been employed fifty years earlier by Vasari in the Palazzo della Signoria's Room of the Elements. There the painter represented a Capricorn (Cosimo's ascendant sign) holding between its hooves a globe that signified both one of the balls of the Medici coat of arms and the cosmos held in check by Cosimo.[65] The Cosimo-as-cosmos theme recurs in other paintings in the Apartment of the Elements as well as in the palazzo's Room of the Geographical Maps.[66] This room contained a large armillary sphere, as well as

---

scientific revolution, see Keith Hutchinson, "Toward a Political Iconology of the Copernican Revolution," in Patrick Curry, ed., *Astrology, Science, and Society* (Woodbridge: Boydell Press, 1987), pp. 95–141.

64. *GO*, vol. 10, no. 199, p. 222. See also Galileo's previous attempt to develop a politically connoted emblem based on the lodestone in a letter to Vinta (ibid., no. 187, pp. 205–9).

65. Allegri and Cecchi, *Palazzo Vecchio e i Medici*, 67; and Vasari, *Ragionamenti*, p. 32.

66. Ibid., p. 22.

a terrestrial globe in the center and maps representing the entire world, all designed and partially executed by the cosmographer Ignazio Danti.[67]

The analogy between Cosimo and cosmos (which Galileo would bring up again a few years later while negotiating the dedication of the *Sidereus nuncius* to Cosimo II) had been an important part of Medici mythology since the mid-sixteenth century. Names incorporating the term "cosmos" proliferated. When in 1548 Cosimo I gained control of Portoferraio, the island of Elba's most important harbor, he had it fully fortified and called "Cosmopoli."[68] This onomastic revisionism found perhaps its strongest expression during the "cultural revolution" that accompanied the constitution of the grand duchy of Tuscany that institutionalized the absolute power of the Medici. At that time Cosimo replaced Florence's old patron saints, Zenobi and Giovanni (who were perceived as emblems of the old republican tradition), with Saints *Cosma* and Damiano, who while on earth were practicing physicians—"medici" being the Italian term for "physicians."[69] The feast day of Saints Cosma and Damiano (September 27) coincided with the birthday of Cosimo il Vecchio (1389–1464)—the *pater patriae*. Like Cosma, both Cosimo I and Cosimo il Vecchio were represented as the physicians of Florence for having saved the city from the deadly plague of political disorder. Even as early as 1513, Leo X, the Medici pope who was instrumental in securing the duchy of Florence for the Medici, had instituted an annual holiday—the *Cosmalia*—allegedly in honor of Saint Cosma. In fact, the holiday was dedicated to the memory of Cosimo il Vecchio and was meant as a tribute to the Medici rule.[70]

In the 1560s the logo ΚΟΣΜΟΣ ΚΟΣΜΟΥ ΚΟΣΜΟΣ—Greek for "The cosmos is Cosimo's world [or domain]" was included in Medici-commissioned works of art.[71] References to Cosimo as cosmos continued to emerge in Medici-related cultural productions, especially when "Cosimo" happened to be the current ruler's name.[72] In his proposal for the impresa of 1608, Galileo reinforced the Cosimo-cosmos theme by suggest-

67. Detlef Heikamp, "L'antica sistemazione degli strumenti scientifici nelle collezioni fiorentine," *Antichità viva* 9 (1970): 3–25; and Allegri and Cecchi, *Palazzo Vecchio e i Medici,* p. 303.

68. Arnaldo Segarizzi, *Relazioni degli ambasciatori veneti al senato* (Bari: Laterza, 1916), 3: 256.

69. Wazbinski, *L'Accademia medicea del Disegno a Firenze nel Cinquecento,* 1: 83. The "medicus"/Medici pun is also found in Erasmus's letters to the Medici popes (I owe this point to an anonymous reviewer of the book in manuscript).

70. Cox-Rearick, *Dynasty and Destiny,* p. 33.

71. Ibid., p. 279.

72. Examples are Gabriello Chiabrera, *La pietà di Cosmo: Dramma musicale rappresentato all'Altezze di Toscana* (Genoa: Pavone, 1622); Giovanni Carlo Coppola, *Cosmo, ovvero l'Italia trionfante* (Florence: Stamperia di SAS, 1650).

ing that *Magnus Magnes Cosmos* should have been the motto on the other side of the medal, which was to contain Cosimo's effigy. "If taken literally [the motto] means only that the world is a great lodestone, but, taken metaphorically, it also confirms the impresa."[73] By substituting "Magnes" for "Dux" in the standard Latin version of Cosimo's title, "Magnus Dux Cosmos" ("Cosimo Grand Duke"), Galileo made the magnet a metaphor for the ruler by reinforcing the analogy between magnetic attraction and the prince's power.

Besides Galileo's remarkable skills in emblematics, this impresa reveals, I think, a turning point in Galileo's patronage strategies.[74] By 1608 he must have realized that the invention of military compasses, however useful, would not help him obtain a high-status position at court. Quite probably the compass brought him a good number of private students interested in fortifications, but it did not make him a very desirable client to a major prince, one preoccupied more with the celebration of his own image than with the quality of his court teacher of mathematics. The Gonzaga appreciated the gift of the compass; the Medici welcomed the dedication of the book which explained its use, but neither prince offered Galileo the type of position he was looking for. I think Galileo realized he needed to produce gifts that were less mechanical in nature than a compass if he wanted to go to court as a gentleman rather than as a teacher of mathematics or a military engineer.

The impresa of 1608 indicates that Galileo understood that marvels such as "mysteriously" behaving lodestones were more rewarding than instruments, especially when they could be represented as an emblematic articulation of the discourse of the court and of the absolute ruler. And indeed the imagery Galileo used in the 1608 impresa had been part of court discourse at least since Baldassarre Castiglione's *Book of the Courtier*. There Castiglione discussed the skills of a good courtier, one able to develop an elaborate presentation of the self that would "attract the eyes of the spectators even as the lodestone attracts iron."[75] The same analogy between the behavior of the lodestone and that of the attractive power of *virtù* occurs in some of the letters Galileo exchanged with Medici courtiers. In December 1605 he received a letter from Cipriano Saracinelli, who concluded by confirming his friendship for and patronage to Galileo: "[But] I would have done the same even if I did not know you, because what is beautiful

---

73. *GO*, vol. 10, no. 199, p. 223.

74. Galileo owned Paolo Giovio's and Ettore Tasso's texts on impresas (Favaro, *La libreria di Galileo Galilei*, 285, 287). One of his sonnets is dedicated to the enigma itself ("Enimma," *GO*, vol. 9, p. 227). As I have mentioned, he was in charge of the imprese of Padua's Intronati (see note 41 above). Finally, he liked to play with enigmas to "communicate" his discoveries, as in the case of the phases of Venus (*GO*, vol. 11, no. 451, p. 12) or of the shape of Saturn (*GO*, vol. 10, no. 427, p. 474; no. 435, p. 483).

75. Castiglione, *Book of the Courtier*, p. 100.

and good—that is, virtue—has the power to attract from far away the soul and the will of even those who can barely recognize it."[76]

Vinta was even more explicit about the attractive force of virtue. In a letter to Galileo in March 1608, concerning the purchase of the lodestone for Cosimo, he concluded with: "And—Your Lord's value being a lodestone that attracts and forces me to love and serve you—I beg you to use me for anything you may desire or need."[77] A week later Galileo returned the courtesy, writing Vinta (whose name within Siena's Accademia dei Filomati was "The Attractive") that:

> I will never admit that the lodestone of my value could attract the affection of Your Most Illustrious Lord, for I know that I do not possess those qualities that would deserve so much favor. Rather, it is my needy status to act as a magnet that moves the pious affection and most courteous attitude of Your Most Illustrious Lord into loving and protecting me.[78]

A month later Galileo presented Vinta with the lodestone-based impresa that Galileo would rework and finally propose to Cristina for Cosimo's wedding medal.[79]

The originality of Galileo's impresa does not lie in the use of technical devices in emblems.[80] Giovio had already discussed them in his emblematics textbook.[81] What was new about Galileo's translation of scientific marvels into the discourse of the court (or of a specific dynasty, as in the

76. *GO*, vol. 10, no. 129, p. 150.

77. Ibid., no. 178, p. 198.

78. On Vinta's academic name, see Giuseppe Fusai, *Belisario Vinta* (Florence: Seeber, 1905), p. 105. The Filomati's choice of that name for Vinta reflected an explicit appreciation for his brokerage skills. As the secretary of the academy put it: "The academic name we have decided to give you is *The Attractive*, because everybody has heard and several know from experience how Your Lordship, because of your great skills in negotiating, is able to win the devotion of their hearts and . . . their affection." Galileo's letter is in *GO*, vol. 10, no. 180, p. 200. Because Vinta was elected to the Filomati in 1603, Galileo's expression might be a pun on Vinta's academic name.

79. Ibid., no. 187, pp. 205–9.

80. The use of emblematics in scientific texts has been studied by William Ashworth in "Iconography of a New Physics," *History and Technology* 4 (1987): 267–97; idem, "Divine Reflections and Profane Refractions," in Irving Lavin, ed., *Gianlorenzo Bernini* (University Park: Pennsylvania State University Press, 1985), pp. 179–95; and idem, "The Habsburg Circle," in Bruce Moran, ed., *Patronage and Institutions* (Rochester, NY: Boydell, 1991), pp. 137–67. See also Ashworth's "Natural History and the Emblematic World View," in David C. Lindberg and Robert S. Westman, eds., *Reappraisals of the Scientific Revolution* (Cambridge: Cambridge University Press, 1990), pp. 303–32.

81. Giovio, *Dialogo dell'imprese militari e amorose*, 37, 66–67. See also Allegri and Cecchi, *Palazzo Vecchio e i Medici*, 113, 149; and Karla Langedijk, *The Portraits of the Medici* (Flor-

case of the satellites of Jupiter) was that he did so both to show that natural philosophy was not necessarily an uncourtly activity, and to legitimize scientific discoveries and theories by linking them to the power image of the prince.[82]

Galileo's claim that the motto *Magnus Magnes Cosmos* meant both that "the world is a great lodestone," as William Gilbert had argued, and that the attractive force of Cosimo's power, as legitimate and "natural," had important implications. It associated Gilbert's theory (one that could be used against the accepted Aristotelian cosmology) with that of the naturalness of the Medici absolute rule.[83] By striking such a medal the Medici would help legitimate Gilbert's theory; at the same time, Galileo's "magnetic" interpretation of the Medici power helped represent their rule as "natural." Quite literally, the medal Galileo proposed to Cristina had two inseparable faces and meanings. Galileo's strategy was aimed at legitimizing scientific theories by including them in the representation of his patrons' power, thus securing both the patrons' involvement and endorsement.[84] This was Galileo's attempt to get out of the deadlock caused by the patrons' noncommittal attitude discussed in the previous chapter.

There was another, more subtle but equally important sense in which Galileo's tactics of socioprofessional self-fashioning were modulated on court discourse. The image of the lodestone attracting pieces of iron presented by Castiglione and adopted by Galileo was an emblem of court life itself. At court, one's status and identity was not safely anchored to one's wealth, academic degree, birth, or professional competence. Status and identity were renegotiated every day in that endless process that Norbert Elias has described so vividly.[85] A courtier's identity depended on how one was perceived by the prince and the other courtiers. As the saying goes, one existed in the eye of the beholder. The way one looked and be-

---

ence: SPES, 1980), vol. 1, p. 212, note 110, on the use of technological and scientific impresas in Medici imagery.

82. Galileo was not alone in trying to present natural philosophy as fitting courtly discourse by using it in emblems. The Jesuits had done the same. As Federico Cesi wrote to Galileo, during one of the usual public debates at the Collegio Romano, the orator had used a fluorescent stone in an impresa as he debated on the sunspots (*GO*, vol. 12, no. 964, p. 12).

83. Galileo's pro-Copernican use of Gilbert is made explicit at the end of the third day of the *Dialogue on the Two Chief World Systems*.

84. For a different but related strategy adopted by Copernicus with Pope Paul III, see Robert S. Westman, "Proof, Poetics and Patronage: Copernicus's Preface to *De revolutionibus*," in David C. Lindberg and Robert S. Westman, eds., *Reappraisals of the Scientific Revolution* (Cambridge: Cambridge University Press, 1990), pp. 167–205.

85. Elias, *Court Society*. On self-fashioning, see Stephen Greenblatt, *Renaissance Self-Fashioning* (Chicago: University of Chicago Press, 1980); and Frank Whigham, *Ambition and Privilege: The Social Tropes of Elizabethan Courtesy Theory* (Berkeley: University of California Press, 1984).

haved was read as a sign of whether or not one had *gratia*.[86] Therefore, one's ability to "attract" the gaze of the courtiers to confirm or increase one's status was a vital skill at court. Nobody was exempt from this, as court etiquette was precisely the framework within which these subtle negotiations of status and identity were performed, usually to the ultimate advantage of the prince.

Castiglione's image of the lodestone was the perfect epitome of the courtier's predicament. It applied to the prince trying to maintain control over his subjects as well as to the courtiers trying to ascend by attracting as many confirming gazes as they could. Galileo's choice of this image is, I think, not accidental. Although the Medici were his direct audience, any courtier (as indicated by Castiglione's text and by Galileo's correspondence) would have probably seen him/herself reflected (either as a subject or as a *terrella*) in his own impresa. Galileo himself could be seen in his impresa as a piece of iron trying to let the lodestone know that he would be happy to be lifted up.

Galileo was trying to show that a natural philosopher could devise gracious impresas like any other literato. Actually, while naturalizing the Medici absolute rule, his lodestone impresa referred also to the courtier's daily life as represented in courtly texts like Castiglione's. In a sense, Galileo was trying to present himself as a competent courtier by showing that he understood the courtly game and, at the same time, by representing it not as a rat race but as an elegant game of sympathies. Quite likely, courtiers cared very little about how Gilbert's views on magnetism could be mobilized against Aristotelian cosmology, but they could probably see and appreciate how these views could provide a gracious metaphor of their daily life. To them, Gilbert's theories may not have explained the processes of the cosmos but they did help to make sense of *le monde*.

However, while Galileo's verbal description of the impresa (like the one he sent to Cristina) allowed him to display his skill in emblematics, the impresa alone was not as self-evident as it needed to be.[87] In fact, who could distinguish a magnet attracting iron fragments from a globe surrounded by irregularly shaped pieces of some unspecified material? Nevertheless, Galileo's attempt was not a total failure but rather an instance of a trial-and-error strategy. What he did two years later by binding the Medici name to the satellites of Jupiter was a successful replay of the same strategy. By turning an astronomical discovery into a dynastic emblem he

86. Randolph Starn, "Seeing Culture in a Room for a Renaissance Prince," in Lynn Hunt, ed., *The New Cultural History* (Berkeley: University of California Press, 1989), pp. 205–32, esp. pp. 210–17.

87. Giovio, *Dialogo dell'imprese militari e amorose*, 37. On the obscurity of impresas, see also Frances Yates, "The Italian Academies," *Collected Essays* (London: Routledge, 1983), 2: 11.

became a very important client, a sort of "cosmic midwife." At the same time he turned Medici power to the legitimation of his discoveries and of his telescope.

From Classified Instruments to Dynastic Horoscopes

After donating his telescope to the Venetian Senate in August 1609 and being rewarded with tenure and a remarkable salary increase, Galileo wrote to his brother-in-law, Benedetto Landucci, that, given the new developments, he perceived his life and career as permanently bound to Padua and its university.[88] However, a few months later he was negotiating with Vinta for his position as "Filosofo e Matematico del Granduca di Toscana," which he formally obtained in July 1610.[89] The four satellites brought about this remarkable change of socioprofessional status and patronage strategies.

For all the exceptional characteristics Galileo perceived in the telescope in August 1609, he still presented it to the doge Leonardo Donà as a *military* instrument. The telescope was a marvel, but one not tailored for any specific patron. Despite its truly exceptional features, the telescope was patronage-generic. It was a gift for everybody and for nobody in particular. Quite correctly, at this point Galileo still perceived the telescope as belonging to the same patronage category as the military compass, the only important difference being that the telescope was much more useful than the compass and therefore could trigger the interest and curiosity of a much wider audience. At this point in Galileo's career, the telescope was still simply an instrument. It was neither a messenger of dynastic destiny nor a ticket to court. From his correspondence of the period we see that until he discovered Jupiter's satellites, Galileo did not make any serious attempt to use the telescope to move to the Medici court. Although Cosimo II asked Galileo for a good telescope, his interest in the instrument did not seem to be essentially different from that he had shown in Sagredo's lodestone a few years before.

Galileo's commitment to Copernicanism seemed to fluctuate with his grasp of possibilities for court patronage. The conditions of his gift of the telescope to the Venetian senate indicates that, at the time, Galileo was not thinking of the telescope as a scientific instrument which could support the Copernican cause but as a sort of classified weapon. In this, Galileo's perception of the telescope's uses was identical to that of his Dutch pre-

88. *GO,* vol. 10, no. 231, pp. 253–54. Favaro questioned the authenticity of this letter, but Edward Rosen has convincingly refuted his argument in his "The Authenticity of Galileo's Letter to Landucci," *Modern Language Quarterly* 12 (1975): 473–86.

89. *GO,* vol. 10, 359, pp. 400–401.

decessor, Hans Lipperhey.[90] In his letter to the doge Leonardo Donà, Galileo claimed that judging the telescope as "worthy of being received and estimated as most useful by Your Lord, I decided to present it to you and have you decide about the future of this invention, ordering and providing according to your prudence whether telescopes *should or should not be built*."[91] This last statement indicates that either Galileo was ready to withhold an effective instrument from the rest of the astronomers or that his commitment to Copernicanism was not strong enough to suggest to him the astronomical potential of the telescope. However, Galileo's Copernican leanings reemerged and his patronage strategies changed abruptly when, four months later, he observed Jupiter's satellites.

The story of the negotiation between Galileo and Cosimo II carried on through Vinta during the first half of 1610 has been told many times.[92] What has not received much attention are Galileo's strategies to gain social status for himself and epistemological legitimation for the Medicean Stars by re-presenting them within the discourse of the Medici mythology, as he had previously tried to incorporate Gilbert's views on magnetism.

Astrological predetermination was a recurrent theme in Galileo's presentation of his discoveries to the Medici. What he had observed, Galileo claimed, was not a discovery but a confirmation of the Medici's destiny, almost a scientific proof of their dynastic horoscope.[93] As he told Cosimo in the dedication of the *Sidereus nuncius,* it was not by chance that the "bright stars offer[ed] themselves in the heavens" right after Cosimo II's enthronement.[94] It was not by chance that such stars were circling around Jupiter (Cosimo's planet) like his offspring and that Jupiter was actually just above the horizon at the time of Prince Cosimo's birth, thus passing on to him the virtues of the founder of the dynasty. And, one might add, it was not by chance that the stars were four in number, like Cosimo II and his brothers.[95] And, given this array of fateful conjunctures, Galileo's role

90. Hans Lipperhey tried to obtain a patent for his telescope in 1608. In his presentation of the instrument to Prince Maurice, he—like Galileo one year later—stressed its military usefulness (Albert Van Helden, "The Invention of the Telescope," *Transactions of the American Philosophical Society* 67 [1977]: 20–21, 26, 36).

91. *GO,* vol. 10, no. 228, p. 251 (emphasis mine).

92. Westfall, *Scientific Patronage,* 16–21; Stillman Drake, ed., *Discoveries and Opinions of Galileo* (Garden City, N.J.: Doubleday, 1957), pp. 1–20; and Galilei, *Sidereus nuncius,* pp. 1–24.

93. For instance, Tommaso Campanella—who did not quite understand Galileo's astrological rhetoric of the dedication—initially saw it as a real horoscope (*GO,* vol. 12, no. 982, p. 32).

94. Galilei, *Sidereus nuncius,* pp. 30–31.

95. Although in the dedication of the *Sidereus nuncius* Galileo did not draw an explicit connection between the four stars and the four brothers but claimed that they were "children

in the appearance of these dynastic signs could not have been a casual one either.

In the dedication, Galileo tended to hide the economic dimensions of the patronage relationship he was trying to establish. As he presented it, he was not trying to sell the Medici a particularly fitting dedication. His relationship with them was a most disinterested one. It was more than completely voluntary: it was *predetermined*.[96] The Medici and Galileo had been brought together by the stars. It could not be by chance that Galileo, a Medici subject and the mathematics tutor of Prince Cosimo II himself, had discovered the stars; *only* he could have discovered them.[97] In a sense, the stars did not need to be dedicated to the Medici: they had always been theirs. As he put it, "four stars were *reserved* for your illustrious name."[98] Like Galileo, they had been assigned to the Medici from the beginning.

Appropriately, Galileo referred not to a *discovery* but to an *encounter* between the Medici and their destiny. His role in this encounter was that of the mediator, and a lowly one at that.[99] As he told Vinta, it was in the best interest of the Medici to "ennoble" him because

> There is only one thing that largely diminishes the greatness of this encounter, and that is the ignobleness and low status of the mediator. Nevertheless . . . the ennoblement of the mediator is no less in the range of possibilities of His Most Serene Highness than the demonstration of my most devout observance was in mine.[100]

If the Medici hesitated, the celestial nature of the encounter might be spoiled by the humble status of the mediator.

However, he was not *asking* the Medici for a title in exchange for a dedication. If the encounter was a predestined one, then his role as mediator was predestined too. Galileo was de facto (or ex Deo) the Medici oracle. The Medici needed only to recognize it. And, with some help from Galileo, they eventually did.

Galileo's tactics go well with the dynamics of the power image of an absolute prince discussed in the previous chapter. As noted, absolute princes behaved as if they had everything. Consequently, nothing could be given to them that was not already theirs. This self-representation le-

---

of the same family" (ibid., 31), he made that analogy in a letter to Vinta (*GO*, vol. 10, no. 265, p. 283).

96. The theme of the predestination of a patronage relationship was not a new one. Vasari used it a few decades earlier when he signed his letters to Cosimo I as "Servitor per fortuna e per istella, Giorgio Vasari" (Frey, *Il carteggio di Giorgio Vasari*, p. 443).

97. Galilei, *Sidereus nuncius*, p. 32.

98. Ibid., p. 31.

99. *GO*, vol. 10, no. 271, p. 289. Galileo played again the theme of the "encounter" a week later (ibid., no. 277, p. 298).

100. Ibid., p. 301.

gitimated the princes' claims that they were not obliged to return gifts from subjects. If they did reciprocate gifts, it was a *favor* they were doing them and was not to be read as the acknowledgment of a debt. In short, subjects could not "challenge" their princes to a potlatch, or, if they did, they had to do so by following a specific "etiquette."

Galileo's dedication of the *Sidereus nuncius* spells out this etiquette. It indicates that a client interested in establishing an exclusive patronage relationship with an absolute prince could try to efface the potlatch qualities of his maneuver by presenting his gift as though it were not really a gift but had already belonged to the prince "since the beginning." By doing so, the client did not present himself as trying to challenge the prince or as asking for something in return. He could pretend to share the same aristocratic ethos of generosity and waste with the prince, to the point of throwing away the most precious thing he had: the authorship of those discoveries. His self-effacement was an extreme gesture of courtly nonchalance. Galileo presented himself as a heroic self-effacer, not as a heroic challenger of the prince. Through this self-inflicted "authorial martyrdom," Galileo could present himself as kin to the prince while stressing that his "heroism" was aimed at celebrating the prince rather than challenging him.

That an absolute prince would not accept being challenged to a potlatch by his would-be clients, but would instead reward those who presented themselves as sharing in his "heroic" ethos (though in a disciplined, if masochistic, fashion) fits quite well with what we have seen about the interaction between great patrons and high-visibility princes. As noted, successful clients were those who were able to present their giftgiving as a fully disinterested act, thus allowing the prince to reward them for apparently equally disinterested reasons. The end result of the process was the mutual legitimation of both client and prince.

Cosimo's "ennoblement" of Galileo by nominating him his philosopher and mathematician reflects these dynamics. In fact, the more the Medici recognized Galileo's disinterestedness and "nobility" in presenting them with the satellites of Jupiter, the more they legitimized their own dynasty by representing the discovery as a preordained celestial encounter with their destiny rather than as an interested gift from a client they had paid. For the discoveries to become an omen from the stars (a *sidereus nuncius*), Galileo *must* be given the status of starry ambassador, that is, of philosopher of the grand duke.[101] By presenting the gift exchange between

101. Similarly, Galileo presented the telescope to the Medici both as a scientific instrument and as a sort of dynastic relic. When, in March 1610, he sent the telescope to Cosimo II together with the presentation copy of the *Sidereus nuncius,* he told Cosimo that the rough-looking and unpolished instrument should be left in its original state, for it was the "instrument through which such a great discovery was achieved." The grand duke, Galileo continued, would receive many and more elegant-looking telescopes, but this was the only one

Galileo and Cosimo as perfectly disinterested, Galileo legitimized his discoveries, his instrument, and his new socioprofessional identity, while Cosimo enhanced his and his house's image.

These dynamics of mutual legitimation of patrons and clients were by no means limited to the case of Galileo and Cosimo II but were in fact typical of the power discourse of absolutism. There are telling homologies between Galileo's tactics for the legitimation of his new socioprofessional identity and Paul Pellisson's attempt to legitimize himself by becoming the historiographer of Louis XIV. As argued by Louis Marin, the message Pellisson sent to Louis through Colbert was that the most effective way to celebrate the king's image and power was by writing *his* history.[102] However, to be legitimate and politically effective this history could not be written by a private historian. Instead, it had to be the king's own history, not a historian's history of the king. But, at the same time, the king could neither write it by himself nor show himself commissioning a self-aggrandizing narrative. Although Louis was absolutely powerful, Pellisson understood that he was also absolutely impotent because he could not sing his own praises nor openly pay somebody else to do it.

According to Marin, Pellisson solved this royal deadlock and, by doing so, he acquired power from the king: "Give me the responsibility of being 'your' historiographer. I will give you a history, 'yours,' but with the precondition that I cannot write your history if I do not receive the office from you."[103] The king's impotent predicament could be overcome only by Pellisson's becoming the king's historiographer. It was only by being connected to Louis that—like an ancient prophet—Pellisson could "speak his voice" and write his history. However, for the king's narrative to be fully credible and effective, it had to come out of Pellisson's pen *naturally*. Pellisson could not be perceived as a hired pen: "It would no doubt be hoped that His Majesty approve and accept this design, which can almost not be well executed without him. But he must not seem to have accepted, known about or ordered it."[104] For Pellisson to become Louis's prophet, it had to happen in an apparently most natural (that is, disinterested) way.

Pellisson's strategy is like Galileo's in his dedication of the *Sidereus nuncius*. Galileo's saying that he and the Medici had been put together by the stars was a perfect tactic to naturalize and legitimize the relationship between patron and client while keeping their "complicity" secret. They

---

that was "there" at "that time" (ibid., pp. 297–98). It alone of all possible telescopes carried that special aura of *hinc et nunc* with it. It alone was not just a telescope but a *nuncius*.

102. Louis Marin, "The King's Narrative, Or How to Write History," *Portrait of the King* (Minneapolis: University of Minnesota Press, 1988), pp. 39–88.

103. Ibid., p. 43.

104. Ibid., p. 44.

did not plan anything. Also, like Pellisson who claimed he could not write the king's history without being (disinterestedly) connected to him, Galileo said that he could not have discovered the Medicean Stars without having been connected to Cosimo since he was a young prince:

> It pleased Almighty God that I was deemed not unworthy by Your serene parents to undertake the task of instructing Your Highness in the mathematical disciplines, which task I fullfilled during the past four years, at that time of the year when it is the custom to rest from more severe studies. Therefore, since I was evidently influenced by divine inspiration to serve Your Highness and to receive from so close the rays of your incredible clemency and kindness, is it any wonder that my soul was so inflamed that day and night it reflected on almost nothing else than how I, most desirous of Your glory (since I am not only by desire but also by origin and nature under Your dominion), might show how very grateful I am toward You. And hence, since under Your auspices, Most Serene Cosimo, I discovered these stars unknown to all previous astronomers, I decided by the highest right to adorn them with the very august name of Your family.[105]

Galileo's self-presentation fits Pellisson's strategy perfectly. He was "naturally" connected to the Medici when he discovered the stars and yet he was not paid by them. In fact, at that time, he was an employee of the Republic of Venice. Quite conveniently, he could represent himself as the unpaid prophet of the Medici glory.[106] The Medici had given him the ability to discover the stars but did not ask him to do so. Becoming the philosopher and mathematician of the grand duke after the publication of the *Sidereus nuncius,* that is, becoming officially and financially connected to the Medici, did not taint his credibility or the Medici's. By then, he had "freely" donated his discoveries to the Medici, and the Medici were giving him back an equally "free" countergift by calling him back to their court. As Galileo reminded them, they were made for each other.

Although we may not really see the discovery of Jupiter's satellites as a celestial encounter between the house of Medici and their destiny, and we may also be skeptical about Galileo's patronage relation with the Medici being written in the sky, Galileo was right (though not for the reasons he put forward) in presenting himself as a "natural" client of the Medici. When he observed the satellites at the beginning of 1610 he probably realized that, given the structure of the Medici's mythology and the patronage

105. Galilei, *Sidereus nuncius,* 32.

106. As noted in the previous chapter, Galileo was not paid cash for his instruction of Prince Cosimo during the summer but was given gifts of various sorts. He was also not paid for either the publication of the *Sidereus nuncius* or the telescopes he sent all over Europe until after the Medici had expressed their intention to invite Galileo back to Florence.

connections he had developed over the years, the Medici were the best patrons he could possibly attract. Quite probably, Jupiter played a role in the political mythologies of other European dynasties, but there is no evidence that Galileo knew of those mythologies or had brokers in those courts who could help him quickly to negotiate a dedication.

Suspicious Stars

Galileo's strategy for the legitimation of both his new instrument and the discoveries it made possible does not seem essentially different from the one he had previously tried out with Cosimo's 1608 *impresa*. By transforming the instrument and the discoveries into Medici fetishes, he tried to tie his patron's image and power to them. But, as we have seen, the use of patrons as legitimizing institutions was not an unproblematic strategy. Usually, patrons did not want to risk their status for that of their clients, even when an important contribution to their own image was at stake. The cautious Cosimo II was not always quick in upholding Galileo against his challengers, and his son Ferdinand II would be even less supportive.

Galileo's strategy seems to have been to try to tie the Medici's image to his discoveries not at once but gradually. In the dedication of the *Sidereus nuncius,* Galileo did not try to say that the Medici had endorsed his discoveries, he simply laid down his credentials: he had discovered the Medicean Stars for Cosimo II because of his particular connection to him. That was why he was dedicating the stars to the house of Medici. He was not doing so to gain credibility. He used the dedication of the *Sidereus nuncius* as a patronage "bait" but did not try to "capture" the Medici immediately; a rush tactic would not have worked.

However, as a result of the Medici acceptance of the dedication, he asked them to distribute, through their diplomatic networks, telescopes and copies of the *Sidereus nuncius* (almost as instruction booklets) to the European nobility. Although he presented this move to the Medici as a way of making sure that their glory would be well publicized among those who counted, he managed to be perceived as a Medici client in the eyes of those who received the telescopes and the *Sidereus nuncius*. Kepler, for instance, given the way he had been approached by the Medici ambassador, thought that Galileo was already in the service of the Medici.

In a sense, Galileo was able to use the extra credibility he derived by being associated to the Medici without them fully realizing it or having officially endorsed his discoveries. Nevertheless, the additional power he obtained through this still quite loose connection with the Medici helped him to succeed in defending his discoveries and, consequently, to gain more recognition from the Medici which, in turn, allowed him to become even more credible and draw further assent to his discoveries from others.

The Medici connection was particularly important in convincing those Galileo could not reach personally.

As he wrote to Matteo Carosio in Paris in May 1610, he had been very successful at convincing people of the truth of his discoveries by going around with a good telescope and showing the satellites to those willing to look at them. However, he had problems in reaching people who were far away.[107] In those cases, having the local prince look through the telescope (one they had received through the Medici ambassador) and see the planets was crucial in disciplining the local astronomers and philosophers.[108] As Martin Hasdale wrote to Galileo in July 1610, the emperor's endorsement of Galileo's discoveries had quenched the opposition at the imperial court: "His August Majesty has been the cause for the decline of the success of the adversaries, because His Majesty has declared himself most happy and most satisfied [with your claims]."[109] What is more interesting in Hasdale's letter is the mention that Galileo's opponents at court had used as a major resource for their criticism the report that Galileo, in his brief visit to Bologna, had not been able to convince Magini and the other mathematicians and philosophers assembled at his house. This report was represented as "an official judgment of the University of Bologna."[110] In short, the emperor's power was enough to overcome the authority of a lesser institution, the University of Bologna—one that, given the received disciplinary hierarchy, was likely to be hostile to Galileo's claims anyway.

Galileo alternated between using those resources he could get from the Medici without putting them on the spot, in order to get credibility from outside, and then using the assent thus gained to strengthen further his link with the Medici and tie his discoveries to their image. By the end of this process, Galileo had slowly tied the Medici to his wagon. More important, in so doing, he had used their own power. A few months later, Galileo was their philosopher and mathematician and was sent to Rome as an official envoy to have the Medici glory and his discoveries endorsed by the greatest of the Italian princes: the pope. As indicated by a letter Vinta wrote to Galileo announcing the Grand Duke's authorization of his trip to Rome, the symbiosis between Galileo's discoveries and the Medici image had finally been achieved:

> About the trip of Your Lordship to Rome, I have told their Highnesses that now is the time [for it] because of the state of the debate

107. *GO,* vol. 10, no. 313, p. 357.

108. Obviously, the first instance in which he adopted this tactic was with Cosimo himself on the occasion of his trip to the Medici court in Pisa on Easter of 1610. That Cosimo had seen the stars gave Galileo the credibility he needed for his subsequent moves.

109. Ibid., no. 360, p. 401.

110. ". . . a definitive judgment of the University of Bologna" (ibid.).

[*speculazione*] and the possibility of observation of those planets, and that, therefore, one should not wait any longer, and that, once this will be cleared in Rome, with the confirmation that has [already] been received from the Mathematician of the Emperor, from Father Clavius and others, as soon as they will be confirmed and established in Rome, it will be possible to say that those claims [*constituzione*] are established for the entire world, and, by sharing them with His Holiness, these new observations and claims will have to be received with universal consensus.[111]

The achievement of this symbiosis was a complicated and delicate process because both the Medici and the Florentine courtiers tended to be unwilling to put their honor at stake for Galileo's discoveries. Just a week after the publication of the *Sidereus nuncius* in March 1610, Galileo wrote to Vinta that:

> because it is most true that our reputation begins with our own self-confidence, and that whoever wants to be esteemed ought to have self-esteem first, when His Most Serene Highness will demonstrate recognition of the importance of this encounter [the discovery of the Medicean Stars], no doubt not only all his subjects but all nations will recognize its importance too, and there will remain no feather in the wings of fame that will not write in praise of the glory of this event.[112]

Galileo then suggested that the distribution of copies of the *Nuncius* and of telescopes to European kings and princes would be most appropriately carried out by the Medici ambassadors in the various Italian and European states.[113] But, while the Medici accepted Galileo's proposal of distributing the books and instruments through their official diplomatic channels, they avoided taking official stands about the reality of the satellites of Jupiter.[114]

Writing again to Vinta on May 7, Galileo went back to the same issue. After reassuring him and the Medici that he had both publicly refuted his challengers at Padua and received a long and very supportive letter from the "Mathematician to the Emperor,"[115] Galileo claimed that the Medici's image in connection to the discoveries had been safely defended. But now "We—especially our Most Serene Lords—have to sustain the importance and reputation of the discovery by demonstrating the esteem such a remarkable novelty deserves, it being so considered by everybody who speaks sincerely."[116] But the Medici maintained their cautious stand. Vincenzo Giugni—the supervisor of the Medici artistic workshops—

111. *GO,* vol. 11, no. 464, pp. 28–29 (emphasis mine).

112. *GO,* vol. 10, no. 277, p. 298.

113. Ibid., no. 277, pp. 298–99.

114. Ibid., no. 311, pp. 355–56.

115. Ibid., no. 306, p. 349.

116. Ibid.

wrote to Galileo on June 5 saying that the production of the dies to strike the medal celebrating the discovery of the Medicean Stars had been put on hold by the grand duke himself. Cosimo II had told Giugni to wait until the debate on the stars was settled.[117]

By this time Galileo had received a long letter from Kepler (soon published as *Dissertatio cum Nuncio sidereo*) in which he confirmed Galileo's observations. Confident of the international credibility provided by Kepler's endorsement, Galileo showed himself annoyed by the grand duke's extreme caution and mentioned to Giugni that even the king of France had intimated his willingness to accept the dedication of whatever planets Galileo might discover in the future. Therefore, Galileo suggested to Giugni that, "whenever possible, please make sure that Your Most Serene Highness would not delay the flight of fame by taking an ambiguous stand about what he has seen many times himself—something that fortune reserved to him and denied to everybody else."[118]

Although by the time Galileo sent this letter he had been already reassured by Vinta of his position at the Medici court, it might not have been by chance that he had not yet received the life contract he had been promised, which in fact reached him only in July.

Cosimo II was not alone in his caution. The Florentine academicians and court poets were not celebrating the Medicean Stars as enthusiastically as Galileo expected or wished them to. Two weeks after the publication of the *Sidereus nuncius,* Alessandro Sertini, a longtime Florentine friend of Galileo's and a member of the Accademia Fiorentina, wrote to him saying that his efforts to mobilize the "Tuscan Muses" had not been very succesful. The Medici court writers seem to be waiting for one of them to give the signal: "The Muses are moving a bit slowly, because nine of them are lagging behind waiting for a tenth one to take the lead. Your Lord should write to him if you want to make sure that he will write something on the Medicean Stars."[119]

In a letter of July 10, Sertini informed Galileo that attacks by Giovanni Magini and Martinus Horky on his discoveries had been widely pub-

---

117. Ibid., no. 326, pp. 368–69.

118. Ibid., no. 339, pp. 381–82. See also pp. 379–80. In addition to the king of France's interest, we know of a number of people who tried to replicate Galileo's patronage strategies. Scheiner dedicated a fifth satellite he thought he had discovered around Jupiter to Welser. It seems that Peiresc planned a "French version" of the *Nuncius* dedicated to Maria de' Medici. The surviving sketch for the frontispiece depicts Maria sitting on Jupiter, surrounded by the four stars that Peiresc had named after the four grand dukes: "Cosmus Major," "Franciscus," "Ferdinandus," and "Cosmus Minor" (*La corte, il mare, i mercanti/ La rinascita della scienza/ Editoria e società/ Astrologia, magia e alchimia* [Florence: Edizioni Medicee, 1980] [an exhibition catalogue], pp. 230–31). Jean Tarde and Charles Malapert thought that the sunspots were congeries of planets and dedicated them to, respectively, the Bourbons (in 1620) and the Austrian Hapsburgs (in 1633).

119. *GO*, vol. 10, no. 282, pp. 305–6.

licized in Florence and that Ludovico delle Colombe seemed to have joined the challengers' side. As a result, Sertini was unsure of the Florentine writers' willingness to publish their sonnets on the stars.[120] In fact, Galileo had proposed to the grand duke to publish a more elegant version of the *Sidereus nuncius* in the Florentine language, one including the sonnets dedicated to the Medicean Stars.[121] Such a version would have been tailored to the Florentine court audience, for the sonnets would spell out the connections between the stars and the Medici mythology. Those connections were not elaborated in the first Latin version of the *Sidereus nuncius* because the European audience to which it was primarily addressed could not have understood them. I suggest that it was because he had a European audience in mind that Vinta, when consulted by Galileo on the name to be assigned to the satellites in the *Sidereus nuncius,* replied that, of the two names proposed by Galileo, "Medicea Sydera" seemed more appropriate because "Cosmica Sydera" might have been misunderstood as referring to "cosmos" rather that to "Cosimo."[122] A Florentine audience would have not made that mistake.

The writers were still unenthusiastic in August, when Sertini wrote to Galileo: "Everybody here is worried because you said you wanted to print [the poems]. [Michelangelo Buonarroti, Jr.] would prefer not to have his name printed but—like Piero de' Bardi—he would be happier if it would say: 'Made by the Impastato, Member of the Academy of the Crusca'."[123] The court writers, knowing that Galileo now wanted to publish not only their sonnets but the challenges to his discoveries, together with his responses, in the new edition of the *Sidereus nuncius,* were uncomfortable with the idea of being perceived as Galileo's allies in his predictably aggressive counterattacks.[124] Sertini went so far as to suggest that Galileo answer all challenges "without mentioning anybody, and by remaining within the specific boundaries of the issue, for it seems the best thing to do, and the one I would prefer."[125]

Although the Medici and the court writers were not Galileo's scientific peers, their caution is like that of a scientist's colleagues in evaluating a discovery claimed by the scientist. At first glance it may seem odd that neither Cosimo nor the court writers seemed to take the opinions of members of the professional elite of astronomers, such as Kepler, as decisive in

120. Ibid., no. 357, pp. 398–99.

121. Ibid., no. 277, p. 299.

122. Ibid., no. 265, p. 283.

123. Ibid., no. 372, pp. 411–13.

124. Ibid., no. 332, pp. 373–74.

125. Ibid., no. 372, p. 412. This new Italian translation was never published. A probably unauthorized reprint of the original Latin version of the *Nuncius* appeared at Frankfurt in the fall of 1610.

determining their own endorsements.[126] This apparent puzzle can be solved by remembering that Cosimo and the writers were actually Galileo's peers (or superiors) by virtue of belonging to the same institution: the court. The court was not a scientific institution but the place where representations of the prince's power were produced; and Galileo was hired there less as an astronomer than as a producer of spectacular dynastic emblems. Therefore, he needed the writers to accept and articulate his discoveries in court cultural productions and representations of the grand duke's power.[127] On the other hand, the Florentine courtiers did not need to believe Kepler or, for that matter, Galileo himself. The opinions of leading astronomers were not binding for the courtiers. The only authority they knew was that of their prince or of their prince's patrons.

Galileo's delicate position in this phase of transition from the university to the court reflects the novelty of the socioprofessional identity he was trying to establish for himself. In a sense, he was a socioprofessional hybrid. He presented himself as a "new philosopher," a role that—given the disciplinary hierarchy structuring the university—could be legitimized only at court. Yet, even though the people who had the professional skills to judge his achievements were mathematicians, not court writers and gentlemen, and even though Galileo might have been in serious trouble had Kepler turned down his claims about the existence of the satellites of Jupiter, Kepler's recognition of his discoveries was not sufficient to win over the courtiers. And Galileo needed the endorsement of courtiers and prince because only at court could he become a philosopher. Schematically put, the mathematicians' endorsement of Galileo's discov-

126. The Medici respect of the Jesuits' scientific authority may be seen as contradicting my point. However, the positive impact that that recognition, in December 1610, of the reliability of the telescopic discoveries had on Galileo's legitimation cannot be interpreted as a sign of their "technical credibility" alone. Their opinion was probably more influential than Kepler's because they were correctly perceived as the mathematicians of the pope. This was particularly true in Florence, where, with the legitimacy of the Medici dynasty precariously dependent on the pope, religious orthodoxy and respect of the Church's positions were crucial. In respecting the Jesuits' views, the Florentine courtiers were thus bowing to the authority of the papal court.

127. Galileo's concern with the "media coverage" of the Medicean Stars or of his discoveries in general was not limited to the Florentine court. For instance, he was very pleased that the Florentine Jesuits to whom he had shown the satellites believed their existence and incorporated those discoveries "in preachings and orations, with very gracious images" (GO, vol. 10, no. 436, p. 484). Similarly, Galileo was quite pleased that Monsignor Giovanni Battista Agucchi—a Roman courtier and future bishop—used the Medicean Stars for an impresa commissioned by a patron who wanted it delivered in a literary academy (GO, vol. 11, pp. 205, 220, 225, 249, 255, 264). The manuscript of Agucchi's impresa—"Del medio"—is at BNCF, "Galileiani 246," fols. 96–110. Similarly, Galileo's friend Cigoli celebrated his discoveries by painting the Madonna standing on the earth-like moon in Rome's Santa Maria Maggiore (GO, vol. 11, no. 814, p. 449).

eries would have been *necessary and sufficient* to establish his credibility *as a mathematician,* but that same endorsement was only *necessary* (and no longer sufficient) in certifying Galileo's credibility *as a court philosopher.* As we will see, this tension between two audiences, two discourses, two different socioprofessional identities, and the different forms and levels of legitimation that went with them, characterized Galileo's entire courtly career.

Shapin's study of the seventeenth-century "house of experiments" suggests that the legitimation of experimental practices in England was caught in an analogous social paradox. Those who had the technical skills to perform experiments (and quite likely to understand them) did not have the high social status needed to be perceived as having "the qualifications to make knowledge."[128] Conversely, many of the gentlemen who had the social qualifications to "make knowledge" did not necessarily have the skills. They could certify, but they often could not figure out how or what to certify.

The Career of the Medicean Stars

Although Galileo was not successful with his first attempts to enlist the support of the court writers, the Medicean Stars eventually became an integral part of the discourse of the court.[129] The medal celebrating Galileo's discovery of the planets was eventually struck. Jupiter sitting on a cloud with the four stars circling about him was presented as an emblem of Cosimo II, whose effigy occupied the other side of the medal (fig. 3). The stars were represented in sonnets, in theatrical machines, in operas, in medals, and in frescoes celebrating the divine pedigree of the house of Medici. We encounter them again in the most important court spectacle of the carnival of 1613—the *barriera* of 17 February.

128. Steven Shapin, "The House of Experiment in Seventeenth-Century England," *Isis* 79 (1988): 395.

129. As noted, the vernacular edition of the *Sidereus nuncius* was never published. Surviving sonnets to the Medicean Stars include those of Buonarroti (*GO,* vol. 10, p. 412), Salvadori (*GO,* vol. 9, pp. 233–72), and Piero Bardi (*GO,* vol. 10, p. 399). Claudio Seripandi's sonnet is lost; Niccolò Arrighetti's was left in manuscript form until it was published in Nunzio Vaccalluzzo, *Galileo Galilei nella poesia del suo tempo* (Milan: Sandron, 1910), pp. 59–60. We do not know whether Chiabrera wrote a sonnet after Sertini's invitation (Galileo had sent him an autographed copy of the *Sidereus nuncius,* which is now at the University of Oklahoma at Norman), but we do know that he included the Medicean Stars in at least one of his compositions. Salvadori's "Per le Stelle Medicee temerariamente oppugnate" makes explicit the use of patronage for the legitimation of Galileo's discoveries. After retracing a mythological history of the Medici family that stresses the link between the Medici and Jupiter (and his tremendous power), Salvadori displays his incredulity at the arrogance of those who, by challenging the existence of the Medicean Stars, were challenging Jupiter's (or Cosimo's) own power (*GO,* vol. 9, p. 272).

Fig. 3. Gaspare Mola, oval medal struck around 1610 to commemorate Cosimo II and the discovery of the Medicean Stars. From Karla Langedijk, *Portraits of the Medici* (Florence: SPES, 1983), vol. 1, p. 579.

It began at two o'clock Florentine time in the theater of the Pitti Palace in front of a selected court audience.[130] After a virtuoso display of
spectacular theatrical machines and effects designed by the court engineer,
Giulio Parigi, the spectacle began to reveal its mythological plot.

Cupid set his own realm over Tuscany, inaugurating a Golden Age.
Unfortunately, peace was soon threatened. Cupid and his knights (six
court pages) were faced by a monstrous dragon (producing flames and
smoke) and twelve Furies led by Nemesis. Although the dragon, Nemesis,
and the Furies were eventually made to disappear into a trap conveniently
connected to hell, Cupid and Tuscany were not safe yet. Sdegno Amoroso
(Disdain of Love) and his five ferocious and barbarous-looking "Egyptian
knights" jumped on stage from hell's mouth.[131] A new combat began, but
peace and Tuscany's Golden Age were reestablished by divine
(Cosimo I's?) intervention.

Thunder was heard and Jupiter arrived on a shimmering cloud (a part
of a very complicated machine which changed in appearance as it moved
across the stage). Jupiter was not alone on arrival:

> down below, among the clouds, appeared the four stars that circle
> Jupiter discovered by Galileo Galilei from Florence, Mathematician
> to His Highness, with the marvelous spyglass, and like the ancients

130. Nagler, *Theatre Festivals of the Medici*, pp. 119–121.
131. Ibid., 122.

who transposed to the sky their greatest heroes, he—having discovered these stars—called them Medicean, and has dedicated the first to His Most Serene Highness, the second to Prince Don Francesco, the third to Prince Don Carlo, the fourth to Prince Don Lorenzo.[132]

The machine brought Jupiter close to the grand duchess, to whom he sang his aria; then it slowly disappeared from the stage. In the process, the four Medicean Stars turned into four flesh-and-blood knights: "After Jupiter finished his song some thunders were heard, the cloud vanished and there appeared four stars which soon turned into four knights who stood up." The Cyclops (who had come on stage right before Jupiter's arrival) handed thunderbolts to the four knights. With such weapons, they were ready to start the new joust in Jupiter's name. The name of the joust was "The Arrival of the Knights of the Medicean Stars." Peace soon followed. The ladies in the audience joined the knights on stage and the final ball began.[133]

The rest of the city had its share of Medicean Stars as well. Two days later, a simpler version of the *barriera* went through the city as a carnival procession. The Medicean Stars, together with the Furies and Nemesis, were in the second troupe of the pageant. However, the stars did not stop moving. Together with Jacopo Cicognini—one of the authors of the *barriera*—they migrated to Rome where, on 9 February 1614, they re-emerged at the wedding of Don Michele Peretti, prince of Venafro, and Princess Anna Maria Cesi—an event recorded in *avvisi* and diaries of contemporaries as the highlight of the Roman carnival of 1614.[134]

That evening, "great was the confusion, the noise, and the excitement of the crowd" near the Palazzo della Cancelleria, where Cicognini's play celebrating the wedding was to be performed. "One could see gathered together all the Roman nobility, . . . the ladies and the princes preceded by a great number of torchbearers and elegantly-dressed servants." The theater was full when the stage curtain was finally raised. A small cloud with a golden chariot on it appeared from the right. Venus was the charioteer. She was searching for her son, Love, who had escaped from Olympus. Love was soon found "with golden hair, all naked, with a most beautiful veil covering only those parts that Nature teaches us to keep hidden." Love's wings were "most delicate and all covered with jewels." He held a bow with his right hand while a quiver full of more precious stones was hanging from his shoulder. More jewels, "even more remarkable for beauty and value, decorated his necklace."

Venus asks Love why he has left Olympus, to which the boy answers

132. Giovanni Villifranchi, *Descrizione della Barriera e della Mascherata fatte in Firenze a' XVII & a' XIX di Febbraio 1613* . . . (Florence, Sermartelli, 1613), pp. 32–33.

133. Ibid., p. 38; and Nagler, *Theatre Festivals of the Medici*, pp. 123–25.

134. Orbaan, *Documenti sul barocco in Roma*, pp. 214–215.

that he has descended to earth to "unite in a holy bond the quality of a great prince and the purity of a great lady." Although pleased by Love's mission, Venus is saddened by the setting in which such a lofty event is taking place: a Rome that is no longer glorious but ruined. Being unable to bear such a view, the goddess immediately restores Rome to its ancient beauty. This instantaneous restoration infuses the bride and the groom with joy and they start dancing to celebrate Love's arrival in a newly beautiful Rome. They are soon joined by the noble audience, with the exception of the cardinals, whose status prevented them from dancing.

At the end of the ball, the scenery is quickly changed. We are now on Olympus, with Venus trying to convince the other gods to help her retrieve her son, Love, who does not want to leave Rome. He likes it down there now. Jupiter intervenes. The usual thunder is heard, the sky opens, and Jupiter appears "from a great distance, covered with gold, and of inconceivable splendor." The clouds now move with slow, circular motion. Jupiter is seated in their midst "on a throne of ivory and ebony, shiny with gold and gems . . . wearing a fiery royal crown." He wears appropriately godly clothes embroidered with stars. And "around him are seen four children with silver armor and golden helmets topped with turquoise plumes and a star emerging from them."[135] A cousin of the bride, Federico Cesi, the founder of the Accademia dei Lincei, was present at the event and reported to Galileo that everyone enjoyed the play and the place of the Medicean Stars in it, with the exception a few "Peripatetic monkeys" who did not appreciate Cicognini's celebration of Galileo's novelties.[136] Cesi apparently felt obliged to enter into a little dispute with them.

This was 1614. Probably as a result of Bellarmine's admonition of Galileo in 1616 and of Cosimo II's declining health and control over cultural and political policies, Galileo's discoveries did not continue the career in the Medici mythology they had begun so brilliantly, and they do not seem to have reappeared in Rome either. Their visibility declined even further after 1621 when, following Cosimo II's death, Grand Duchess Cristina and her counselors took over the government of Tuscany and the

135. Jacopo Cicognini, *Amor pudico* (Viterbo: Discepolo, 1614), contains the text of the play. I have not been able to locate a copy of it. On the party, see Filippo Clementi, *Il Carnevale romano nelle cronache contemporanee* (Città di Castello: Unione Arti Grafiche, 1939), pp. 396–411. My description of the play is based on Clementi. The quotes are from the contemporary *Avvisi di Roma* cited therein. The *Avviso* reproduced in Orbaan sketches a more intricate plot.

136. "Cicognini certainly satisfied me. Finding myself at the party and scenic soirée [pageant] of the wedding of Princess Peretti, my cousin, I saw that among the other planets he had, with much propriety, placed the Medicean ones around Jupiter. Everybody loved the spectacle and the discoveries [novelties] included in their proper place. Well, it is true that I made myself heard by some Peripatetic monkeys who could not stop themselves from snarling like old men hostile to any novelty" (*GO*, vol. 12, no. 980, p. 29).

management of court culture. Carnival festivals were played down, and sacred comedies became the dominant genre.[137] Moreover, the lack of an actual prince (Ferdinand II would reach his majority only in 1628) made it difficult to develop new prince-centered cultural productions. Jupiter was unemployed. When, in 1628, Ferdinand II finally took power, Galileo had already developed his patronage niche in Rome.

References to the Medicean Stars, however, were still included in the work of writers connected to the Medici court. Alessandro Tassoni, in his (then) famous *Secchia rapita,* had Jupiter enter the scene "with those stars that have been found around his head," while Chiabrera—a poet on the Medici payroll—praised Galileo for having put "the name of our great Medici among the eternal stars, a name so powerful that it even improves the value of stars."[138] Although less conspicuous than during Cosimo II's reign, the presence of the Medicean Stars in Florentine court culture continued, as shown by a large painting (listed in the 1638 inventory of the Pitti Palace but now lost) of Jupiter riding an eagle surrounded by four *putti* on the Medicean Stars.[139]

As a result of the court moving from the Palazzo della Signoria to the Pitti Palace, a new Medici Olympus was painted in the new palace's Planetary Rooms. The context in which the pictorial program of the Planetary Rooms was developed (about ten years after Galileo's condemnation) posed serious problems to the design of the representation of Jupiter and the Medicean Stars. Emblematics offered the way out of the dilemma.

Just as Galileo linked the Medicean Stars to Jupiter-Cosimo I's virtues in the dedication of the *Sidereus nuncius,* the Pitti Palace's Room of Jupiter (one of the Planetary Rooms)[140] presented the god surrounded by the Medicean Stars as the four cardinal virtues (fig. 4).[141] Such an emblematic representation of the Medicean Stars was repeated—this time much more explicitly—in a large engraving of 1664 (fig. 5). Cosimo III was there represented as Augustus.[142] Above him, we find Jupiter (resembling

137. Ludovico Zorzi, *Il luogo teatrale a Firenze* (Milan: Electa, 1975), p. 88.

138. Alessandro Tassoni, *La secchia rapita* (Ronciglione, 1624), in Alberto Asor Rosa, ed., *I poeti giocosi dell'età barocca* (Bari: Laterza, 1975), p. 28; and Gabriello Chiabrera in his "Sermone a Gio. Francesco Geri," in Alberto Asor Rosa, ed., *La lirica del Seicento* (Bari: Laterza, 1975), p. 134. For a comprehensive compilation of poems referring to Galileo and to the stars, see Vaccalluzzo, *Galileo Galilei nella poesia del suo tempo.*

139. The same theme would be adopted later for Cosimo III's medals. See Figures 10, 11, 12.

140. ". . . canvas depicting Jupiter riding the eagle with four putti representing the Medicean Stars, $3^1/_3 \times 2^1/_2$ braccia" (*ASF,* "Guardaroba medicea 535," fol. 143).

141. The frescoes were begun by Pietro da Cortona and completed around 1665 by his pupil Ciro Ferri (Langedijk, *Portraits of the Medici,* 1: 210).

142. Ibid., pp. 211–12.

Fig. 4. Pietro da Cortona, *Jupiter Accompanied by the Cardinal Virtues*, Room of Jupiter, detail of ceiling, Palazzo Pitti, Florence. From Karla Langedijk, *Portraits of the Medici,* vol. 1, p. 209.

Cosimo's father, Ferdinand II). On the clouds around Jupiter/Ferdinand II we find the four cardinal virtues (embodied by the former Medici grand dukes) with the four Medicean Stars shining over their heads.[143]

The Medicean Stars reemerged conspicuously in the Medici mythology during the reign of Cosimo III (1670–1723). The grand duke's name lent itself to references to the Medicean Stars especially because, having five ancestors, he could be portrayed as directly related to Jupiter and the four stars. The revival of the Medicean Stars was most evident in 1661 on the occasion of the politically important wedding of Prince Cosimo and Marguerite-Louise d'Orléans—the cousin of Louis XIV.[144] The *Mondo Festeggiante,* an equestrian ballet, was the highlight of a long series of ceremonies, pageants, and spectacles.[145] According to the official descrip-

143. Ibid., pp. 215–16. Ciro Ferri also was the painter who completed the Planetary Rooms. Spierre engraved the frontispiece of the memoirs of the Cimento. See also Filippo Baldinucci, *Cominciamento e progresso dell'arte dell'intagliare in rame* (Florence: Stecchi, 1767), pp. 215–16.

144. Langedijk, *Portraits of the Medici,* 1: 216–17.

145. *Memorie delle feste fatte in Firenze per le reali nozze de' Serenissimi Sposi Cosimo Principe di Toscana e Margherita Luisa d'Orleans* (Florence: Stamperia di SAS, 1662).

Fig. 5. Frans Spierre, engraving after Ciro Ferri, *The Medici Stars Protecting Cosimo III*, 1664. From Karla Langedijk, *Portraits of the Medici*, vol. 1, p. 208.

Fig. 6. Stefano della Bella, engraving of Hercules carrying the cosmos on his shoulders during a festival celebrating the wedding of Prince Cosimo de' Medici and Marguerite-Louise d'Orléans. From Alessandro Carducci, *Il mondo festeggiante*. Courtesy of the Harvard Theater Collection, Harvard University.

tion of the event, twenty thousand spectators were present at the ballet.[146]

The spectacle began with the entrance of an exceptionally large theatrical machine representing Hercules carrying the cosmos on his shoulders (fig. 6). Once Hercules reached the center of the stage, the machine slowly transformed itself into Mount Atlas. Numerous knights representing the earth's four continents entered the stage paying homage to Hercules and—implicitly—to the new "Herculean" couple being celebrated there. But while the knights of Europe and America were happy about the wedding, those of Asia and Africa felt threatened by such a powerful union. An elegant duel-ballet between the two factions began but did not last long.[147]

Powerful thunder was heard, announcing Jupiter's arrival on a very tall theatrical machine surrounded by clouds (fig. 7). Immediately, all knights stopped dueling. As soon as the machine had lowered Jupiter to the level of the stage, the clouds disappeared and "Four knights riding four elegant horses appeared very close to Jupiter. They symbolized the four Medicean Stars which [this is a quotation from the *Nuncius*] never depart from his side."[148] Jupiter then sang a song celebrating the wedding, one which would make Cosimo's Medicean Stars even more beautiful and shining because of the new splendor contributed by the golden lilies of Marguerite-Louise.[149] Apollo joined Jupiter praising the wedding as the union of the "French Sun and the Medicean Stars."[150] As the spectacle continued, "Four Medicean Stars reached His Highness and took their places around him, that is, around the Tuscan Jupiter, and they never left him during the remaining part of the ceremony, but they always accompanied him and remained orderly and close to him throughout his pageants."[151]

The Medicean Stars also appeared in a medal struck on the occasion of Cosimo's wedding. His impresa was a ship at sea guided by the Medicean Stars with the motto: *Certa Fulgent Sidera* (fig. 8). They were also represented in the cycle of frescoes called the "Medici Apotheosis" painted by Luca Giordano on the ceiling of the Medici-Riccardi Palace[152] as well as in other official medals (figs. 9, 10, 11).[153] When Cosimo died in 1723, a

146. Alessandro Carducci, *Il mondo festeggiante, balletto a cavallo fatto nel teatro congiunto al palazzo del Sereniss. Gran Duca per le reali nozze de' Serenissimi Principi Cosimo Terzo di Toscana e Margherita Luisa d'Orleans* (Florence: Stamperia di SAS, 1661).

147. *Memorie delle feste*, p. 106.

148. Carducci, *Il mondo festeggiante*, p. 46.

149. Ibid., p. 49.

150. Ibid., pp. 51, 53.

151. Ibid., p. 53.

152. Langedijk, *Portraits of the Medici*, 1: 215; 2: 639.

153. Ibid., 2: 630–32, 637, 639. At pp. 630–32, Langedijk has not noticed either the sign of Jupiter or the four stars in these medals.

Fig. 7. Stefano della Bella, engraving of Jupiter arriving among clouds (center) during a festival celebrating the wedding of Prince Cosimo de' Medici and Marguerite-Louise d'Orléans. The four "Medicean Stars" knights are at the base of the theatrical machine. From Alessandro Carducci, *Il mondo festeggiante*.

Fig. 8. Francesco Tavani, later copy (1666) of a medal Tavani made on the occasion of the marriage of Prince Cosimo and Marguerite-Louise d'Or-léans in 1661. From Karla Langedijk, *Portraits of the Medici,* vol. 1, p. 640.

medal with the Medicean Stars was placed on his chest (fig. 12). The Medici dynasty survived him by only fourteen years.

Court Culture, Absolutism, and the Legitimation of Science

Even as the Medicean Stars began to reappear in court mythology during the reign of Ferdinand II, their association with Galileo was on the wane. His condemnation of 1633 hastened the process. Galileo's role in the stars' discovery was mentioned in the *barriera* of the carnival of 1613, but no such reference is to be found in the *Mondo Festeggiante* of 1661. By that time, Medici court culture had severed the Medicean Stars not only from their discoverer but from astronomy as well. As shown by the *Mondo Festeggiante,* the Medicean Stars were stars no longer. All that was left of them was a dynastic fetish, a name assigned to Jupiter-Cosimo's knights. Analysis of this process of fetishization uncovers both the avenues and structural limits of Medici patronage for the legitimation of science.

Medici patronage did not reward authors of scientific theories or proponents of research programs but appreciated marvels that fit the discourse of the court and contributed to legitimizing the Medici image. Consequently, Galileo could be rewarded as a celestial ambassador of the Medici glory but not as a Copernican astronomer. Galileo understood well the discourse of the court and presented the satellites of Jupiter to the Medici not as Copernicus-supporting astronomical discoveries but as dynastic emblems.[154]

154. Galileo's awareness of the codes of Medici patronage can be found not only in his representation of the satellites of Jupiter as dynastic emblems. In fact, while negotiating with

Fig. 9. Luca Giordano, *Medici Apotheosis,* galleria ceiling, Palazzo Medici-Riccardi, Florence. From Karla Langedijk, *Portraits of the Medici,* vol. 3, p. 1513.

29,107 var.   *29,107 rev. variant*

Fig. 10. Anonymous, no date. Medal celebrating Cosimo III. On the reverse, surrounded by the motto "FAMAM EXTENDERE FACTIS," is Fame floating among clouds over the globe. Directly above Fame's trumpet is Jupiter surrounded by the Medicean Stars. From Karla Langedijk, *Portraits of the Medici,* vol. 1, p. 631.

The interesting paradox of Galileo's successful patronage strategy is that he had to efface his authorship in the discovery in order to become a more legitimate author, that is, a philosopher. As we have seen, this ritual effacement was rooted in the dynamics of absolute power and the way it framed authorship. In his dialogue on the court, Torquato Tasso presented the court as a gathering through which courtiers increase the prince's reputation and honor because it is only by doing so that—like a stream becoming such by being fed from a spring—they could gain honor for themselves.[155] Similarly, as shown by Galileo and Pellisson (but also by the Cimento's dedication of the *Saggi*), a subject could become a legitimate author not by presenting himself as an arrogant producer (a "challenger") but by presenting himself as the prince's "agent." In this way, the client could gain legitimation while the prince remained, so to speak, the ultimate, absolute author.[156]

Vinta for his position at court, he tried to be perceived as an appealing client by presenting himself as literally swamped by marvels (*GO,* vol. 10, no. 307, p. 351).

155. Torquato Tasso, *Il malpiglio, o vero de la corte,* reprinted in Cesare Guasti, ed., *I dialoghi di Torquato Tasso* (Florence: Le Monnier, 1901), 3: 13.

156. Differently from potlatches, where the competitors engage in conspicuous consumption in order to challenge (and possibly ruin) each other, we have here a more disciplined format in which one party (the prince) is assumed to be the winner from the start. Consequently, the other party can gain status not through challenging the prince (who is ax-

Fig. 11. Giovanni Battista Foggini, ca. 1683, reverse of a bronze medal celebrating Vittoria della Rovere, Cosimo III's mother. Above Fame is Jupiter
and the four Medicean Stars. From Fiorenza Vannel and Giuseppe Toderi,
La medaglia barocca in Toscana, table 1.

However, the prince could not celebrate his image by himself. He
needed the client to do so. But patrons and clients could not openly trade
legitimation for celebration. This would have destroyed the image of
princely power that the client was supposed to celebrate and that, in turn,
was supposed to legitimize the client himself. As Marin puts it, "The only
outlet is for both sides to keep the secret of complicity."[157]

It was the client who had to propose and deploy the trick, and it was
precisely for doing what the prince could not do that the client was rewarded. This may explain why Galileo and Pellisson presented themselves

iomatically unchallengeable) but by "wasting himself" in ways that enhance the image of the
prince. Because the prince is not challenged (but actually benefits from the client's "wasting
himself") he can then "recognize" the tribute to his image by rewarding the client. But this
reward is not costly to the prince. It is more a recognition than a real reward. The link between these processes and the ways in which political absolutism developed out of the do-
mesticization of the aristocrats through court culture is, I think, telling. The aristocratic
challenging drive is turned onto itself. The "wasting" of the clients illuminates the prince's
power.

157. Marin, *Portrait of the King*, p. 44.

Fig. 12. Antonio Selvi, bronze medal celebrating Cosimo III. On the reverse, under the motto "CERTA FULGENT SIDERA," is a ship guided by the Medicean Stars. From Fiorenza Vannel and Giuseppe Toderi, *La medaglia barocca in Toscana,* table 115.

as not giving anything to their prince that was not already his. It was only by doing so that the rewards the prince would bestow on them would not appear as remuneration for a service they had performed. The clients had to efface themselves as authors in order to keep the trick hidden. Only thus could the prince be represented as the ultimate author of whatever his clients produced and, therefore, have his image celebrated while, at the same time, legitimizing his clients as the "agents" through which the celebration was produced.[158] Quite literally, Galileo could not be an independent philosopher; he could only be the philosopher of the grand duke.

Consequently, the complete alienation of the Medicean Stars from their discoverer displayed in the *Mondo festeggiante* and in the other later representations of the stars was already inscribed in the patronage strat-

---

158. However, because the client was in charge of the trick, in a sense, he could trick the patron himself. Nonetheless, the patron was not losing anything by being "tricked" because the trick ended up confirming his power image. The prince is an "omnipotent puppet." See ibid., p. 44.

egy Galileo had adopted fifty years before. In the long run, his extraneous-
ness to the discovery of the stars, which Galileo had claimed rhetorically,
became reality. The Medicean Stars became nothing but Medici fetishes
and were celebrated as such within Medici court culture until the very end
of the dynasty. Galileo left the stage much earlier.[159]

Galileo obtained the title of philosopher by presenting his discoveries
as having been made possible by the prince himself. However, he did not
gain Medici support for the legitimization of Copernican astronomy and
the mathematical analysis of nature. That was something that did not fit
the codes of the prince's power image.[160]

Although the codes of Medici court patronage were both a blessing
and a curse for Galileo, they represented an opportunity he could not ignore. Although the Medici's patronage agenda may have overlapped only
locally or temporarily with Galileo's strategies for social and cognitive legitimation, the overlap was of great historical significance. Besides its obvious importance for Galileo's own career, his being hired at the Medici
court with the title of philosopher may indicate the intersection between
two more general historical processes: the formation of court culture associated with the emergence of the absolute state, and the process of social
legitimation of science. Let me briefly outline how the strategies for the
social and cognitive legitimation of science that emerge from the analysis
of Galileo's career may be compared to other patterns of socioprofessional
legitimation also connected to the formation of court society and culture.

Recent works on early modern courts suggest that, although baroque
courts differed in specific ways, the fundamental features of their culture
were closely associated with the discourse of increasingly absolute princes
and displayed a number of similarities across national boundaries.[161] One

159. Interestingly, Galileo was resurrected later on by Prince Leopold, Cosimo II's son, as
part of his attempt to celebrate the Medici image—this time by presenting the Medici as
having patronized (in his eyes) not only European art but European science as well. Again,
Leopold was not celebrating Galileo *per se*, but rather because he happened to fit very well
into a narrative of Medici self-celebration that Leopold had developed.

160. The paradoxes inherent in Galileo's patronage-bound representation of the Medicean Stars were connected with the other paradox embodied in his moving to court—that is,
to an institution that could legitimize the new socioprofessional role he was seeking, but that
could not understand or care about the technical dimensions of his work.

161. See, for instance, Elias, *Court Society;* idem, The History of Manners (New York:
Pantheon, 1982); idem, *Power and Civility,* (New York: Pantheon, 1982); Marin, *Portrait of the
King;* Jean-Marie Apostolides, *Le prince sacrifié* (Paris: Minuit, 1985); idem, *Le roi machine*
(Paris: Minuit, 1981); Sergio Bertelli and Giuliano Crifò, eds., *Rituale, cerimoniale, etichetta*
(Milan: Bompiano, 1985); Amedeo Quandam and Marzio Achille Romani, eds., *Le corti farnesiane di Parma e Piacenza,* (Rome: Bulzoni, 1978), 2 vols.; Adriano Prosperi, ed., *La corte e
il "cortegiano": Un modello europeo* (Rome: Bulzoni, 1980); Hubert Ch. Ehalt, *Ausdrucksformen Absolutischer Herrschaft* (Munich: Oldenbourg, 1980); Frank Whigham, Jr., *Ambition
and Privilege: The Social Tropes of Elizabethian Courtesy Theory* (Berkeley: University of California Press, 1984); Jean-Francois Solnon, *La Cour de France* (Paris: Fayard, 1987); and Ran-

common feature was self-referentiality. Especially since the end of the sixteenth century, court society tended to close itself off (both culturally and geographically) from surrounding society to focus on and refer exclusively to itself, to the prince, or to the culture of other courts. It is to this process that we can relate the development of closed theatrical court spaces which then replaced public spectacles.[162] Similarly, if we look at court literature and poetry, we soon notice that their subject matter was a more or less subtle mix of the ruling family's mythologies with contemporary events (ceremonies, military exploits, public works and monuments) and the lives and works of living courtiers. The works of the writers courted by Galileo to write about the Medicean Stars (Gabriello Chiabrera, Michelangelo Buonarroti the Younger, Andrea Salvadori; or his friend Salvadore Coppola) are full of references to actual court life. A similar pattern can be found in court paintings.[163]

The descriptions of some of the court spectacles of the time indicate another aspect of this self-referentiality: the courtiers acted themselves. Together with professional performers, courtiers and the prince himself went on stage and performed roles commensurable with those they had in actual life. In the *barriera* of 1613, Cosimo II landed on stage from a galley coming from Elba's Cosmopoli and, crossing the stage, sang a song to his grand duchess in the audience.[164] In the *Mondo Festeggiante* of 1661, Cosimo III was on stage (surrounded by the Knights of the Medicean Stars) leading his courtiers in the equestrian ballet.[165] Quite literally, the court represented itself and its mythologies through court spectacles.[166]

The effect was a cultural closure which sometimes accompanied the geographical isolation of the court from the rest of society. Versailles is probably the most visible example of this process, but the various Medici *Ville* in the countryside near Florence shared Versailles' political function.[167] They were princely gardens of Eden. Together with this cultural-geographical isolation of the court from the city and the "crowds" which populated it, we find the formation of a new social group, that of the court society, out of (in the Florentine case) the former patriciate of commercial origins. This closure gave the would-be courtiers a sense of differentiation

dolph Starn and Loren Partridge, *Arts of Power* (Berkeley: University of California Press, 1992).

162. See note 23 above.

163. See, for instance, Allegri and Cecchi, *Palazzo Vecchio e i Medici,* pp. 145–47.

164. Nagler, *Theatre Festivals of the Medici,* p. 123.

165. Carducci, *Il mondo festeggiante,* pp. 60–66.

166. Elias, *Court Society,* p. 112; for a Spanish example, see J. E. Varey, "The Audience and the Play at Court Spectacles: The Role of the King," *Bulletin of Hispanic Studies* 61 (1984): 399–406.

167. Apostolides, *Le roi machine,* especially "Les plaisirs de l'île enchantée," pp. 93–113.

from the urban crowds and helped shape their new social identity. If Louis XIV used Versailles to control a politically restless aristocracy, the Medici used the court to create an aristocracy out of their former fellow merchants. Contemporary treatises on the court refer to its culture with a specific term: *civiltà*. As Matteo Pellegrini put it in 1624, "The Prince is the heart and the court the limbs of civilized living [*vita civile*]." Courtly lifestyle is civility itself.[168]

But the formation of court society and its increasing isolation from the lower classes did not affect the status only of the upper classes that it included or controlled. The development of court society required more than the formation of court aristocracy, that is, of a competent and collusive audience for the representations of the prince's power. As indicated by the development of official academies of fine arts as institutions that controlled the codes of those representations, competent producers of the prince's images were needed as well. Although artists have always celebrated the image of the powerful, we find that with the emergence of the baroque court and the centralized state the artistic representations of the prince's power began to be controlled by specialized institutions. As a result of their incorporation in this sort of artistic bureaucracy, academic *artists* obtained a much higher social status than that of the nonacademic *craftsmen* practicing the visual arts.[169]

It is here that the development of court society and culture intersects with the process of social legitimation of science. While princes like the Medici were trying to develop absolute states and needed legitimizing representations of their power, university mathematicians like Galileo were facing a status gap between them and the philosophers. As mentioned earlier, this gap in status delegitimized the use of mathematics as a tool for the study of the physical dimensions of natural phenomena. Therefore, in the same way artisans had become academic artists by representing the prince's mythologies of power in painting, sculpture, and architecture, Galileo turned himself from a mathematician into a philosopher by representing the satellites of Jupiter as Medici dynastic emblems. Although the court was not a scientific academy, it was an institution that could offer social legitimation which, in turn, could help establish the credibility of mathematicians-turned-philosophers. Given these disciplinary hierarchies, existing social institutions, and patterns of sociocultural change, the court represented Galileo's most promising option, although a problematic one.

My concern here is not with presenting Galileo's career and strategies

168. Pellegrini, *Che al savio é convenevole il corteggiare,* pp. 82, 171.

169. For a general treatment of the topic, see Nikolaus Pevsner, *Academies of Art* (Cambridge: Cambridge University Press, 1940). For the Accademia del Disegno, see note 22 above.

of social legitimation as *determined* by the court and its forms of patronage. Galileo did not need to move from the university to the court and did not discover the satellites of Jupiter because he was a client of the Medici. However, the historical processes, institutions, and patronage dynamics that made *possible* Galileo's career were not unique to him. Similarly, the fundamental aspects of baroque court culture and patronage related to the discourse of the absolute ruler and the low epistemological status assigned to mathematics by a disciplinary hierarchy that privileged theology and philosophy were by no means exclusive to the Florentine context.[170]

To say that Galileo was simply lucky with his patronage strategies—or to say that he was just an exceptional scientist—is to ignore the broader historical dynamics that made possible his unusual career and informed his strategies for the legitimation of Copernicanism and mathematical physics. Rather, I would say that Galileo was a great *bricoleur*. Many of the ingredients of his career (from telescopes to courts) were already there. The bricolage was not.

170. Westman, "Astronomer's Role in the Sixteenth Century"; and Biagioli, "Social Status of Italian Mathematicians."

THREE

Anatomy of a Court Dispute

LTHOUGH GALILEO BECAME THE philosopher and mathematician of the grand duke, he did not spend much time at
court. Instead, he tried to adopt a life-style comparable to that
of many Florentine patricians, who—like his friend Salviati—
would pay visits to the grand duke regularly yet would not live, work, or
spend their days at court. Moreover, there was no real job for Galileo at
court. His position was quite peculiar. In the *ruoli* of the Medici court,
Galileo was not listed in the category of artists, engineers, and cosmogra-
phers (where his predecessor Ricci can be found).[1] Also, he was not paid
as a courtier. His stipend did not come from the Medici's treasury (the
Depositeria Generale) but rather from the *Decime Ecclesiastiche* (the taxes on
Church properties in the grand duchy), which provided for the funds of
the University of Pisa.[2]

Therefore, Galileo was a courtier listed among those gentlemen who
had free access to court but did not have to work (and therefore was not
paid) there. At the same time, he was paid by an institution where he never
worked and which (as far as I can tell) he never even visited after returning
to Florence in September of 1610. Thereafter, his salary was sent to him
twice a year in Florence through the Pisan branch of his friend Salviati's
banco.[3]

1. ASF, "Guardaroba medicea 309," under "Familiari a ruolo senza provvisione a godere
di privilegi", we find Galileo at fol. 38v. In other ruoli, the same category is called "Gen-
tilhuomini a ruolo senza provvisione." Mathematicians and engineers are found in much
different categories, either among the "Maestri de' Sigg. Paggi" (like Ricci, Pieroni, and
Cantagallina), or among the "Architetti, pittori, et altri manifattori" (like Buontalenti,
Parigi, and Neroni).

2. *GO*, vol. 19, pp. 233–64.

3. *GO*, vol. II, no. 671, p. 292.

The fact that Galileo did not fit the court taxonomy is a confirmation of the novelty of his socioprofessional identity. There was no established category in which he could fit. However, the sort of privileged marginality Galileo experienced at the Medici court was a necessary phase in his attempt to shape a new and previously unknown socioprofessional identity. He wished to be neither a university mathematics professor nor a court mathematician busy compiling astronomical tables for his prince's horoscopes. His position at Florence combined the advantages of those two professional identities while avoiding many of their drawbacks. Galileo was an honorary university professor at Pisa and an honorary courtier in Florence. He operated in a privileged yet fairly unarticulated socioprofessional space.

The peculiar status he enjoyed in Florence around the time of the dispute on buoyancy can be interpreted from his pattern of residence. In January 1611, just a few months after his return to Florence, Galileo was already at Salviati's Villa delle Selve.[4] He did not follow the court in its winter trip to Pisa and Livorno. As he wrote to Sarpi, the "air" of Salviati's villa suited his health much better than that of the court.[5] The grand duke understood this and (maybe also to ward off the competition from a lesser patron such as Salviati) offered Galileo a stay at any of his villas around Florence.[6] Although we do not know whether Galileo took Cosimo's offer seriously, a year later we find him at Don Antonio de' Medici's (the grand duke's cousin) Villa di Marigniolle. It is there that Don Antonio sent him a gift of a wild boar and other game.[7]

Between January and October, Galileo did not stay at court for any extended period of time. At the beginning of February he was still at Le Selve and a month later he was traveling to Rome, where he stayed until the beginning of June.[8] A month after his return from Rome, Galileo was again at Le Selve at the time the dispute on buoyancy started. He was there again in late August or early September, when the last meeting of the first phase of the dispute took place. Then, as noted, Galileo moved to the Villa di Marigniolle at the end of October. Shortly after, we find him at Le Selve again, working on the conclusion of the dispute on buoyancy and on the first phases of the debate on the sunspots. He would stay there from the end of December through March (probably with a brief intermission to go to Florence to receive the Grand Duke returning with his court from

4. On the few traces of Galileo's presence at Le Selve, see Mario Biagioli, "New Documents on Galileo," *Nuncius* 6 (1991): 157–69.

5. *GO*, vol. 11, no. 461, p. 27.

6. Ibid., no. 476, pp. 46–47.

7. Ibid., no. 600, p. 227.

8. Ibid., no. 476, pp. 46–47; no. 504, pp. 78–79; no. 538, p. 121.

Pisa).[9] But his stay in Florence must have been a short one, for at the beginning of April Salviati wrote him in Florence asking why—contrary to what he had promised—he was not back at Le Selve yet. They could not discuss Ruzante without him.[10]

Galileo maintained this shifting pattern of residence until the departure of Salviati from Florence at the end of 1613.[11] But even after that date Galileo did not spend much time at court. Although he could not afford a villa in the country, Galileo succeeded in maintaining a sort of suburban life-style. Also, he participated very rarely in the meetings of the academies to which he was elected.

This was not the life-style of a court mathematician or of a scientist seeking a productive setting away from mundane distractions. Le Selve and Marigniolle were neither research centers nor monasteries. Galileo lived not as a secluded scientist but as a patrician. Through his association with Salviati, and by stressing the ancient (and nearly forgotten) patrician origins of his family, Galileo tried to represent himself as a noble.[12] Consequently, he lived in a country estate and came to Florence or went to Livorno to visit the grand duke whenever court etiquette required him (and Salviati) to do so.[13] In those years, his house in Florence was not his

9. Ibid., no. 633, p. 254; no. 640, pp. 258–59; no. 647, p. 265; no. 648, p. 266; no. 659, p. 278; no. 833, p. 468. Although the last letter dates from 1613, it refers to a habitual seasonal transfer of the court from Florence to Pisa and Livorno.

10. Ibid., no. 668, p. 290.

11. In May 1612 Galileo was still at Le Selve (ibid., no. 672, p. 293; no. 674, p. 294; no. 675, p. 295). At the very end of May and during June he was mostly in Florence (ibid., no. 681, p. 301; no. 684, p. 304). In August he was in Florence (ibid., no. 741, p. 374), but we do not have any letters from him in the period June-August to track his movements. When we begin to find his letters again it was October and he was at Le Selve (ibid., no. 787, p. 419). He seems to have remained there from October to February (or maybe even March) 1613 (ibid., no. 792, p. 426; no. 806, p. 440; no. 827, p. 459; no. 833, p. 465; no. 842, p. 477; no. 850, p. 485).

12. The most comprehensive source of biographical information about Salviati is Niccolò Arrighetti, *Delle lodi del Sig. Filippo Salviati* (Florence: Giunti, 1614). See also Mario Biagioli, "Filippo Salviati: A Baroque Virtuoso," *Nuncius* 7 (1992).

13. It is also to the anomaly of Galileo's role at court that we can attribute his absence from the court diaries. Basically, court diaries were etiquette data bases; bodies of information about the treatment of foreign dignitaries to be consulted to avoid future etiquette gaffes. Consequently, as a literary genre, the court diary is a very selective one. People mentioned in it had to have some political role, otherwise they would not be involved in public receptions or ceremonies. For instance, artists are never mentioned—except Giambologna or Bernini, who were *cavalieri*—for they were not nobles or military people. Even major functionaries like Vinta (himself only a *cavaliere*) were rarely mentioned in court diaries. Although Galileo had a high visibility in Florence and was visited by many important princes, aristocrats, and cardinals, he was not seen as a nobleman (despite his attempts to present himself as such) and could not cut through the barriers court etiquette posed to non-aristocrats. He tried act as a noble, but in the court taxonomies he was only a gentleman.

residence but a sort of urban hotel where he stayed while taking care of his business in town before going back to the country. Actually, Targioni Tozzetti believed that Galileo did not have any house in downtown Florence in this period.[14] By 1617, he moved to a small villa in Bellosguardo and, from there, to another villa at Arcetri.[15]

As stated in his contract, Galileo needed neither to work at court nor to teach at Pisa. Simply, his presence (and performance) was required at court whenever the grand duke wished to entertain himself, his family, or important visitors. Even in this, Galileo's role could be compared to that of Tuscan nobles whose daily presence at court was appreciated but not required; however, they were expected to show up (and display *magnificentia*) at major ceremonies. As explicitly laid out in his contract with the Medici, one kind of ceremony Galileo was expected to participate in were court disputes.[16]

Science at a Cleared Table

Given the dynamics of Galileo's social system, it is not difficult to understand why a good portion of his scientific production was either of the topical type (like the debate on buoyancy, on the Bologna stone, or on sunspots) or related to accidental happenings (like the comets of 1618 or the new star of 1604). Simply, his patrons perceived marvels, unusual events, and discoveries as occasions or materials for *questions*.[17] These questions were of the type: "What are comets about?" "Why does ice float on water?" "Why is Saturn three-bodied?" "Why does the Bologna stone shine of its own light?" "What are sunspots?"[18]

14. Giovanni Targioni Tozzetti, *Notizie degli aggrandimenti delle scienze fisiche accaduti in Toscana nel corso di anni LX del secolo XVII* (Florence: Bouchard, 1780; reprint Bologna: Forni, 1967), 1: 67.

15. On Galileo's houses in Florence, see Maria Luisa Righini Bonelli and William Shea, *Galileo's Florentine Residences* (Florence: Istituto e Museo di Storia della Scienza, n.d.).

16. ". . . and without the obligation to live in Pisa, nor to teach there except as an honor when it pleases you, or when we expressly and extraordinarily want you to, *for our pleasure or for that of Princes and foreign gentlemen in visit.* You will ordinarily reside here in Florence pursuing the perfection of your studies and your work, however with obligation to come to us wherever we will be, also outside Florence, whenever we will call you" (*GO*, vol. 10, no. 359, pp. 400–401 [emphasis mine]).

17. With some qualification, we may also include discoveries in this latter category. Although it may be tempting to see Galileo's discoveries as units of a Copernican "research program," we may alternatively try to consider them as the quasi-accidental results of a patronage strategy aimed at maximizing the "patronage capital" provided by the telescope.

18. Confirming a pattern observed in so-called traditional societies, we find that Galileo's answers to problems posed by his patrons were perceived less as "solutions" and more as "gifts" to the patrons who asked them. That enigmas were forms of challenges or duels is clear enough (for instance, the sphinx in Sophocles's *Oedipus Rex*). And, as we have seen, challenges and gifts have an analogous role in the economy of honor, status, and credibility

This type of environment did not structure only Galileo's courtly life but was typical for anyone working within a patronage framework. In his 1604 *Ad Vitellionem paralipomena*, Kepler thanked one of his patrons for having asked him about the behavior of eyeglasses, a question that was central to his treatise on optics.[19] Patrons tended to ask questions or pass reports of discoveries to their (or to their friends') astronomers and philosophers. On some occasions, university philosophers were also caught in these dynamics. Galileo's answer to a query on Hero's lamp that he received from Alvise Mocenigo in 1594, or the Pisan philosophers' report to Cosimo II on the meteorite that fell somewhere in the Tuscan countryside in 1613 and was then sent to Pisa are instances of these customary rituals of patronage.[20]

Contrary to what Galileo might have hoped, the court was not precisely the place in which systematic research could be best accomplished.[21] In contrast to a modern scientist, Galileo had little control over the questions that were asked him. Nevertheless, he had to answer them somehow, and in a witty manner fitting the codes of court culture. Moreover, he was expected to deliver his responses promptly—sometimes too promptly, given the difficulty of the topic at hand. Descartes was probably right in thinking that Galileo was not an orderly or consistent philosopher (a judgment shared by several modern historians and philosophers of science), but his lack of system was probably more a result of the court reward system than of his own intellectual attitudes.[22] I would say that, given the courtly environment in which he operated, Galileo's science was "performative." Its performance-like quality was then preserved in

in a number of traditional cultures (see Pierre Bourdieu, "The Sentiment of Honour in Kabyle Society," in J. G. Peristiany, ed., *Honour and Shame* [Chicago: University of Chicago Press, 1966], p. 215; and Marcel Mauss, *The Gift* [New York: Norton, 1967]). As we have seen, in the case of Galileo critiques (i.e., questions) were presented as honorable gifts or challenges, and Galileo's answers were received as counter-gifts. Moreover, enigmas were also literally exchanged as "challenging" gifts, as in the case of Galileo's sending out ciphers representing his latest discoveries. From the reaction of Welser, Kepler, Giuliano de' Medici, and Rudolph II upon receiving Galileo's enigmas, it seems that they were quite addictive gifts (*GO*, vol. 10, nos. 384, 378, 385, 417, 432, 435, 443, 445; vol. 11, nos. 451, 454, 455, 471).

19. Johannes Kepler, *Ad vitellionem paralipomena* (Frankfurt: Marnium, 1604), p. 201.

20. *GO*, vol. 10, no. 53, pp. 64–65; vol. 11, no. 922, p. 562.

21. The issue I am addressing here is that questions from patrons tended to be of the nontechnical type, although they might well have had very complex technical and methodological implications. One would not get questions about sizes and periods of the epicycles of a certain planet, or about the demonstration of the law of free fall. The questions from nonscientific patrons were broader and needed to be answered with the fewest technicalities possible. The *Assayer* and the *Dialogue* reflect these requirements very well. That the *Discourse* on buoyancy did not receive the same attention as other of his works may have to do not only with its less glamorous topic but also with its somewhat technical style.

22. William R. Shea, "Descartes as a Critic of Galileo," in Robert E. Butts and Joseph C. Pitt, eds., *New Perspectives on Galileo* (Dordrecht: Reidel, 1978), pp. 139–59.

Galileo's literary genres: the dialogue, the letter, and the discourse. As Campanella was quick to notice, Galileo's *Dialogue* was a "philosophical comedy," though one with a serious agenda.

Not only did Galileo's patronage system contribute to the topical character of some of his scientific production, it also put him in difficult situations because of the irresolvable tension between the patronage system's requirement for quick, witty, and preferably nontechnical answers, and his attempt to foresee and control the full implications (cosmological, methodological, theological) of his statements.

Disputes were a common dimension of the life of the court and academies.[23] In Florence we encounter at least two kinds of court disputes. Some were aimed at the private education and entertainment of the grand dukes and princes; others were spectacles the Medici offered to their guests. In his *Elogio* of Cosimo II, Michelangelo Buonarroti the Younger praised the deceased grand duke for having organized frequent academic disputes at court.[24] The later experimental academies of Ferdinand II and Prince Leopold's Accademia del Cimento can be seen as the development of this tradition. The Medici court diary reports that on many occasions the grand duke entertained himself with *virtuosi*.[25] On occasions the diary gives us some detail. In July 1603 during the reign of Cosimo's father, Ferdinand I, we find that:

> His Most Serene Highness being in Florence and desiring that the Most Serene Prince would grow very virtuous, he ordered that many Florentine doctors and academicians gather every other day in the ground-floor apartments at Pitti Palace to dispute in vernacular on humanistic and *pleasant* subjects. His Highness, the Grand Duchess Cristina, the Most Serene Prince, and the Duchess of Bracciano with all her children were present. Among the many doctors there were: Signor Mercuriale, Signor Bonciani, Signor Rucellai, Signor Adriani, Father Civitella, . . . and many others.[26]

23. In a letter to Galileo, Cesi mentioned a public dispute at the Collegio Romano on a Sunday. It seems to have been a custom (*GO*, vol. 11, no. 761, p. 395). Disputes were a popular genre in private salons as well. We have evidence that Galileo participated in some of them during his visit in Rome in 1616 (*GO*, vol. 12, no. 1156, p. 212; no. 1170, pp. 226–27). For instance, his 1624 "Reply to Ingoli" resulted from a disputation he had in Rome in 1616 with Ingoli in the presence of Monsignor (later Cardinal) Magalotti. The 1611 disputation on the lunar mountains, which involved the Jesuit mathematician Giuseppe Biancani (and, marginally, Galileo), also developed from a court disputation, this time at Mantua in the presence of Cardinal Gonzaga.

24. Michelangelo Buonarroti il Giovane, *Elogio di Cosimo II* (Florence, 1621), quoted in Targioni Tozzetti, *Notizie*, pp. 10–11. A very similar statement is found in an anonymous *Elogio di Cosimo II*, ASF, "Miscellanea medicea 359," insert 9, p. 19.

25. Targioni Tozzetti, *Notizie*, p. 73.

26. "Diario di corte di Cesare Tinghi" (21 July 1603), BNCF, "Fondo Capponi 1," fol. 68v (emphasis mine). Also mentioned in Targioni Tozzetti, *Notizie*, p. 12. A very similar descrip-

This academy continued to meet for at least another year, as the court diary mentions it twice more.[27] In both cases, young Cosimo is reported as participating.[28]

But there were court disputes that were not primarily aimed at the education of the young princes. These (like the impromptu debate on Copernicanism and the Scriptures, which caught Castelli off guard and led Galileo to write the "Letter to the Grand Duchess") took place after lunch at the table of the grand duke in Pisa in December 1613.[29] Antonio de' Medici, Prince Orsini, Cosimo II, and Cristina sided with Castelli, while Cosimo's wife, Maria Maddalena, and the philosopher Boscaglia (who was probably responsible for triggering the debate) were critical of his position.

In engaging in these forms of entertainment, the Medici were evidently conforming to the customs of the time; in his dialogue on the court, Torquato Tasso has one of his interlocutors claim that disputations "can be seen every day at the princes' tables."[30] Games to be played "after the table was cleared" were so common as to be discussed in classic textbooks of polite and courtly behavior like Stefano Guazzo's *La civil conversazione*.[31] Many of these games aimed at finding some pretext for the

tion is found in *ASF*, "Diari di etichetta di guardaroba 4," fol. 42. Father Civitella was a Dominican. He is mentioned as a teacher of humanities in Attavanti's 1615 deposition to the Florentine Inquisition on Caccini's denunciation of Galileo (*GO*, vol. 19, pp. 318–20).

27. "During the day [after lunch] His Highness with all his sons and Madame went to the usual Lecture at the Academy held downstairs in the ground floor rooms of the Pitti [Palace], recited by the usual doctors" (31 August 1604); "The next day His Highness with the Most Serene family were at the usual academy given by the usual doctors" (9 September 1604; both quotes are in "Diario di corte di Cesare Tinghi," *BNCF*, "Fondo Capponi 1," fol. 103r, fol. 104v). Also mentioned in Targioni Tozzetti, *Notizie*, 12. This academy must have met regularly, because it is mentioned among Ferdinand's cultural accomplishments in G. Giraldi, *Delle lodi di D. Ferdinando G. D. di Toscana* (Florence: Giunti, 1609), p. 29.

28. This custom must have been quite common in aristocratic households, as indicated by something very similar taking place at the same time in Spain. In fact, part of the humanistic training of the future Count-Duke Olivares—the favorite of Philip IV—came from engaging every fortnight in set debates with members of his household (J. H. Elliott, *Richelieu and Olivares* [Cambridge: Cambridge University Press, 1984], p. 30).

29. I am aware that the lunch at court in Pisa, in which Castelli participated and was questioned by Cristina on the religious orthodoxy of the Copernican theory, is not usually presented as a dispute. However, it was. Castelli was not directly questioned by the grand duchess, but by Boscagli through her. Therefore, it was a dispute between Castelli and Boscagli managed by Cristina (*GO*, vol. 11, no. 956, pp. 605–6). Galileo's "Letter to Castelli" is dated one week later, 21 December (*GO*, vol. 5, pp. 281–88).

30. Torquato Tasso, *Il malpiglio, o vero de la corte,* reprinted in Cesare Guasti, ed., *I dialoghi di Torquato Tasso* (Florence: Le Monnier, 1901), 3: 18. See also Michel Jeanneret, *A Feast of Works* (Chicago: University of Chicago Press, 1991).

31. Thomas Frederick Crane, *Italian Social Customs of the Sixteenth Century* (New Haven: Yale University Press, 1920), p. 410.

display of the guests' skills in polite and witty conversation. As shown by Jay Tribby, natural philosophy lent itself to these types of performances.[32] These convivial debates were common at the table of cardinals as well. In July 1613, Cardinal Cesi invited his nephew Prince Federico (the founder of the Lincei), the philosopher Giulio Cesare Lagalla, and other Lincei for a lunch at which Lagalla discussed some of his cosmological views.[33] In his 1611 description of the Roman court, Lunadoro mentions that Cardinal San Giorgio's lunch was always "a public academy."[34] A few decades later, the Jesuit mathematician Honoré Fabri, then living in Rome, was asked to write his *Dialogi physici* by Cardinal Facchinetti and his friends as a result of a discussion at the cardinal's table.[35]

Although we find other cases of lunchtime disputes for the private entertainment of the granducal family, it seems that the most spectacular were those offered in honor of important guests.[36] When Cardinal Perron visited Florence in September 1607,

> he had lunch with the ducal family, and, once the tablecloth was removed, some very beautiful disputes took place between the cardinal, Father Civitella, Doctor Libri, and the physician of His Highness—Signor Biagio Bernardi. They disputed on philosophy and mathematics. Later on, the cardinal joined the academy of the prince and of all the sons of Don Virginio Orsini.[37]

32. Jay Tribby, "Of Conversational Dispositions and the *Saggi*'s Proem," in Elizabeth Cropper, ed., *Documentary Culture: Florence and Rome from Grand Duke Ferdinand I to Pope Alexander VII* (Florence: Olschki, forthcoming); idem, "Stalking Civility: Conversing and Collecting in Early Modern Europe," *Rhetorica*, forthcoming; and idem, "Cooking (with) Clio and Cleo: Eloquence and Experiment in Seventeenth-Century Florence," *Journal of the History of Ideas* 52 (1991): 417–39.

33. "1613, Julii 8. Pransi sumus in palatio Cardinalis Caesii, ubi Lagalla habuit lectionem de Animabus Caeli" (Giuseppe Gabrieli, "Verbali delle adunanze e cronaca della prima Accademia Lincea [1603–1630]," *Memorie della R. Accademia Nazionale dei Lincei*, Classe di Scienze morali, storiche e filologiche, series 6, 2 (1927): 490.

34. Girolamo Lunadoro, *Relatione della corte di Roma* (Rome: Frambotto, 1635), p. 13.

35. W. E. Knowles Middleton, "Science in Rome, 1675–1700, and the Accademia Fisicomatematica of Giovanni Giustino Ciampini," *The British Journal for the History of Science* 8 (1975): 140.

36. Targioni Tozzetti, *Notizie*, p. 17.

37. *ASF*, "Diari di etichetta di guardaroba 4," fol. 109, (20 September 1607). Other disputes were arranged by the court but did not take place within its physical space. They were part of the "grand tour" usually given to important visitors. After lunch, visiting cardinals were taken around Florence and shown the "Galleria," the "Guardaroba," the "Botteghe degli Uffizi," the "Libreria di S. Lorenzo," the "Sagrestia di Michelagnolo," the "Capella di S. Lorenzo," and the Medici theater—the "Stanzone delle Commedie in Via della Pergola" (*ASF*, "Miscellanea medicea 438," fol. 24). Disputes were sometimes added to this entertainment program. Similarly, when Cardinal Barberini came to Florence in September 1626, he was taken to meetings and disputes at the Accademia degli Alterati and at the Accademia della Crusca (*ASF*, "Miscellanea medicea 441," fol. 88).

The lunchtime dispute on buoyancy at the table of the grand duke in the fall of 1611 between Galileo and the peripatetic Papazzoni, with Cardinal Barberini supporting Galileo and Cardinal Gonzaga siding with the philosopher, falls right in this genre of "science at a cleared table."[38]

From the available evidence it seems that court disputes were not usually open to mathematicians. Although mathematics was a subject of the dispute before Cardinal Perron, those who disputed were not mathematicians. Libri was a philosopher, Bernardi a physician, and Civitella a Dominican man of letters. Similarly, when Castelli debated at the grand duke's table in Pisa, the argument became a theological one and, as Castelli said in a letter to Galileo, he was interrogated (and answered) as an astronomically literate theologian.[39] I have not found any evidence of Ricci, Santucci, Neroni, or other court mathematicians debating in front of the grand duke. On the other hand, we find that philosophers and physicians were welcome at the table of princes and cardinals. Mazzoni was often at court in Florence, the Aristotelian Boscagli was a frequent guest of Cosimo II in Pisa, and Papazzoni was often at the table of Cardinal Barberini in Bologna.[40] Mercuriale, the first physician to the grand duke's family, spent almost more time at the court in Florence than at the university in Pisa.[41]

This seems to confirm the sharp difference in social status and skills between mathematicians and philosophers discussed in the previous chapters. Given the different status attributed to the liberal disciplines, it was probably improper to have a mathematician challenge a philosopher. Moreover, it would have been difficult to find a mathematician with the rhetorical and courtly skills that were needed to engage in these disputes and succeed in pleasing a sophisticated audience such as that of the Medici court. The social skills of Galileo, Guidobaldo, Commandino, or Raimondi were not usual among mathematicians.

Court disputes were dangerous performances. Through them a vir-

38. *GO*, vol. 4, p. 6 and note 1.

39. *GO*, vol. 11, pp. 605–6.

40. On Mazzoni's visits, see *ASF*, "Diari di etichetta di guardaroba 3," p. 86: "When the philosopher Signor Mazzoni comes to court, he is given a room and two servants in the dining room" (1595). On Boscagli, see *GO*, vol. 20, p. 398. On Papazzoni, see *GO*, vol. 10, no. 820, p. 455.

41. Like Galileo, Mercuriale was not paid at court, but as "Archiatra" he was very frequently at court, and not always to take care of the Medici health. His presence at court is recorded in many documents, among them *ASF*, "Diari di etichetta di guardaroba 3," p. 47. At p. 214: "On 9 May 1604 Signor Mercuriale the Physician arrived from Pisa with Her Most Serene Ladyship. He was lodged in the Pitti in the attic of the wine storerooms, where he stayed with two of his servants. He left on the thirteenth in the litter after lunch, for dinner at the Ambrosiana [a Medici villa], and by boat to Pisa without cost." But he was soon back in Florence: "Signor Girolamo Mercuriale on 4 June [1604] came from Pisa to Florence and stayed in the Pitti palace with three of his servants" (p. 217).

tuoso could either advance or seriously damage his career. Galileo's friend Ciampoli built a remarkable career through performances of this kind, beginning when he was still a teenager in Florence and Rome. Apparently, he impressed his potential patrons with his ability to compose elegant poems on the spot about any subject proposed to him.[42] Similarly, Galileo and Castelli were expected to answer on the spot the difficult or delicate questions of natural philosophy that were posed to them.

Not only did this topical discussion format frame several of Galileo's works and performances, it may have been directly responsible for some of Galileo's troubles. As a thought experiment, we may consider what might have happened if Castelli did *not* have to engage without warning in a discussion on Copernicus and the Scriptures at the grand duke's table in Pisa, and did *not* have to answer on the spot. I would suggest that probably Galileo would not have needed to write the "Letter to the Grand Duchess" and, consequently, the admonition of 1616 might never have taken place.

The potential dangers of these table disputations were a matter of common knowledge. In April 1616, the Medici secretary Curzio Picchena cautioned Galileo (who was still in Rome after the admonition):

> I understand that you are thinking about staying in Rome as long as Cardinal de' Medici will be there. In this regard I recall what Their Highnesses told me at some point, that is, that I should advise you that when you find yourself at the table of the Lord Cardinal, where other learned people are likely to be as well, Your Lordship should not get into disputations about those matters that have triggered the friars' persecution against you.[43]

The dangers of impromptu questions were felt not only by courtly virtuosi but also by anatomists performing public dissections. One of the doubts concerning the usefulness of public anatomy lessons like those that took place in Bologna and other Italian cities during carnival was precisely that the students could not learn much from such an episodic event, while the professors were put in the dangerous position of having to answer impromptu questions for which answers were not possible or could not be compressed to fit such a format.[44] As shown by the later develop-

---

42. On Ciampoli's after-lunch performances in Rome, see J. A. F. Orbaan, *Documenti sul barocco in Roma* (Rome: Società Romana di Storia Patria, 1920), vol. 2: 218; and Guido Bentivoglio, *Memorie e Lettere,* ed. Costantino Panigada (Bari: Laterza, 1934), pp. 74–75.

43. *GO,* vol. 18, no. 1198bis, p. 422.

44. Giovanna Ferrari, "Public Anatomy Lessons and the Carnival: The Anatomy Theater of Bologna," *Past and Present* 117 (1987): 91.

ment of science, the tendency was to move away from disputational formats toward more constructive (if less spectacular and entertaining) ways of doing science. Boyle's experimental philosophy can be seen as the epitome of this trend.[45]

However, public scientific disputes (whether at court or in a public anatomy theater) should not be seen as a silly, archaic format of scientific production. With all their constraints, they provided a mathematician such as Galileo with the resources that were most important for his social and scientific career: status and credibility.[46] They were an offer Galileo could not refuse. Disputes at the prince's cleared table fit perfectly the protocols of patronage discussed before. To the patron, they were the epitome of "good sport." It was by performing well in these courtly events or by writing "performative" texts that Galileo could present himself not as a lowly mathematician but as a true philosopher.

Galileo knew very well that he did not have a formal philosophy degree but that he could get that title by making an impression on an absolute prince and also by displaying his knowledge and skills in public disputes. Writing to Vinta in the spring of 1610, asking for the title of philosopher in addition to that of mathematician of the grand duke, Galileo added "and that I may or should deserve this title, I can show to their Highnesses, whenever they might like to give me the chance to discuss these matters in their presence with the most distinguished practitioners of that discipline."[47]

It was not crucial that Galileo succeed in winning over his adversaries during these disputes. All that counted was the audience's appreciation of his skills. When, after the publication of the *Sidereus nuncius,* Galileo visited the Medici court at Easter, at Pisa, he was immediately made to dispute with the Pisan philosopher Giulio Libri (whom we have already encountered at one of the court disputes in Florence) on the existence of Jupiter's satellites. The dispute must have pleased the grand duke, since it was at this time that Galileo's position at the Florentine court became a serious option for him.[48] However, Libri remained quite unconvinced by Galileo's arguments. A few months later, when the philosopher died, Galileo told a friend that he hoped Libri might get a chance to see the Medicean Stars on his way to heaven.

---

45. Steven Shapin and Simon Schaffer, *Leviathan and the Air Pump* (Princeton: Princeton University Press, 1985), pp. 22–109.

46. On this issue, see Mario Biagioli, "Scientific Revolution, Social Bricolage, and Etiquette," in Roy Porter and Mikulas Teich, eds., *The Scientific Revolution in National Context* (Cambridge: Cambridge University Press, 1992), pp. 11–54.

47. *GO,* vol. 10, no. 307, p. 353.

48. Ibid., no. 379, p. 423; vol. 11, no. 820, p. 453.

CHAPTER THREE

Conflicting Narratives

When Galileo finally returned to Florence as the philosopher and mathe-matician of the grand duke in September 1610, the status gap between him and the philosophers had been bridged, at least theoretically. He could now address and challenge the Aristotelian philosophers as his equals, and they were no longer able to dismiss his claims just because he was a practi-tioner of a lower discipline. Yet he could not dismiss their claims either. Their status was now comparable and they shared the same patron—since the Studio Pisano was the university of the grand duchy. Galileo and the Pisan Aristotelians were two species with comparable power competing for the same patronage niche.

Although the dispute on buoyancy ended up at the grand duke's table, it did not develop there. Rather, it was a link in a long chain of disputes Galileo had been involved in since his astronomical discoveries. While writing his *Discourse on Bodies in Water,* Galileo was also responding to Welser and Scheiner's queries on sunspots. And one of his adversaries on buoyancy—Ludovico delle Colombe—had just tried unsuccessfully to engage Galileo (and Clavius) in a debate on the irregularities of the moon's surface, after not having received any reply from Galileo on a piece he wrote against the *Sidereus nuncius.*[49]

The dispute on buoyancy started as a discussion on the nature of cold and took place at Salviati's in the summer of 1611.[50] Vincenzo di Grazia

49. Delle Colombe wrote Clavius in May 1611 trying to involve him in a dispute with Galileo on the irregularities of the moon's surface (*GO*, vol. 11, no. 534, p. 118). The topic attracted some attention in Rome in the summer and early fall of 1611 (ibid., no. 587, p. 212), and the philosopher Lagalla joined the dispute. Although delle Colombe perceived his views as similar to those of Clavius, the Jesuit was not responsive, and by November it became clear that he would not answer to, or ally himself with, delle Colombe (ibid., no. 602, pp. 228–29). Nevertheless, delle Colombe continued to try to interest important Roman patrons so that Galileo would be forced to stop dismissing his critiques (as he had done with his "Contro il moto della terra"). I tend to see delle Colombe's participation in the dispute on buoyancy also as a result of his failure to engage Galileo on Copernicanism and on the surface of the moon. It was while he was trying to find some supporter in Rome that he entered the dispute on buoyancy.

50. The history of the debate is outlined in Stillman Drake, "The Dispute Over Bodies in Water," *Galileo Studies* (Ann Arbor: University of Michigan Press, 1970), pp. 159–76; and in Drake's preface to the reprint of the first English translation of Galileo's *Discorso* (Galileo Galilei, *Discourse on Bodies in Water* [Urbana: Illinois University Press, 1960], pp. ix–xxvi). Drake has included a new English translation of the *Discorso* in his Galilean-style dialogue *Cause, Experiment, and Science* (Chicago: University of Chicago Press, 1981). Different con-ceptual dimensions of the dispute have been dealt with by William R. Shea, "Galileo's Dis-course on Floating Bodies: Archimedean and Aristotelian Elements," *Actes du XIIᵉ Congrès International d'Histoire des Sciences, Paris, 1968* (Paris, 1971), 4: 149–53; idem, *Galileo's Intellec-tual Revolution* (New York: Science History Publications, 1972), pp. 14–48; idem, "Galileo's Atomic Hypothesis," *Ambix* 17 (1970): 13–27; Thomas B. Settle, "Galilean Science: Essays in the Mechanics and Dynamics of the *Discorsi*" (Ph.D. diss., Cornell University, 1966),

and Giorgio Coresio, two Aristotelians who taught at Pisa, participated in the discussion. On that occasion, Galileo stirred the philosophers by claiming that ice was rarified water instead of condensed water, as Aristotle had maintained. To Galileo, the fact that ice floated on water proved its being less dense than water. The admission that cold could be a cause of rarefaction instead of condensation would have implied an important anomaly in Aristotle's element-based explanation of sublunary phenomena. Not surprisingly, the Aristotelians refused to entertain that the density of ice had anything to do with its floating in water. Instead, they claimed that ice floated because of its shape, because it was relatively flat and thin.

Although the Aristotelians' interpretation of the flotation of ice was very much improvised, since their master had left them only a page and a half on buoyancy to comment upon, they felt compelled to attack Galileo's statement on the rarefaction of water in order to defend the overall coherence of their world view. Similarly, the Aristotelians' views on buoyancy flew in the face of Galileo's mathematical approach to the problem (as well as to motion in general) which he had modeled after Archimedes' *On Floating Bodies*. To Galileo (as to Archimedes), the shape of the object had nothing to do with buoyancy, which instead was a direct result of the difference between the specific weight of the body and that of the surrounding medium. All that shape influenced was the speed at which the body would sink or surface in the medium. He summarized his position in a statement he delivered to his adversaries at the end of the first meeting: "a solid body which falls to the bottom in water when reduced to a spherical shape, will also fall there if given any other shape. Therefore, in general, the difference of shape in bodies of the same material does not alter their sinking or not sinking, rising or not rising, in that given medium."[51]

Galileo's statement was soon seriously challenged by the Pisan Aristotelians. A few days after the first meeting, the Aristotelians' ranks were swelled by a longtime opponent of Galileo's, the Florentine "independent" philosopher Ludovico delle Colombe.[52] The newcomer was able to

pp. 226–34; Paolo Galluzzi, *Momento* (Rome: Edizioni dell'Ateneo, 1979), pp. 227–46; Raffaello Caverni, *Storia del metodo sperimentale in Italia* (Florence, 1900; reprint New York: Johnson Reprint, 1972), 4: esp. pp. 89–146; Richard S. Westfall, "The Problem of Force in Galileo's Physics," in Carlo Golino, ed., *Galileo Reappraised* (Berkeley: University of California Press, 1966), pp. 86–88; and William Wallace, *Galileo and His Sources* (Princeton: Princeton University Press, 1984), pp. 284–88.

51. *GO*, vol. 4, p. 34. The English translation is adapted from Drake, "Dispute Over Bodies in Water," p. 166.

52. As noted, Ludovico delle Colombe had been a longtime opponent of Galileo. But his frustrated interaction with Galileo antedates his attack on the *Sidereus nuncius* (*GO*, vol. 3, pp. 251–90). Before that, he wrote a short treatise on the new star of 1604, which was mocked in a printed response by a certain Alimberto Mauri. It seems quite probable that Galileo hid behind that fictitious name. Delle Colombe had been systematically dismissed or ridiculed

produce a powerful experiment which seemed to refute Galileo's views on buoyancy as represented in his first statement. Delle Colombe showed that a sphere of ebony (a material with a specific weight superior to that of water) placed on water would sink, while a thin piece of the same material would remain afloat. From this he concluded that, contrary to Galileo, buoyancy was not a matter of difference in specific weight, but depended on shape.

The evidence put forward by delle Colombe was based on what we call surface tension, a phenomenon that, as we will see, did not fit Galileo's conceptual repertoire. The entire debate that followed revolved about the contestants' contrasting interpretations of this phenomenon. Without such a conceptual gap on Galileo's side, the Aristotelians would have been refuted easily (at least experimentally). Instead, with the help of delle Colombe's experiment they could produce a crucial anomaly to the Archimedeo-Galilean theory of buoyancy. Moreover, the experiment allowed them to criticize Galileo not by deploying a barrage of citations from Aristotle but by meeting him on the ground he claimed to prize the most: empirical evidence. In fact, delle Colombe's experiment was self-evident. It did not require any special experimental expertise to be performed or interpreted. One did not need to know much about either Aristotle or Archimedes to see that delle Colombe had a very good point.

Because of delle Colombe's intervention, the debate became characterized by an attempt on both sides to set the terms of the dispute and of the experimental conditions in such a way that the behavior of the water surface would be either circumvented (Galileo) or turned into the main focus of the dispute (Aristotelians). As a result, most of the intellectual energies of the two parties were spent on subtle legalistic arguments aimed at defining the subject matter of the dispute in ways favorable to one side while declaring unacceptable those proposed by the adversary. The dispute turned almost totally into a confrontation over the very rules

by Galileo, who never, however, addressed his responses to him in person—hence delle Colombe's aggressive style in his *Discorso apologetico*. Even in Galileo's *Discourse* he was not able to find the satisfaction of being addressed by name. And he would never get it, for even his *Discorso apologetico* would be responded to by Castelli and not by Galileo. Delle Colombe's inability to obtain a personal response from Galileo was, I think, a problem of patronage. As we have seen in the first chapter, authors without authoritative patrons could be easily dismissed. Delle Colombe never managed to have his patrons force a direct response from Galileo. He did succeed in bringing Giovanni de' Medici on his side. That was not difficult, given Giovanni's cold feelings toward Galileo. Delle Colombe dedicated his *Discorso apologetico* to Giovanni de' Medici and accompanied him to Salviati's for the last meeting—the one at which Galileo refused to be engaged in a verbal dispute. Yet Galileo, with the support of Cosimo, could dismiss delle Colombe and his patron because Giovanni (the illegitimate son of Cosimo I) was not a member of the main branch of the house of Medici.

of the game. As we will see, this was not an arbitrary power game. It resulted from the incommensurable notions of bouyancy, causality, method, and structure of matter held by the two parties.

Given these circumstances it is not surprising to find that Galileo and his opponents presented sharply different descriptions of the dispute for the period following the arrival on stage of delle Colombe and his ebony spheres and splinters. Galileo wrote in the *Discourse* that his Aristotelian adversaries had informed him of their partner's experiment and that he consequently agreed to meet delle Colombe publicly. But, according to Galileo, delle Colombe never showed up at that meeting. Instead, he went around the city "in the squares, in the churches, and in other public spaces" displaying his experiment and claiming that he had defeated Galileo.[53] This move was presented by Galileo and Castelli as implying that delle Colombe, unable to convince a qualified audience, went for a vulgar and impressionable audience. Quite probably, Galileo was worried by the support such an easily graspable experiment was likely to bring to his adversaries' camp. Galileo tried to counter delle Colombe's popular success by putting his views in writing and printing the *Discourse*—a move that, according to him, would have upgraded the dispute and kept it within the boundaries of a nonvulgar audience. In doing so, he presented himself as following the desires of the grand duke. But, he added, while writing his treatise he sent out a second statement concerning his adversary's new experiment:

> Every kind of shape, of any size, when wetted, goes to the bottom in water. But if a small portion of the same shape is not wetted, it will sit on the water without sinking. Therefore, the cause of sinking or not sinking is not the shape and not the size, but the complete or incomplete wetting.[54]

It is important to notice that Galileo included the clause about experimentation with wet bodies in this second statement. And he did so in order to avoid the problem posed by delle Colombe's experiment based on the behavior of the water surface. Although Galileo's earlier statement tacitly implied that a body which sank if placed on the water surface would not emerge if placed at the bottom of the water container, it had not said so explicitly. Actually, the first part of the statement mentioned sinking bodies only, and it was only in the second part that Galileo equated the conditions of buoyancy with those of emergence through water. By reframing his first statement, Galileo was trying to emphasize that the causes of

---

53. *GO*, vol. 4, pp. 31 and 34.

54. Ibid., p. 35. The translation is adapted from Drake, "Dispute Over Bodies in Water," p. 167.

bodies rising *in* water are the same that make them rest *on* it. Consequently, one should not focus exclusively on evidence concerning bodies *on* water. However, given the nature of the evidence supporting their claims, the Aristotelians were not willing to accept the symmetry proposed by Galileo between "in water" and "on water." If they accepted having their thin piece of ebony put at the bottom of the water container, they would be defeated. If they were to win, it had to be placed *on* the water surface.

Therefore, the Aristotelians strove for a very literal interpretation of Galileo's first statement in order to refute it and to say that, in his second statement, Galileo was changing the rules of the game as a result of seeing himself defeated. At the same time, Galileo needed to introduce a qualified version of his earlier statement to neutralize the devastating effects of delle Colombe's experiment. This is the reason for Galileo's pushing the "wetness clause" which—experimentally speaking—was equivalent to saying that the body should be placed at the bottom of the water container.

The account of the dispute delle Colombe presented in his later *Discorso apologetico* is much different from Galileo's. The two narratives coincide only in the sections describing the very first discussion at Salviati's on the nature of cold—one in which delle Colombe did not participate. Delle Colombe claimed that, after he entered the dispute, a number of written agreements between him and Galileo were drafted in order to circumscribe the argument of the dispute and the set of experiments acceptable to both parties. But, to his surprise, Galileo did not reproduce or mention those agreements in his *Discourse*.[55] According to delle Colombe, at first Galileo issued a statement that resembled quite closely the one he gave to the Pisan Aristotelians at the end of the first meeting:

> Signor Lodovico delle Colombe being of the opinion that shape affects solid bodies with regard to their descending or not descending and emerging or not emerging in a given medium, such as water, and in such a manner that for example a solid of spherical shape which would go to the bottom would not do so if shaped differently; and I, Galileo Galilei, on the contrary deeming this not to be true, but rather affirming that a solid body which sinks to the bottom in spherical or any other shape will also sink no matter what its shape is, being opposed to Signor Colombe in this particular, am content that we proceed to make experiments of it. And since these experiences might be made in various ways, I am content that the Very Reverend Canon Nori, as our common friend, shall choose among

55. Ludovico delle Colombe, *Discorso apologetico d'intorno al discorso di Galileo Galilei circa le cose che stanno sull'acqua o che in quella si muovono* (Florence: Pignoni, 1612; reprinted in *GO*, vol. 4, pp. 313–69).

the experiments that we shall submit, selecting those that may seem to him best suited to reveal the truth, as I also defer to his judgment the decision and the settling of controversies that may arise between the parties in making the said experiences.[56]

Delle Colombe added to this agreement the statement (which he claimed was accepted by Galileo):

That the body is to be of the same material and the same weight, but the different shapes are at the choice of Lodovico; and the choices of bodies (which shall be chosen as nearly equal as possible in density) at the election of Signor Galileo; and the shapes at the election of Ludovico: and the experience shall be tried out four times, with the same material but with as many pieces of that material as the number of times the experience shall be made.[57]

Subsequently, Galileo and delle Colombe agreed to have Filippo Arrighetti join Francesco Nori as a judge of the experiments.[58] Both Nori and Arrighetti were members of the Florentine upper clergy and were therefore represented as being impartial in the dispute.

Although delle Colombe's representation of the agreements between him and Galileo seems fair, for it was not challenged in Castelli's later refutation of his *Discorso apologetico,* delle Colombe seemed to forget something. He presented the dispute he was talking about as having happened at Salviati's and claimed that, contrary to the agreements,

Signor Galileo could neither be brought to dispute, nor did he wish to perform the experiment with materials of a suitable size, shape, and quantity of material. Rather, he was resolved (and let everyone judge the reason for himself) to publish a treatise of his on this subject, hoping to make others believe by arguing that which he could not show to the senses.[59]

---

56. *GO*, vol. 4, p. 318. The translation is adapted from Drake, "Dispute Over Bodies in Water," p. 173.

57. *GO*, vol. 4, pp. 318–19. The translation of this segment that is included by Drake in his "Dispute Over Bodies in Water," p. 173, is inaccurate.

58. Francesco Nori was "Canonico della metropolitana fiorentina" and a member of the Florentine College of Theologians since 1620. A member of the Accademia Fiorentina, Nori became its consul twice, in 1598 and 1613. In 1624, Urban VIII appointed him bishop of S. Miniato. Filippo Arrighetti was twenty years old at the time of the dispute. He became a member of the College of Theologians in Florence and a courtier. He was in the household of Cardinal Carlo de' Medici, and on friendly terms with Urban VIII. In 1631, he took what had been Nori's post as "Canonico." In November 1608, Arrighetti traveled from Florence to Pisa with Galileo and stayed at his house until the following spring (*GO*, vol. 19, p. 165).

59. *GO*, vol. 4, p. 319. The translation is from Drake, "Dispute Over Bodies in Water," p. 167.

But from Galileo's narrative (which is confirmed by a letter of Cigoli's) we find that these agreements and the choice of judges were not for the dispute at Salviati's but for a previous one that was supposed to have been held at Nori's.[60] Therefore, according to Galileo and Cigoli's narrative, delle Colombe avoided mentioning the meeting he deserted, and jumped to the last dispute of the series—the one at which Galileo refused to perform. Galileo too glossed over something. In the *Discourse* he stated that he decided to put his views in writing because the tone of the debate had degenerated and because the grand duke found it inappropriate for his philosopher to be engaged in such a noisy enterprise. However, he did not say that he also refused to participate in an already scheduled debate at Salviati's.[61]

Even in those cases in which delle Colombe and Galileo's descriptions of the dispute reported the same facts, they offered radically different interpretations. Delle Colombe acknowledged that, after Galileo's refusal to dispute, in order to put everything in writing, Galileo nevertheless sent delle Colombe an additional statement containing the "wetness clause" which coincides with what Galileo wrote in the preliminary version of the *Discourse*.[62] But while Galileo represented his move as an attempt to reestablish an orderly framework for an increasingly messy dispute, delle Colombe perceived these addenda as Galileo's tricky attempt to defend his position once he had realized that it was only by wetting delle Colombe's piece of ebony that he could safeguard his Achilles' heel.[63]

Delle Colombe was particularly abrasive on this issue because, according to him, this was not Galileo's first attempt to dictate the rules of the game. As he reported, Galileo had already raised hell (*faceste si gran schiamazzo*) when delle Colombe proposed to experiment with a very large body—as he was entitled to do by the mutual agreement presented above.[64] The strategic use of memory and of sophistical skills displayed by both parties in arguing on the experimental conditions suggests the presence of a deadlock which could not be broken by additional experimental evidence but only by setting the rules of the game to the advantage of one party or the other.

It is in this context that we can understand both delle Colombe and Galileo's "desertions." From the agreements between the two, we see that

60. "I have heard how the Pippione [delle Colombe] was to get into a fight with you at the house of Signor Nori, but that he did not show up" (*GO*, vol. 11, no. 573, p. 176).

61. *GO*, vol. 4, pp. 34–35, 65–66.

62. Ibid., p. 35.

63. ". . . believing that to wet [the thin piece of ebony] was your Achilles' heel" (ibid., p. 319).

64. Ibid.

judges Nori and Arrighetti were given remarkable power. Not only were they in charge of settling arguments, they were also supposed to screen the experiments proposed by the two adversaries and "chose those which they considered most conducive to the certification of truth."[65] Arrighetti and Nori were also friends of Galileo. Arrighetti was cousin to two of Galileo's strongest supporters in Florence, Andrea and Niccolò Arrighetti, and had been a guest of Galileo in Padua for a few months in 1608–9, while Nori was Galileo's friend and fellow academician at the Accademia Fiorentina.[66] Galileo was by then an international star and a distinguished client who had the same patron as Arrighetti and Nori—Cosimo II. Although delle Colombe had some visibility on the Florentine cultural and political scene since he had become a member of the Consiglio de' Duecento (a political position Galileo would achieve almost ten years after delle Colombe obtained it),[67] he may have been better known as a "character" than he was respected as a philosopher, to judge from the fact that the popular vernacular writer Ruspoli made him the target of one of his satirical sonnets.[68]

Consequently, it is hard to believe that the judges would have given the same credibility to Galileo and to delle Colombe. If we are to believe delle Colombe, Galileo had actually been able to turn the tables in his favor once before, preventing delle Colombe from performing an experiment that was legitimate according to their agreements. Therefore, it should not be difficult to understand delle Colombe's worries about enter-

---

65. Ibid., p. 318.

66. Drake has probably mistaken Filippo Arrighetti for one of his cousins (Andrea or Niccolò), for he claims that he had been a former pupil of Galileo (Drake, "Dispute Over Bodies in Water," 160). On Arrighetti's trip to, and stay in, Padua with Galileo, see *GO*, vol. 19, p. 165. On Nori's connection to Galileo, see *GO*, vol. 10, no. 282, p. 305; no. 409, p. 447.

67. In a list of Florentine citizens elected to the Consiglio dei Duecento in 1623 we find Ludovico di Zanobi delle Colombe and his brother Corso (*ASF*, "Manoscritti 133," fol. 215v., fol. 216r). Galileo would be elected to that office only in 1631 (*GO*, vol. 19, pp. 484–86. The delle Colombe family must have been moderately influential: Corso held minor political roles in the late 1620s (*ASF*, "Tratte 645," fol. 153).

68. Almost nothing is known about the life of delle Colombe. The "Diario fiorentino del Settimanni," at the date 3 December 1625, reports the death of Francesco Ruspoli and gives a brief biography of him. In it, Settimanni mentions the "Colombaia," a spot in Florence "he had so named because it was the ordinary residence of Colombo, as the philosopher Ludovico delle Colombe was commonly called. [He was] an erudite man who wrote a book of pleasurable responses against Galileo Galilei, the esteemed man of letters of our century. This Ludovico was a solitary and melancholy man, large in stature, thin (actually, extremely lean) with a long and very white beard, a small and completely bald head, and sunken eyes; he looked precisely like a ghost, and because of this Ruspoli used to call him the 'Manager of Limbo'" (*ASF*, "Manoscritti 133," fol. 301). I have not been able to locate Ruspoli's sonnet in the available works by and on him. In a letter by Galileo, we find that delle Colombe was more than fifty years old in 1611 (*GO*, vol. 11, no. 555, p. 153).

ing a judged competition. Unlike Galileo, who had a comprehensive, if incomplete, theory of buoyancy which allowed for a range of experimentally testable predictions, delle Colombe held views on buoyancy whose empirical support was concentrated in a single experiment. Delle Colombe had much more at risk than Galileo. If Nori and Arrighetti judged his experiment irrelevant or pushed the thin piece of ebony underwater, delle Colombe and his fellow Aristotelians would have been defeated publicly. Probably for these reasons delle Colombe tried to find a different audience (and a safer one) by showing his experiment around town. In such a context, he could fully control the experimental setting. Delle Colombe's strategic desertion and search for an alternative audience must have proven effective if Galileo felt he had to try to arrange a second meeting with him. Delle Colombe's plain experiment must have convinced enough important Florentines to put Galileo on the defensive.

From Performance to Text, from the City to the Court

The fact that Cosimo II rebuked Galileo for letting himself be caught in a noisy and improperly managed dispute may have been also a result of delle Colombe's strategy.[69] But the grand duke's firm advice to Galileo to put his views in writing rather than to sustain verbal disputations helped Galileo get out of a serious deadlock.[70] It offered him a perfect excuse to withdraw from the last dispute at Salviati's—a dispute in which he would have been confronted again with delle Colombe's powerful experiment.

More important, the writing of the *Discourse* was crucial to Galileo's attempt to establish his own rules of the game. The book allowed him to move away from a situation in which he had to defend his position by refuting delle Colombe's very specific experience to one in which Galileo could set the boundaries of his adversaries' future critiques by putting forward his theory as a coherent whole. Moreover, Galileo's distancing himself from the dispute by shifting from an aggressive and narrowly focused ad hominem line of argument to a more general and systematic approach to the problem of buoyancy could be easily represented as fitting the "scientific demeanor" of a client of a great prince.[71]

Galileo knew how to turn these etiquette protocols to his advantage. For instance, although he devoted much of his treatise to the careful refu-

69. Ibid., vol. 4, p. 66. Galileo mentions that many experiments had been carried out and that Cosimo saw some of them. It is not clear whether Cosimo saw experiments performed by Galileo or delle Colombe.

70. Ibid., pp. 30, 34–35, 65.

71. This relationship between the patron's high social status and the "loftiness" of his client's style of argumentation fits the sociology of taste presented by Pierre Bourdieu in his *Distinction* (Cambridge, Mass.: Harvard University Press, 1984).

tation of the views of his adversaries, he never referred to them by name.[72] Although delle Colombe's experiment remained a relevant anomaly to Galileo's theory of buoyancy even after the publication of the *Discourse,* it could hardly aspire to be considered the crucial experiment after Galileo's publication of an articulated theory of buoyancy. The publication of the *Discourse* raised the threshold for the possible refutation of Galileo's positions. Now his adversaries, too, were expected to come up with a holistic evaluation of his theory of buoyancy.

Castelli's response to delle Colombe's attack on Galileo for not having included their agreements in his *Discourse* confirms this reading of Galileo's tactics.[73] Castelli took delle Colombe's accusation to be beside the point: "In no way should these agreements have been reproduced in the *Discourse,* for I am sure that it never crossed Signor Galileo's mind to write his treatise in response to Signor Ludovico. Instead, he simply wanted to find the truth about this matter without antagonizing anybody."[74]

The switch from an ad hominem style of argumentation to a more polite and systematic one resulting from the grand duke's intervention helped Galileo in other ways was well. Not only could he gloss over the dangers of delle Colombe's local but devastating anomaly and set the rules of the game (something that had escaped him in the earlier, more performative phase of the dispute), it also allowed him to define his audience and, therefore, whose questions to answer and whose to ignore. This message was spelled out in Castelli's claim that the *Discourse* was not written as an answer to Galileo's critics.

Such a patronizing attitude (which Galileo adopted in the *Discourse* as well) was also a rhetorical strategy aimed at lowering the status of his critics so as to facilitate their dismissal. If Galileo had been unable previously to dismiss their rules of the game, now he could try to ignore them altogether and address his work to an audience sympathetic to his discourse. In a passage that reminds us of the comment by Copernicus, that "mathematics is for mathematicians," Castelli claimed that:

> I am sure that Signor Galileo did not write with Signor Colombo in mind, and that this book is not directed to him. And Signor Colombo could have figured that out by himself by noticing that he is never mentioned in it, and that the *Discourse* treats and proves most of its arguments geometrically. This alone should have sufficed to convince Signor Colombo that *this text was written for those who know*

72. While the lack of reference to specific people could be presented as a way of lifting the debate above the *ad hominem* level, it was also a way for Galileo to insult his adversaries by representing them as unworthy of being named.

73. However, Galileo contributed substantially to Castelli's replies.

74. *GO,* vol. 4, pp. 465–66.

mathematics, and not for those who are totally ignorant in that discipline.[75]

It is worth noticing that the transition from a noisy dispute Galileo could not control to a textual and legitimate presentation of his theory was made possible by the intervention of his patron, Cosimo II. Galileo could get out of the dispute without losing his honor precisely because his patron (who happened to be the Aristotelians' prince as well) "ordered" him to do so.

A client was obliged to answer the challenges from clients of his or other patrons. At the beginning of the dispute, Galileo could not dismiss his interlocutors. Some of them were legitimate philosophers from the University of Pisa and were, therefore, clients of his own patron. Others, like delle Colombe, were not clients of the Medici in a direct way, but—besides being subjects of the Medici—they were considered legitimate challengers by other Florentine patrons of Galileo like Salviati. Galileo represented Cosimo's intervention as meaning that his patron had ordered him to drop the fight because the challengers were unworthy. By doing so, the Florentine gentlemen who followed the dispute or contributed to keeping it alive could not accuse Galileo of having run away from the challenge. They too were subjects and clients of the Medici. Galileo was able to deflect the grand duke's disdain for his behavior in that dispute onto his adversaries. Once they were indirectly "disgraced" by the grand duke, he too could legitimately dismiss them.

However, Cosimo II was not necessarily trying to get Galileo out of trouble by asking him to put his views in writing. As we have seen, the client's welfare per se was not an item very high on the list of a patron's priorities. Rather, I believe that Cosimo II may have been annoyed by seeing his personal philosopher enmeshed in what he may have perceived as messy bargainings worthy of a shopkeeper. By telling Galileo to put things in writing, Cosimo was not trying to save Galileo's honor but his own.

By writing a treatise that was probably longer and more systematic than the grand duke had expected (and using it to put in print all his astronomical discoveries since the *Sidereus nuncius*) Galileo maintained the grand duke's favor and controlled his adversaries. Although they were able to produce four books against his *Discourse* (each of them dedicated to a member of the Medici family), Galileo's position at court must have been strong enough that he did not feel compelled to answer them. Not until two years after the publication of the last of the attacks did Castelli—one of Galileo's clients—take charge of refuting them. Although Galileo still did not know how to explain the water's surface tension that had given him so many headaches, he won (at least temporarily) against the Pisan

75. Ibid., p. 467 (emphasis mine).

philosophers and their supporters in the specific sense that he obtained the power to ignore their attacks.

Cosimo II's rebuke to Galileo does not mean that he did not appreciate disputes in general. Quite to the contrary, Galileo's obligation to participate in disputes was explicitly included in his contract with the Medici.[76] Cosimo II wanted to see his philosopher engage in properly framed disputes, in a proper setting, against proper challengers, and in front of a proper audience. The occasion for such a dispute came up in the fall of 1611 when Cardinals Gonzaga and Barberini happened to be in Florence at the same time. This event marked the dispute's shift from the city to the court, that is, to a site where Galileo could exercise better control.

We do not have a full description of the dispute for it is not recorded in Cesare Tinghi's official court diary, but we find several references to it and partial descriptions of it in Galileo's correspondence and in the various texts of the dispute on buoyancy.[77] As with the other court disputes mentioned before, this one too took place at the table of the grand duke after lunch, and—as during the visit of Cardinal Perron—the visiting cardinals participated in the dispute. Barberini sided with Galileo, while Gonzaga supported his opponent, the Aristotelian philosopher Papazzoni. Judging from the content of the texts they would publish later, delle Colombe, Coresio, and di Grazia did not participate in the dispute at court. The grand duke's decision to have Galileo dispute against the newly hired philosophical "star" of the University of Pisa rather than against his early adversaries legitimized Galileo's claims that the later *Discourse* was not a response to their critiques. Delle Colombe and his cohorts were not granted direct access to the new "higher" phase of the dispute. As we will see, they could try to reach the dispute's new courtly site with books but not in person.

It is difficult to understand who was perceived to have won the dispute at court. Ciampoli—already a good friend of Galileo's—was present at the dispute as a member of the entourage of Cardinal Barberini and remained convinced of Galileo's superiority over Papazzoni. To him the dispute seemed characterized by a "great inequality" (*gran disuguaglianza*), "since one of them presented brilliant and empirically sound arguments, to which the other answered with very narrow and dry distinctions such as *per accidens* or *secundum potentiam* or *secundum quid*."[78] Cardinal

---

76. *GO*, vol. 10, no. 359, pp. 400–401.

77. Ibid., pp. 298, 329, 331; vol. 11, no. 820, pp. 453–55. In addition to delle Colombe's *Discorso apologetico* mentioned above, the other responses to the discorso were: *Considerazioni di Accademico Ignoto sopra il Discorso del Sig. Galilei* (Pisa: Boschetti, 1612); Giorgio Coresio, *Operetta intorno al galleggiare de' corpi solidi* (Florence: Sermartelli, 1612); and Vincenzo di Grazia, *Considerazioni sopra il Discorso di Galileo Galilei,* (Florence: Pignoni, 1613). All four works are reprinted in *GO*, vol. 4.

78. *GO*, vol. 11, no. 820, p. 453.

Barberini also thought that Galileo performed well, but Cardinal Gonzaga remained unconvinced by his arguments.[79] Galileo's adversaries, who seemed to be well informed about what was said in front of the grand duke, were not at all convinced of Galileo's victory.

Given the state of the debate and the interpretive deadlock on what we call surface tension, it is not surprising that different judgments about the outcome of the dispute were possible. But more important to Galileo's legitimation of his views of buoyancy (and of his mathematical method in general) was the fact that the debate had moved to court, where Galileo's audience and patrons appreciated (and digested) a "good show" better than a demonstrative argument. Given the patrons' noncommittal attitude, an undecidable debate was the perfect and safest topic for a courtly dispute. Moreover, Papazzoni was a client of Galileo and, being indebted to him for a major and very recent favor, it is likely that the philosopher was not as aggressive with Galileo as delle Colombe may have been.

The quite favorable outcome this dispute had for Galileo was also a result of patronage dynamics. By becoming the philosopher and mathematician of the grand duke, Galileo became an important patron of mathematicians and philosophers, as we can see from the many letters from clients he received after 1610.[80] He was consulted by the Medici for the hiring of mathematicians and philosophers at the University of Pisa, but he was also very influential at the Universities of Bologna and Padua as well as at Rome's Sapienza. It was through Galileo's influence that Cavalieri obtained his post at Bologna, Castelli his chairs first at Pisa and then at Rome, and Aggiunti and Peri theirs at Pisa. Then, after Libri's death, Galileo also succeeded in securing Papazzoni the prestigious chair of philosophy at Pisa—a favor Papazzoni could not easily forget.[81]

Although Galileo was not president of an important scientific institution and could not mobilize it against other mathematicians or philosophers as Newton would later do with Leibniz, he nevertheless had somewhat docile clients willing to undergo "friendly" abuses.[82] The dis-

79. Ibid., no. 684, pp. 304–5; no. 690, pp. 317–18; no. 698, p. 325; no. 711, p. 338.

80. *GO,* vol. 10, nos. 98, 100, 106, 112, 115, 119, 179, 217, 229, 281, 282, 386, 441, 444, 445; vol. 11, nos. 469, 471, 473, 474, 480, 482, 483, 488, 490, 577; vol. 14, no. 1973.

81. *GO,* vol. 11, no. 464, pp. 28–29.

82. In February 1611 Papazzoni wrote Galileo a letter thanking him in advance for whatever he could do to help him get the job at Pisa. Papazzoni concluded the letter, which is an impressive exercise in flattery, by expressing his "moral duty" to "exalt my Signor Galileo" (ibid., no. 483, p. 59). Four days later, Papazzoni repeated the praise, claiming: "I assure you that you will have a trumpeter of your deserved praises. Please love me, give me orders, and stay healthy" (ibid., no. 487, p. 63). Together with Papazzoni's first letter, Galileo received one from Roffeni, the mediator between him and Papazzoni. Roffeni wrote him: "And I assure you that this gentleman is a faithful servant, and if he ends up selected [for the chair] *Your very Illustrious Lordship will notice the many signs [of his dedication] on every occasion he will have to [help] your honor*" (ibid., no. 482, p. 58 [emphasis mine]).

pute on buoyancy might not have ended happily for Galileo had he not
been the philosopher of the grand duke and had Papazzoni not been a du-
tiful client of his.

Discourse on Bodies in Water

The structure of the *Discourse* reflects the ambiguities of Galileo's tactics
and perceptions of the "battlefield."[83] It displays Galileo's attempt to present his treatment of buoyancy as the paradigmatic framework for any future discussion on the topic among competent discussants. At the same
time, it also indicates that Galileo was not quite sure of having the power
necessary to ignore completely the dispute between him and the Aristotelians and to pretend that the *Discourse* was not a response to delle
Colombe and his cohorts.

These tensions are also mirrored in the content of the *Discourse* which,
after starting out as a Euclidean-style treatise, turns very soon into a dialectical point-by-point attack on the positions of his Aristotelian adversaries. I say "positions" because Galileo did not mention any of his
adversaries by name. Actually, at one point he even claimed that—given
the ridiculousness of their views—he was doing them a favor by not putting their names in print.[84] Wanting to dismiss the connection between
the 1612 *Discourse* and the dispute, while feeling obliged to respond his adversaries, Galileo adopted an interesting rhetorical strategy: not only did
he avoid mentioning his real adversaries, he turned the then-deceased
Pisan Aristotelian philosopher Francesco Bonamico into his principal in-
terlocutor in the *Discourse*.

Quite conveniently, Galileo could make Bonamico represent his live
critics while avoiding the drawbacks of associating his *Discourse* with
them. Having been a well-known commentator of Aristotle, Bonamico
was a proper match for Galileo. By challenging him, Galileo was not
lowering the level of the dispute (and his own prestige) as he would have
done by challenging an undistinguished philosopher such as delle
Colombe. In contrast to delle Colombe and his cohorts, Bonamico was
mathematically literate and, in the fifth book of his *De motu*, he had pro-

---

83. While Galileo was engaged in the dispute on buoyancy in Florence, his Roman
friends and patron were putting pressure on him to publish a revised version of the *Sidereus*
in order to prevent probable priority disputes about his more recent astronomical discoveries (ibid., no. 572, p. 175; no. 573, p. 176). At some point Galileo decided to publish the *Discourse* rather than a revised version of the *Sidereus nuncius*, and to include his latest
discoveries in it. But the writing took much longer than he or his friends and patron expected. In December he told Cesi that the *Discourse* was ready (ibid., no. 652, p. 248), but it
came out only six months later, when Galileo was already fully involved in the dispute on the
sunspots.

84. *GO*, vol. 4, p. 73.

posed a refutation of Archimedes' theory of buoyancy. Also, Bonamico's views on buoyancy were a coherent *summa* of the positions fragmentarily presented by delle Colombe and his colleagues.

Bonamico's resurrection was not the *Discourse*'s only rhetorical device. Although there is no question that Galileo had a deep admiration for Archimedes, the emphasis he put on the continuity between Archimedes' *On Floating Bodies* and his own *Discourse* was not totally justified by the content and methodological assumptions of the two texts. It seems that Galileo presented his work as more Archimedean than it actually was in order to back his analysis of buoyancy with Archimedes' authority. Also, Galileo's work was much less systematic than Archimedes' and contained only a fraction of the propositions in *On Floating Bodies*. In a sense, the *Discourse* was more of a polemical attack on Galileo's adversaries dressed up as a systematic mathematico-deductive treatment of buoyancy. As Sagredo was quick to notice, Galileo was pretending to deal with demonstrable matters but was frequently sliding into extensive dialectical arguments.[85]

In fact, the identity of Archimedes' and Galileo's views of buoyancy did not go much beyond the claim that the cause by which some solid bodies descend in water is the excess of their specific weight over the specific weight of the water, while conversely, the excess of the specific weight of the water over the specific weight of the solid body is the cause of others not descending, even of rising from the bottom and surmounting the surface.[86]

In fact, after presenting the definitions of specific and absolute weight, Galileo began to move beyond the boundaries of hydrostatics (and Archimedes' treatment of buoyancy) and to trespass into dynamics—the philosophers' domain. He did so by introducing the notion of *momento*—a principle he claimed to have adopted from the science of mechanics.[87] He justified this novelty as a way to account for an aspect of buoyancy that had escaped Archimedes himself: the relationship between the size of a container and the depth to which a floating body sinks.

85. "You know that I am neither Peripatetic nor crazy, and that I rather take the occasion to tell you with my usual liberty that I am surprised that you have written about such a subject in the form of a discourse and that, by responding to those who do not understand anything about it, you have almost put in difficulty the patent and demonstrated truth, thus lending credence to the philosophical errors of the present time" (*GO*, vol. II, no. 701, p. 330).

86. Ibid., p. 67. The translation is from Drake, *Cause, Experiment, and Science*, p. 26.

87. Galileo's two principles are: (1) "Absolutely equal weights, moved with equal speeds, are of equal force and *momenti* in their operation"; and (2) "The *momento* and the power of heaviness is increased by speed of motion, so that absolutely equal weights conjoined with unequal speeds are of unequal power, *momento*, and force, in the ratio of one speed to the other" (*GO*, vol. 4, p. 68. The English translation is adapted from Drake, *Cause, Experiment, and Science*, pp. 29–31).

As indicated by recent studies of his mechanics, Galileo tried consistently to cross from statics into dynamics (and from mathematics into philosophy) by using *momento* as the hinge between those two branches of mechanics.[88] However, his attempts to connect through *momento* the dynamical notion of *momentum velocitatis* (or momentum) and the static *momentum gravitatis* (or moment) failed, and in his last and major work on mechanics—the 1638 *Discourse on the Two New Sciences*—Galileo retreated into a more kinematic approach. Nevertheless, his final kinematic approach did not result from his methodological decision of not seeking the causes of natural motion, but rather from his failure to find a way to connect them to his mathematical framework. The kinematic approach of the *Two New Sciences* (and maybe the fact that he waited for so long before publishing that work) indicates Galileo's lack of success in living up to the title of "philosopher" he had prized so highly. The mathematical investigation of the real causes of motion escaped him.

However, at the time he wrote the *Discourse on Bodies in Water* in 1612 Galileo had not yet given up on his dynamical agenda. On the contrary, as the grand duke's philosopher he had the opportunity to legitimize it. When his Aristotelian adversaries saw the two principles regarding *momento* Galileo had included in the *Discourse,* they quickly realized that, behind the screen of buoyancy, Galileo was trying to live up to the standards of his newly acquired title by presenting the core of a more general theory of sublunary motion.[89] In particular, both the Anonymous Academician and di Grazia questioned Galileo's attribution of a new meaning to *momento*—a semantic unorthodoxy that reflected Galileo's unconven-

88. On Galileo's notion of *momento* and on his dynamical connotations, see Galluzzi, *Momento,* pp. 153–259, 287, 343, 353; Shea, *Galileo's Intellectual Revolution,* 23; Winifred L. Wisan, "The New Science of Motion: A Study of Galileo's 'De Motu Locali,'" *Archive for History of Exact Sciences* 13 (1974): pp. 222–29, 292, 297; Settle, "Galilean Science," pp. 157–247, esp. pp. 226–34; Adriano Carugo and Ludovico Geymonat, "Note," in Galileo Galilei, *Discorsi e dimostrazioni matematiche intorno a due nuove scienze* (Turin: Boeringhieri, 1958), pp. 724–26; Westfall, "Problem of Force in Galileo's Physics," pp. 67–95; and Edith Dudley Sylla, "Galileo and the Oxford Calculatores," in William A. Wallace, ed., *Reinterpreting Galileo* (Washington: Catholic University of America Press, 1986), pp. 53–108, esp. p. 88.

89. *GO,* vol. 4, pp. 156–57. That Galileo was not talking just about buoyancy can be seen by this: "I would not distrust myself to be able to sustain as most truthful the sentence of Plato and of some others, which absolutely negates lightness, and affirms that in elementary bodies there is no intrinsic principle of movement except one directed toward the center of the earth, and that there is no cause for upward movement (I mean that which has semblance of natural motion) except the displacement of the fluid medium produced by the heavier [downward] moving body. I think one can fully answer Aristotle's contrary opinion, and I would do so if it were necessary at this point, or if it would not entail too long a digression in this brief treatise" (ibid., pp. 85–86). Actually, in an earlier draft of the *Discourse,* Galileo thought of presenting this as a physical assumption: "I assume one of the following two axioms and suppose . . . that the natural order required that the heavier bodies stand under those less heavy, under which they will go unless restrained" (ibid., p. 36).

tional philosophical agenda.[90] In fact, the debate which followed the pub-lication of the *Discourse* was about Aristotle's and Galileo's theories of movement in general as well as about the more narrow issue of the conditions of equilibrium of bodies in water.

As mentioned, Galileo introduced the notion of *momento* as a way to amend what he saw as a faulty aspect of Archimedes' theory of buoyancy. While he agreed with Archimedes that buoyancy was fundamentally connected to the different specific weights of the body and the surrounding medium, he claimed that it was wrong to say that a body specifically lighter than water sank in it until it displaced an amount of water of equal weight. According to Galileo, bodies sank less and displaced less water than Archimedes had claimed. In fact, the sinking would stop as soon as the water—pushed *up* by the sinking body—reached the height at which the weight of the volume of water equivalent to the volume of the immersed solid weighed as much as the entire solid. In the case of a container only slightly larger than the sinking body, one could see that by the time equilibrium (that is, buoyancy) was reached the body had displaced much less water than its own weight because the water had moved *up* considerably. Galileo compared this example to that of two connected cylindrical containers of different diameter, as shown in fig. 13. Depression of the water level in the large cylinder would cause a steep increase of the water level in the smaller connected cylinder.

Because of the effect of the size of the container, a heavy object could float in a volume of water of absolute weight much smaller than the body's. As Galileo put it, a ship could float in a barrel of water, provided one constructed a suitable container. Conversely, in the case of a very large container (like an ocean) the body's sinking in it would not raise the water's level by any significant amount while—if we take the sea level as reference—the body would have markedly sunk in it to reach the condition of equilibrium. In short, Archimedes' problem was that he had not considered the interaction between the sinking body and the rising level of water, a phenomenon that becomes particularly conspicuous when one considers relatively small containers.

By extending the analysis to the effect of containers, Galileo managed to reduce buoyancy to one of his favorite models of motion: the lever.[91] Accordingly, he could compare the vertical displacements of body and water to those of two weights attached to the arms of a lever. In particular, he demonstrated that the ratio between the fall of the body and the rise of water was equal to the ratio of the surface of the water containers and the base of the sinking solid. For instance, when the ratio between the con-

90. Ibid., pp. 159, 385, 387–88. See also Galluzzi, *Momento*, pp. 240–46.
91. Galluzzi, *Momento*, pp. 70–79.

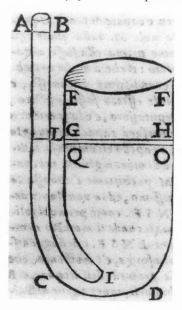

Fig. 13. The depression of the water level from GH to QO in the larger cylinder, to the right, causes the water in the smaller cylinder to rise from L to AB. From *GO*, vol. 4, p. 78.

tainer's and the body's sections is much less than two, a body sinking for, say, two inches would propel the water upward much more than two inches, to stop only when the condition of equilibrium had been reached. In short, a relatively small quantity of water is able to balance a much heavier body—provided the container is small enough. Such a case is comparable to that in which a heavy body is attached to a relatively short lever arm and balanced by a much lighter body attached to a much longer arm (fig. 14).

This explains how Galileo could consider speed (and therefore *momento*) as an essential dimension of both buoyancy and of the equilibrium of a lever. In the example above, the water moves up faster than the body moves down. Basically, Galileo transferred the notion of virtual velocities from the lever's arms to the vertical displacement of body and water. Consequently, he could argue that water and body were in equilibrium despite their different weights because—as in a balance with unequal arms that begins to move because of an infinitesimal weight or external force applied to one of the arms—the lighter weight would equilibrate the heavier one because (its arm being longer) it would move at higher tangential speed. A lever was in equilibrium when the *momenti* (that is the product of the weight by the speed) on the two sides of the lever were equal.

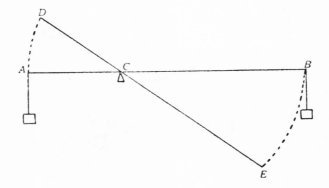

Fig. 14. Although the lever's arms have different lengths, the two bodies at A and B move through the arcs AD and BE in the same amount of time as it takes for the lever to make its swing. Any change in the lever's position makes the body attached to the longer arm move faster than its counterweight, as happens to the water in the narrower part of the water container in fig. 13. From *GO*, vol. 2, p. 163.

Bringing speed and *momento* to bear on buoyancy through considerations of virtual velocity in the "hydraulic lever" allowed Galileo to live up to his title of philosopher. It allowed him to revise his early theory of free fall—that of the *De motu antiquiora* manuscript—which he never published, probably because of its serious shortcomings in accounting for uniformly accelerated motion.[92] Because the dispute had focused on the conditions of equilibrium rather than on the acceleration and speed of bodies sinking in a medium, the shortcomings of Galileo's theory in accounting for acceleration were not foregrounded. Moreover, the water's density and viscosity drastically reduced the sinking body's velocity and therefore the visibility of its acceleration. The writing of the *Discourse* gave Galileo the occasion to rework and bring together some thoughts on mechanics that he had been developing since the early 1590s and to present them in a more favorable environment that minimized their problems while allowing him to present himself as a philosopher and his theory of buoyancy as a philosophical one.[93]

92. On this work see Raymond Fredette, "Galileo's 'De Motu Antiquiora,'" *Physis* 14 (1972): 321–48.

93. However, it is not completely clear how Galileo's view of buoyancy related to the thoughts he was having on free fall in that same period. This may account for the little serious interest in this text shown by specialists of Galileo's mechanics (with the exceptions of Settle and Galluzzi). What is, instead, more clear is the relationship (through the notion of *momento*) between the *Discourse* and his "Trattato delle mechaniche," an unpublished piece he wrote around 1599. As is known, Galileo presented a first (but flawed) enunciation of the law

However, in so doing, Galileo escalated the confrontation with the philosophers. Now, his adversaries did not have to confront only a very specific claim about buoyancy (and one that remained within the boundaries of the mathematical disciplines) but something that could look like a theory of sublunar motion in disguise. Moreover, Galileo's articulation of the philosophical implications of his theory (something he did mostly by arguing against Bonamico) led him to enter into more areas—such as the structure of matter—that the philosophers considered as their professional jurisdiction.

Galileo's articulation of his views on buoyancy highlighted the fundamental differences between his worldview and that of the philosophers and brought about a communication deadlock between the two parties. If the earlier exchange had already ended in a deadlock, the transition from performance to text did not solve but rather worsened that predicament, making the outcome of the dispute even more dependent on the grand duke's verdict. Next, I shall briefly map out the irresolvable differences between Galileo and the Aristotelians on issues such as locomotion, the classification of causes, the status of empirical evidence, the structure of matter, and the phenomenon of surface tension.

Articulating a Deadlock

At the very end of *On the Heavens,* Aristotle wrote that "The shapes of bodies are not responsible for the actual downward or upward direction of their motion, but for making this motion faster or slower."[94] This apparently Galilean statement was quickly qualified. Aristotle observed that heavy bodies such as lead can float if they are flattened enough to cover a wide area of the water surface. Instead, lighter but smaller bodies may sink. He concluded that buoyancy depended on the body's weight not exceeding the water's resistance.

Although his example covered only the case of bodies on water, it reflected the structure of Aristotle's general view of locomotion. A body moved naturally in a medium when the tendency of its elements to reach their natural place overcame the resistance of the medium. Speed was thought of as proportional to the ratio of the force causing motion and the resistance of the medium. The omnipresence of matter in the universe and the assumption of a resistance in the medium were fundamental to the Ar-

---

of free fall and kept working on that problem until the 1638 *Two New Sciences.* Despite the reexamination of previously overlooked manuscripts and a lively debate among specialists of Galileo's mechanics, the development of his thought between 1604 and 1638 is still a matter of well-informed and argued hypotheses.

94. Aristotle, *On the Heavens,* trans. W. K. C. Guthrie (London: Heinemann, 1939), pp. 366–69.

istotelian system. Among other things, they ruled out the possibility of infinite speed (a threatening concept in a finite cosmos such as Aristotle's). The resistance of the medium was a necessary assumption for preserving some of the fundamental principles of that system and, consequently, the Aristotelians could not give it up easily.[95]

Galileo's views on buoyancy asked the Aristotelians to do precisely that. As he argued against Bonamico, the cause of buoyancy had nothing to do with the resistance of the medium.[96] It was the difference in specific weight (*gravità per ispecie*) between the body and the medium that determined whether it would float or not. Resistance of the medium influenced only the speed with which the body would sink or emerge.[97] Galileo's view of buoyancy entailed the complete denial of the role of the resistance of the medium in determining the conditions of movement. Although Galileo could admit a relation between resistance and speed while denying one between resistance and movement, the Aristotelians could not. According to their views on locomotion, both the possibility of movement and its speed were tied to resistance. That is why, I think, they systematically disregarded Galileo's distinction and attacked him for (allegedly) saying that the medium did not resist movement. Galileo's two distinct claims on the role of resistance in relation to the conditions of movement and to speed were collapsed into one by the Aristotelians' categories.[98]

Unlike Galileo, Aristotle and the League (as the Aristotelians came to call themselves) referred the fundamental cause of buoyancy to the elements that made up the bodies. If a body was predominantly made up of the earthy element, then it would sink in water toward the center of the earth. If a body contained much air element, it would float on the water's surface. If it were fiery, it would fly up toward the sphere of the moon.[99] According to the Aristotelian terminology, a body's elemental makeup was the cause *per se* (or *simpliciter*) of motion. The shape of the body was not the main cause of its motion but could become the cause *per accidens* or *secundum quid* of its rest. As stated by Aristotle in *On the Heavens,* in certain cases the shape of the body together with its weight and the resistance of that specific medium caused the body to float. Shape was a secondary cause of buoyancy, which became important only under certain circumstances.

Judging from the responses to Galileo's *Discourse,* the core of the Aris-

95. *GO,* vol. 4, p. 415.
96. Ibid., pp. 81, 86–87, 103, 126.
97. Ibid., pp. 33–34, 44–45, 50, 91–92, 96.
98. Ibid., pp. 412–13.
99. Ibid., pp. 85–86.

totelians' thesis consisted of this: "the shape of a body is a cause *secundum quid* of buoyancy" (that is, of a body's *rest* on water).[100] To them, shape was not a cause for a body's rising from the bottom of a water container. That motion was determined solely by the body's elemental makeup. This was the case because, as we will see in a moment, the water surface was credited with properties different from those of its inner parts. The behavior displayed by delle Colombe's sphere and thin piece of ebony provided all the evidence the Aristotelians needed to defend such a view. Those bodies were made up of the same material and had the same weight but differed in shape. As a result, one floated while the other sank. As Galileo himself admitted, delle Colombe's experiment fit the Aristotelian doctrine very well.[101]

Contrary to the reading of the dispute presented by a modern historian, the Aristotelians were not quibbling.[102] Delle Colombe's interpretation of his experiment was logically correct and empirically sound *on its own terms*. The League did not claim to be able to predict conditions of buoyancy for different media, weights, and shapes. They claimed only that, if a body heavier than water floated, it was because of its shape.

What the Aristotelians feared now was not a refutation of delle Colombe's experiment (which was virtually unfalsifiable if they stuck to their rules of the game), but the cosmological implications of Galileo's claims. In fact, if buoyancy was not a direct result of the body's innate "levity" resulting from its elemental makeup, but a result of the specifically heavier medium falling toward the center of the earth, then the very notion of natural motion would be questioned. As the Anonymous Academician put it, if Galileo were right then all upward motions would become violent.[103] Buoyancy would no longer be the result of a body's natural tendency but, as Galileo put it, of other bodies pushing it up as they were moving down. In short, there would be only one type of natural motion: downwards—a motion that was shared by all objects independently of their elemental makeup.

Galileo's claim was devastating not only for the Aristotelians' element-based view of motion and change but for their notion of demonstration as well. In fact, while the Aristotelians saw buoyancy as an instance of natural motion (sometimes brought to an accidental rest by the shape of the body), Galileo restricted the Aristotelians' notion of natural motion and therefore questioned the taxonomy of causes the Aristotelians had developed within that cosmological framework. Given the close connection be-

100. Ibid., pp. 28, 43–45, 86, 96, 174, 212, 329, 337, 403, 420.
101. Ibid., p. 90.
102. Drake, *Cause, Experiment, and Science*, pp. xix–xx.
103. *GO*, vol. 4, p. 157.

tween natural causality and logical demonstration in Aristotle, Galileo was perceived by his adversaries not just as putting forward causes different from theirs but as questioning the very notion of cause. Galileo was quite aware of this. As he wrote to one of his supporters,

> there is only one, true, and proper cause of buoyancy—the one known to me and to others. Distinctions such as *per se* or *per accidens*, *proprie vel improprie, absolute vel respective* cannot be applied to it. Those distinctions are brought only to help those who cannot grasp the true, proper, and immediate causes of the philosophical problems they are confronting.[104]

Along the same lines, Galileo accused Bonamico of producing, not causes, but *causes of causes* of the phenomenon of buoyancy when he referred to the properties of the four elements.[105]

Galileo's critique of the Aristotelians' modes of reasoning appeared tenable only to those who accepted his notion of causality and the cosmology that went with it. Basically, he reversed the causal link established by the Aristotelians between "element" (that is, elementary species) and buoyancy. To him, all that was observable was a body floating or sinking in a given medium. The inference that "this body contains more air than water or earth therefore it will float on water" was replaced by "this body floats on that fluid therefore it has a lesser specific weight than that fluid." Buoyancy ceased to be an *effect* of the elements' properties and—as shown by Galileo's "Bilancetta"—became a sort of *instrument* through which information about the body's density (a mathematical feature) could be obtained.[106] To Galileo, his explanation of buoyancy had more the status of an ontological principle than of an effect. Therefore, even the very term "buoyancy" had sharply different meanings for the two parties.

Di Grazia was well aware of the incompatibility of their methodological views:

> concerning the things that fall within the domain of the senses and that we see all the time, he wants to demonstrate them mathe-

104. Ibid., p. 299. A similar critique of the Aristotelian classification of causes is put forward in Castelli's response: "He [Galileo], without ever needing to resort to causes that are primary, secondary, instrumental, essential, accidental, related to figure, dryness, resistance of continuous mediums, fluidity, hardness, uncovered surfaces, antipathies, unctuosity, circumstances, qualified matters, skillful terms, and other hundred chimeras (that are your excuses), explains everything with only one simple and clear conclusion exempt from all those limitations and distinctions" (ibid., p. 580).

105. Ibid., p. 87.

106. "But who does not know that the real cause is the immediate rather than the mediated one? Moreover, gravity provides a cause that is very clear to the senses, so that we can determine very easily if ebony, for example, or pine is heavier or lighter than water. But who

matically. Instead, concerning those things that cannot be grasped through the senses or, if so, only poorly, he insists on explaining them through the senses, as with the cavities of the moon, the sunspots, and a thousand more things like that. Rather, he should go the other way around. In fact, it is superfluous to argue about things that can be directly grasped through experience. Instead, it is in those cases in which sensorial experience is inadequate that we need to correct and help it through reason.[107]

The views on buoyancy of both the Aristotelians and Galileo were closely tied to their more general views about motion, causality, the structure of the cosmos, and mathematics' place in it. Galileo strove to make sure that his theory would be perceived as having fundamental cosmological implications as well as dynamical dimensions. It is interesting that di Grazia, being unwilling to give cognitive legitimacy to Galileo's mathematics-informed experience, claimed that Galileo's theory was intimating cosmological principles (and a cosmology in which they were rooted) but failed to present either.[108]

The incompatibility of the categories of Galileo and the Aristotelians emerged again in their views on the structure of matter. Since the preliminary draft of the *Discourse,* Galileo had considered water as made up of particles.[109] Such a view was then presented in the *Discourse* and articulated in some detail in a later letter to Tolomeo Nozzolini.[110] To Galileo, the structure of water was analogous to that of a metal reduced to fluidity by the action of the thin *ignicoli* (the atoms of fire) which, like thin knives, were able to penetrate the microscopic interstices between the body's particles and destroy the cohesion that had been maintained among those particles by the *horror vacui*—nature's abhorrence of the vacuum.[111]

Galileo's views on the structure of matter were not exhausted by the treatment they received in the *Discourse* but reemerged modified in the *As-*

---

could tell us whether they are predominantly earthy or airy? Certainly no other experience is better than to check whether they float or sink to the bottom" (ibid., p. 87).

107. Ibid., p. 436.

108. He concluded his *Considerazioni* by criticizing Galileo for not yet having published the "System of the World" he had already announced. Ibid., p. 439.

109. Ibid., pp. 27–28.

110. Ibid., pp. 26–27, 103, 105–6, 301 note 1. Galileo's letter to Nozzolini confirms the patronage dynamics discussed before. Nozzolini had been a tutor of young Prince Cosimo in the 1590s (*ASF,* "Depositeria generale 389," fol. 47v) and was probably still an intimate client of his. Consequently, Galileo could not ignore a query coming from him because if he failed to respond, Nozzolini might have complained to the grand duke.

111. Galileo Galilei, *Two New Sciences,* trans. Stillman Drake (Madison: University of Wisconsin Press, 1974), pp. 27–28.

sayer and especially in the *Two New Sciences*.[112] At the time he wrote the *Discourse,* Galileo thought that the atoms of fire were indivisible and *quanti*. In contrast, he thought of water and other fluids as made up of particles that were still divisible.[113] As a result of this corpuscularism, Galileo had to confront the issue of cohesion, one that would trouble Descartes as well. If fluids were not continuous bodies but constituted by contiguous particles, how was it that—unlike vapors—they still had some level of cohesion? He tried to answer this question by comparing water to a configuration of smooth, spherical magnets. Although magnets so shaped would still be difficult to separate, they could easily be given different configurations by sliding their points of contact around.[114]

The corpuscular view of water was crucial for the tenability of Galileo's theory of buoyancy. More precisely, there was a fundamental relationship between Galileo's corpuscularism and his view of movement based on the notion of *momento*. If—as Galileo claimed—movement in fluids could be triggered by an infinitesimal *momento* (generated either by a minimal difference in specific weight or by the application of an infinitesimal external force), then he needed to think of the medium as something that—when moved through at infinitesimal speed—would offer only an infinitesimal resistance to motion.[115] All that an infinitesimal *momento* had to cause was the displacement of a medium's particles one at a time.[116] Conversely, Galileo's view of motion could not be tenable if the medium had some *finite* resistance for that would have neutralized the effect of infinitesimal *momenti*.[117]

While the Aristotelians presented a view of motion in terms of finite

112. See Ugo Baldini, "La struttura della materia nel pensiero di Galileo," *De homine* 57 (1976): 91–164; William R. Shea, "Galileo's Atomic Hypothesis," *Ambix* 17 (1970): 13–27; and idem, *Galileo's Intellectual Revolution*, pp. 27–31, 98–106. Galileo's atomistic beliefs are at the center of Pietro Redondi, *Galileo Heretic* (Princeton: Princeton University Press, 1986). For a general survey of the Greek development and later reemergence of atomistic theories, see Andrew G. Van Melsen, *From Atomos to Atom* (Pittsburgh: Duquesne University Press, 1952).

113. As with any other type of physical body, Galileo theorized that particles needed a "knife" to be cut with and, consequently, one could not think of particles smaller than the edge of the smallest knife. According to him, fire atoms were the smallest "knife" around.

114. The reason why he could not follow the Greek atomists in thinking of cohesion as a result of the atoms hooking onto each other will become clear later in this text.

115. *GO*, vol. 4, p. 86.

116. This, I think, also would have worked if water's internal cohesion could be explained in terms of the attraction between smooth magnets sliding against each other. In fact, one could assume that the body's effort to separate two atoms opposing its motion was equal to the "kick" the body would receive from two other atoms rejoining at the opposite extremity of the body.

117. For a different interpretation of the emergence of Galileo's atomism during the dispute on buoyancy, see Shea, "Galileo's Atomic Hypothesis," 14–15; and idem, *Galileo's Intellectual Revolution*, p. 29.

forces and equally finite hindrances, Galileo dismissed the existence of thresholds both of force and resistance. As he stated, one could pull a ship with a hair provided the speed of translation was *very low.*[118] I would suggest that Galileo saw motion on a fluid's surface (or the motion of a body in a medium whose specific weight was like its own) as something related to his concept of "indifferent motion."

As is known, Galileo distinguished among three basic types of motion: natural (toward the center of the earth, driven by weight), violent (upward, caused by an external force), and indifferent (horizontal, caused by no natural or violent force). It has often been argued that indifferent motion (something he discussed in several of his works) was a proto-inertia notion. In the second letter on sunspots (a text he wrote a few months after the *Discourse*), Galileo described indifferent motion:

> All external impediments removed, a heavy body on a spherical surface concentric with the earth will be indifferent to rest and to movements towards any part of the horizon. And it will maintain itself in that state in which it has once been placed, that is, if placed in a state of rest, it will conserve that; and if placed in movement toward the west (for example), it will maintain itself in that movement. Thus a ship, for instance, having once received some impetus through the tranquil sea, would move continually around our globe without ever stopping. . . . [119]

However, the ship would travel indefinitely only on a very special sea, one that would not present any "extrinsic impediments" to the ship's motion. But in the *Discourse* Galileo also claimed that a real ship in real water could be pulled by a real hair provided the speed of translation was very low. Therefore, speaking in terms of resistance to motion, Galileo seemed to perceive fluids as occupying a position between solid bodies and the vacuum. A fluid did hinder motions with *finite* speed, but did not do so when—as in the case of speeds produced by an infinitesimal *momento* acting on a body at rest in a fluid—the resulting speeds were *infinitesimal*. Instead, unlike a fluid, a solid obstacle may well stop even an infinitesimally slow motion. To put it differently, a solid may prevent motion from taking place altogether; a fluid would not prevent motion but would influence its speed; the vacuum would hinder nothing—neither the possibility of motion nor its speed.[120]

118. *GO*, vol. 4, pp. 104, 107.

119. *GO*, vol. 5, pp. 134–35. The translation is from Stillman Drake, *Discoveries and Opinions of Galileo* (New York: Anchor Books, 1957), pp. 113–14.

120. What I want to indicate here is that Galileo theorized a close connection between motion and the structure of matter. It is not by accident that Galileo's changing notions of *momento* are accompanied by different notions of the structure of matter and of the continuum. For instance, in the *Two New Sciences,* we find that the articulation of the new notion

Given the relationship Galileo saw between movement and the structure of matter, he could not accept the Aristotelians' view of fluids as something that presented a finite resistance to motion. Therefore, the behavior of the water's surface proved to be a major headache for Galileo's theory of buoyancy and his corpuscular view of matter. In fact, that phenomenon could be interpreted as indicating that the resistance opposed by water to the floating body was by no means infinitesimal and that, on its surface, water seemed to behave like a continuous rather than a contiguous entity. Interestingly enough, Archimedes' only postulate on his *On Floating Bodies* put forward a notion of fluids as "continuous" (by which he probably meant isotropic).[121] Being concerned with hydrostatics rather than with the cause of motion of bodies through water, Archimedes—unlike Galileo—did not need additional assumptions about the structure of matter.

That we cannot separate the debate on the causes of buoyancy from that on the structure of matter becomes clearer when we notice that the Aristotelians' view of buoyancy also assumed a specific structure of matter—one opposite to Galileo's. The League attacked Galileo's corpuscular view of water with determination and somewhat hysterical repetitiveness.[122] Not only did they understand the symbiosis between Galileo's theory of buoyancy and his corpuscularism, but they felt they had to stop the threat corpuscularism (and the vacuum with it) posed to their worldview.[123] Symmetrically, a view of matter as continuous fit the Aristotelians' explanation of buoyancy very well. First, the idea of a continuous medium went hand in hand with their need for the medium's noninfinitesimal resistance, in order to maintain their explanation of motion in general. Second, the idea of some sort of continuous "skin" en-

of *momentum velocitatis* goes hand in hand with the introduction of Galileo's theory about the composition of the continuum (Galluzzi, *Momento*, pp. 331–62). Similarly, in the *Assayer,* Galileo presents an interesting fantasy about what would happen in breaking down the smallest finite particles constituting bodies to the ultimately indivisible components of matter: "And perhaps when the thinning and attrition stop at, or are confined within, the tiniest particles [*minimi quanti*], their motion is temporal and their action is calorific only, but when their ultimate and highest resolution into truly indivisible atoms is reached, light is created which has instantaneous motion—or let us say instantaneous expansion and diffusion—and is capable of occupying immense spaces" (translation from Stillman Drake and C. D. O'Malley, *The Controversy on the Comets of 1618* [Philadelphia: University of Pennsylvania Press, 1960], p. 313). On this passage, see Shea, "Galileo's Atomic Hypothesis," 20. Differently from a finite body in equilibrium in a fluid, which—when acted upon by an infinitesimal *momento*—moves with an infinitesimal speed, or a body in equilibrium in a vacuum, which would move at finite speed, the effect of a finite *momento* acting on an infinitesimal particle would be to generate an infinitely fast motion.

121. Archimedes, *On Floating Bodies,* in *The Works of Archimedes,* trans. T. L. Heath (Cambridge: Cambridge University Press, 1912), p. 253.

122. *GO,* vol. 4, pp. 329, 416, 430.

123. Ibid., pp. 258, 329, 416, 430.

veloping water helped explain the different behavior of the thin piece of ebony that floated on the water surface but did not emerge once it was placed at the bottom of the water container.

The Aristotelian system made much better sense of the phenomenon of surface tension than Galileo's corpuscularism did. Surface tension could be presented as a result of the water element's need to preserve its cohesion and natural place by preventing foreign bodies from splitting and displacing it.[124] Given the teleological character of the Aristotelian system, it was not difficult to understand that such a resistance to displacement would be strongest at the boundary between water and another element such as air. Surface tension could be related conceptually to the water element's "place" and to its boundary. It could be seen as a "natural" effect of the properties of the water element.

A view of matter as continuous gave conceptual consistency to the Aristotelians' theory of motion and offered a coherent explanation (in their eyes) of the ebony's inability to float back up to the water surface after it had been submerged. To the Aristotelians, Galileo's request to place the body at the bottom of the water container did not make sense. If to Galileo specific weight (and therefore buoyancy) did not depend upon the body's position within the medium, to the Aristotelians buoyancy depended on the medium's resistance, and they had good arguments to claim that the water's surface had different properties than the rest of the fluid.[125] Therefore, they did not simply sneak away from the experimental setting proposed by Galileo. They could justify their refusal on grounds that—if judged within their framework—were not ad hoc. Once it is perceived against this conceptual background, not only was delle Colombe's experiment locally correct; it also tied together a number of crucial com-

124. Di Grazia claimed that the behavior of the water surface reflected the "desire to preserve itself" (ibid., p. 418). Also, he quoted Aristotle as claiming: "Continuous bodies have the property of resisting division" (p. 434). Similarly, delle Colombe claimed that it was the continuity of water that allowed for the "small banks" (p. 330). Why, he asked Galileo, can we make bubbles of continuous bodies such as water, but we cannot do so with a contiguous body like sand? (Galileo had taken sand as a model for contiguous bodies like water at p. 103.)

125. If the surface tension was a result of the water's natural tendency to "keep to itself," that is, to prevent bodies made up of different elements (which should belong somewhere else) from taking over its place, then the Aristotelians had good reason to refuse Galileo's rule that bodies should be wetted or put at the bottom of the container. To delle Colombe, it was the water's reaction against the dryness (a property of another element) of the object to prevent it from sinking. Therefore, Galileo's pressure to wet it was unacceptable. If the body were wet, then it would no longer be perceived as "alien" by the water, and would be allowed entrance into it: ". . . being heavier than water, if it dived [into the water] what can [possibly] make it come back up to the surface?" (ibid., p. 337). For delle Colombe's treatment of the issue, see pp. 338–41. Di Grazia was also clear about the fact that the interior of the water behaved differently than did its surface: "Signor Galileo's little walnut strips do not remain at the bottom because there is not that resistance that can be found at the surface, that is, the one produced by the water's desire to preserve itself" (ibid., p. 418).

ponents of the Aristotelian system. Moreover, while surface tension seemed to prove their system, it put in crisis Galileo's corpuscularism which, in turn, was a fundamental element of his own theory of motion and buoyancy. In short, both Galileo and the Aristotelians had their own "systems" with their strong and weak points. Delle Colombe's experiment was particularly powerful because it highlighted the tightness of the Aristotelian system while exposing the one local but devastating weakness of Galileo's.

It was only halfway into the *Discourse* that Galileo went back to the dispute (still without naming his adversaries) and proposed an Archimedean interpretation of delle Colombe's experiment—something he considered "the principal issue of the present debate."[126] He prepared his ground by refuting Bonamico's attack on Archimedes. According to Bonamico, the fact that an empty earthenware vase could float on water went against Archimedes' theory by providing an instance in which a body specifically heavier than water floated in it. Moreover, Bonamico continued, if we fill up that same vase with water, we see it sink. This too went against Archimedes: water did not have any weight in water and should not have changed the buoyancy of the vase.[127]

Against Bonamico, Galileo argued that what floated was not the earthenware but the earthenware-air compound. Because the specific weight of this compound was less than that of water, the floating of the vase happened in accordance with Archimedes' principle. When, a few pages later, Galileo faced delle Colombe's experiment, he applied the same line of reasoning:

> Without doubt the cause of the ebony chip or the gold leaf going to the bottom when they do so is that their heaviness is greater than that of water. It follows that when they stay afloat, the cause of this is their lightness, which in this case, *perhaps through some accident not previously observed,* comes to be attached to them, rendering them no longer heavier than water, as they were before they sank, but instead less heavy.[128]

The "accident never previously observed" was another "discovery" by Galileo which—according to him—turned the situation to his advantage. If you look carefully at the thin piece of ebony floating on the water, Galileo said, you will see that the object is not at the same level as the water surface but a bit lower. It is as if small banks (*arginetti*) prevented the water from closing over the object (fig. 15). As in the case of the earthen-

126. Ibid., p. 88.
127. Ibid., pp. 80–81.
128. Ibid., p. 97 (emphasis mine). The translation is adapted from Drake, *Cause, Experiment, and Science,* p. 94.

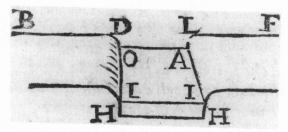

Fig. 15. According to Galileo, the ebony plate does not sink because the specific weight of the composite body constituted by the ebony plate HIOAIH and by the air volume included between the water surface BDLF and the surface of the plate IOAI is less than that of the water. The curved lines DO and AL are the "small banks." From *GO*, vol. 4, p. 98.

ware vase, what floated was not the piece of ebony by itself but an ebony-air composite. And that was in perfect accord with Archimedes' theory of buoyancy. The Aristotelians should stop saying that delle Colombe's experiment refuted Archimedes: quite to the contrary, it confirmed it.

However, Galileo seemed to suffer from an acute strategic amnesia. As noticed by the Anonymous Academician, with the vase it was its external surface that acted as a containing wall, preventing the water from closing over it, but with the thin piece of ebony Galileo could not point to any comparable device.[129] The consistency with which Galileo avoided talking about the cause of the small banks exposed the severity of his struggle with delle Colombe's experiment. When pressed on this issue, Galileo took a "positivist" position: whatever the cause of the containing walls may be, they are there, we can see them clearly, and we understand that they make buoyancy possible according to Archimedes' basic principle.[130] Instead of dwelling on the cause of the small banks, Galileo pretended to take their existence as a fact and articulated it in a series of geometrical propositions in which he showed that, independently of the body's geometrical figure, the small banks allowed for buoyancy according to Archimedes' principles.[131] However, these propositions seemed to be produced deductively and without experimental confirmation. He

129. "Since the walls of the vase are preventing the water from naturally flowing in, it easily maintains its unity and cannot expell the air inside. But these walls that keep the water out are not to be found in the case of the flat small piece [of ebony]" (*GO*, vol. 4, p. 170).

130. In replying to the Anonymous Academician's critique of his explanation of the "small banks," Galileo was not able to present a counterargument but wrote that "that's the way it is" (ibid., p. 166, notes 45 and 46). In the letter to Nozzolini (who had previously expressed "positivistic" views on the "small banks"), Galileo wrote: "I will then care very little [to explain] that these watery banks do not break up, [letting] the water flow to occupy that hole and cavity" (ibid., p. 301).

131. Ibid., pp. 98–120.

simply took for granted that there was a maximum admissible depth to the small banks and that such a depth was *not* dependent on the shape of the body impinging on the water.[132] Moreover, a smooth flat body or a spiky object did not seem to make any difference in terms of how much weight the water surface could stand.[133] Consequently, these propositions looked very much like a cloud of "geometrical smoke" ejected to efface his inability to deal with the banks' cause.[134]

As one can imagine, the Aristotelians were not at all convinced by Galileo's treatment of the small banks. To them, he had introduced some sort of occult affinity, a "magnetic virtue," between the object and the air above it in order to explain why the water did not close over the object.[135] Their claim had some validity, for Galileo explained buoyancy due to the small banks in these terms:

> in submerging until its [upper] surface arrives at the water level, it loses part of its weight, and it goes on then losing the rest of it in going deeper down and lowering itself beneath the surface of the water, while that makes a ridge and a bank around it. It suffers this loss [of weight] by drawing down and making descend with it the air above itself by adherent contact, which air follows in to fill the cavity surrounded by the little ridges of water. . . . [136]

Therefore, Galileo was forced to rely on the adhesion between the body and the air above it because he could not explain why water would resist and allow for the formation of the small banks.

One explanation of this peculiar interpretation of the cause of the

132. For instance, at p. 111 he claims that "the height of the bank AI is the maximum allowed by the nature of water and air" (ibid., p. 111). Similarly, at p. 114 he claims: "Let the line DB be the maximum height of the bank."

133. Actually, he tried to "domesticate" the Aristotelian claims about the role of the shape by showing that, when shape is relevant to buoyancy, it is because of its interplay with the Archimedean principle of buoyancy. For instance, a cone or pyramid that is put in the water with its vertex down is more likely to float than the same cone or pyramid laid on the water with its base down. This is because the former would sink more into the water and therefore become relatively lighter, so that the small banks would not need to be stretched beyond their "natural" limits. The latter would instead sink past the maximum depth of the "small banks" well before it could become buoyant.

134. William Shea has also noticed the peculiar nature of these propositions and has called them a "mathematical holiday" (Shea, *Galileo's Intellectual Revolution*, p. 27).

135. *GO*, vol. 4, pp. 163, 166, 172, 213, 335, 416. Nozzolini shared the same opinion as the Aristotelians (ibid., pp. 290–91).

136. Ibid., p. 98. The translation is from Drake, *Cause, Experiment, and Science*, p. 97. In his response to Nozzolini, Galileo disowned the notion of "magnetic virtue." According to him, such a term was introduced by a courtier adversary of Galileo during the dispute at court. But it seems that Galileo wanted to get rid of a problematic notion by attributing it to his adversaries. With Nozzolini, Galileo stressed an extreme adhesion (*fine contatto*) between the body and the air as the cause of the small banks (*GO*, vol. 4, p. 299).

small banks can be found in Galileo's denial that shape had anything to do with the conditions of buoyancy. Furthermore, he took water to be a contiguous body, a view that did not give him any clue about why water behaved differently at the surface and in its interior.[137] Therefore, he may have felt compelled to rule out any explanation of the small banks that could be referred to the interaction between the water and the body's shape, for this could have opened the door to an indirect acknowledgment of the role of the body's shape in buoyancy. Being opposed to giving any role to shape and unable to find an explanation for the small banks in his corpuscular view of water, Galileo could only blame them on air and its "magnetic virtue."

Such an interpretation was not a difficult one for Galileo to develop. Actually, what he presented here was something analogous to his explanation of the impossibility of pumping water more than eighteen *braccia* (about ten meters) above its original level.[138] According to Galileo, an eighteen-*braccia* water column simply broke under its own weight. The *horror vacui* which held the water particles together was not strong enough to hold them together once the column reached that weight. When the column of water dropped ("sank") a "blade of air" had inserted itself between the water column and the pump's piston. Analogously, in the buoyancy case, Galileo claimed that the sinking following the thin piece of ebony getting wet was caused by the water "cutting" the contact between the ebony and the air above it.[139] Therefore, in both cases, "sinking" was the result of the breakdown of the *horror vacui*. The only difference from the cause of the water pump in which the water column was "cut" by a blade of air was that here a column of air was "cut" by a blade of water.[140]

137. ". . . because, if it [water's resistance] was there, it would be no less in the interior parts than in those close to the surface" (ibid., p. 103).

138. Galilei, *Two New Sciences*, pp. 23–26.

139. ". . . the height of the bank . . . is the highest allowed by the nature of water and air *without water expelling the air that adheres to the surface of the body*" (*GO*, vol. 4, p. 111 [emphasis mine]).

140. Galileo may have perceived pumping and buoyancy as different aspects of the same scenario involving the displacement of two differently dense bodies into each other. The explanation of the behavior of a body (water) being *sucked up* into a specifically lighter body (air) was probably seen by Galileo as symmetrical to that of a specifically lighter body (air) being *sucked down* into a specifically heavier body (water). Ebony's weight could have been perceived as playing the role of the pump. In fact, if the body was too heavy (or the pump too strong), the column of either air or water would break down (or up). This kind of reasoning was probably made possible by the fact that Galileo thought of buoyancy as a three-body rather than a two-body system. The scenario he confronted was a water-ebony-air system. If something floated on the water it meant that it did not fly up in the air; consequently, it meant that it was at the same time specifically lighter than the body below it and specifically heavier than that above it.

This interpretation fits Galileo's Archimedean paradigm. As shown in the *Two New Sci-*

Whatever the intellectual origin of Galileo's interpretation of the small banks and their eventual breakdown may have been, he must have been aware that it could not be empirically tested because it was derived directly from his assumptions about the structure of matter. If the "magnetic virtue" may have convinced already sympathetic audiences, it definitely left the Aristotelians cold. It is likely that Galileo expected a negative response from his adversaries since he simultaneously deployed other strategies to bypass delle Colombe's experiment. One of them was to push for a holistic evaluation of the two competing theories so that the Aristotelians' very local emphasis on delle Colombe's experiment would sound like the result of simpleminded (and unethical) pickiness and stubbornness:

> My adversaries are not silenced by this, but say that what I have said thus far matters little to them. They are content that in one particular case, in material and shapes suitable to them (that is, a chip and a ball of ebony), they have shown that the latter placed in water goes down, and the former remains afloat. The material being the same, and the two bodies differing in nothing but shape, they believe they have completely demonstrated and made palpable all they needed to, and have finally carried out their intent.[141]

In opposition to the Aristotelians' attitude, he emphasized the generality and wide range of different empirical evidence which supported his views.[142] He also tried to hide his inability to explain the anomaly by focusing exclusively on those aspects that could be accounted for by his theory. When that did not work, he assumed what could be called a "positivistic" stand about the phenomenon of the small banks by claiming that their unexplained cause was not a problem. All that mattered was the fact that—for some reason—small banks happened to be produced. He seemed to intimate that seeking their cause would be a pointless search for final causes. That Galileo expressed this attitude only in notes to himself or in letters to supporters indicates that his demarcation between causes that were worth pursuing and those that were beyond the scope of his investigation was a statement of power—a power Galileo was not sure he had.

However, the most conspicuous sign of deadlock in the debate on buoyancy was shown by the strategic interpretation of the agreements on the experimental setting.

---

ences, Galileo thought of bodies falling in air as falling in a fluid specifically lighter than water. The letter by Antonio de' Medici dealt with a sphere suspended between two liquids of very similar density. Similar situations were studied by Galileo in the earlier drafts of the *Discourse* (*GO,* vol. 4, p. 37).

141. *GO,* vol. 4, p. 94. The translation is from Drake, *Cause, Experiment, and Science,* p. 88.
142. *GO,* vol. 4, p. 91.

A Matter of Principles

As we have seen, the debate on the terms of the agreement had charac-terized the dispute since its beginning. The very legitimacy of delle Colombe's experiment stemmed directly from the lack of specificity in the statement Galileo gave his adversaries after the first meeting at Salviati's.[143] In later statements, Galileo tried to dismiss delle Colombe's experiment by insisting that buoyancy and emergibility were the same thing. The "wetness clause" which he later tried to push on delle Colombe reflected this strategy.[144]

In the *Discourse,* Galileo kept upholding his interpretation of the agreements while stressing the remarkable importance words had as-sumed in that dispute.[145] His forensic skills reached a peak in discussing the small banks.[146] The Aristotelians saw the small banks not as confirm-ing Galileo's views but as proving that—as they had claimed—the body's shape could be a *secundum quid* cause of buoyancy. As shown by the thin piece of ebony, the body did not sink; it did not overcome the resistance of the medium and did not *divide* it. Galileo tried to undermine the favorable evidence the small banks had for the Aristotelians' thesis by claiming that, in this case, one should not use the verb "to divide." Assuming his view of the structure of matter as granted, he claimed that water was not a con-tinuous body but rather a contiguous one and, therefore, there was noth-ing there to divide. Instead, the verb that best described the process under scrutiny was "to move" (the particles of water).[147]

Then, Galileo said, if one looked carefully, the thin piece of ebony was actually below the water surface, therefore it had "already penetrated and won the continuation [*continuazione*] of the water" (fig. 15).[148] Notice that here Galileo did not talk about water *surface* but *continuazione,* that is, the *geometrical line* representing the water surface before it was touched by the ebony. Galileo would not allow himself to say that the body had actu-ally divided the water surface but—by claiming the verb "to divide" inapplicable here because of the contiguous structure of water—he tried to show that according to geometry (that is, his own language) he was right. To him, the body had sunk (without dividing anything, for there was nothing to divide) below the water's *continuazione.* Consequently,

143. Ibid., p. 34.

144. Ibid., p. 35.

145. Ibid., p. 94.

146. Ibid., pp. 35, 44, 91, 94–95, 99, 101.

147. Ibid., p. 106: "The bodies that are placed in water move only and do not divide." However, he does not always use *muovere,* but frequently uses *dividere*—although he meant "moving apart" (pp. 91–93).

148. Ibid., p. 98.

delle Colombe's experiment should not be seen as an anomaly to Galileo's theory of buoyancy. The splinter of ebony was *in* water and not *on* water.

Although Galileo's exegesis of the agreements may seem quite one-sided here, the Aristotelians replied accordingly. To di Grazia "to create a cavity" (*far cavità*) did not mean "to cut the surface" or "to penetrate."[149] Similarly, the Anonymous Academician stated that "Here it is necessary to notice that floating does not exclude entering to some extent into the water. It only excludes sinking to the bottom. The dispute hinges upon these two contrary readings [*punti*]."[150]

Both parties were concerned with preserving the coherence of their claims within their own systems. Also, if Galileo tried to efface delle Colombe's puzzle, the Aristotelians avoided confronting many of the positive features of Galileo's theory of buoyancy as well as his proposed Archimedean interpretation of delle Colombe's experiment. That Galileo's theory could predict measurable phenomena such as how much a body would sink in a fluid or how much the level of that fluid would rise in a given container did not seem to impress his adversaries at all. Actually, as we will see in a moment, they dismissed the philosophical validity of his mathematical method altogether.

Galileo was in an interesting deadlock, one that resembles the predicament often experienced by mathematicians and Copernicans during the scientific revolution. By assuming a presentist point of view, we would say that Galileo (or Copernicus) were "basically right" in the sense that their claims are genealogically connected to those we hold today. However, at the stage of articulation at which they were published, Copernicus's and Galileo's theories contained anomalies and unanswered questions which problematized their acceptance. Delle Colombe's experiment could be construed as a refutation of Galileo's views even after Galileo had tried—with mild success—to introduce auxiliary hypotheses (like the magnetic virtue of air) to bypass that anomaly. I would say that the danger of early mortality (a common one among new and yet unarticulated paradigms) cannot be countered by dialoguing with adversaries but rather through a range of tactics aimed at gaining time so as to allow for the further articulation of one's claims.[151]

In fact, it is not at all evident that Galileo tried to dialogue with the Aristotelians—and certainly not on their own grounds. Rather, he tried to raise the stakes of the discussion by attaching all sorts of philosophical, methodological, and cosmological issues to his initial treatment of buoy-

149. Ibid., p. 405.

150. Ibid., p. 162.

151. While several philosophers have discussed the need of devices aimed at preventing the early refutation of a theory, paradigm, or research program, to my knowledge this issue has been best articulated by Paul Feyerabend in *Against Method* (London: Verso, 1975), esp. pp. 145–61.

ancy. In doing so, he did not quite hope to convince his adversaries but to present and establish his own alternative philosophical package.

The Aristotelians adopted similar tactics. By linking delle Colombe's experiment as tightly as possible to the Aristotelian worldview, they tried to confront Galileo's package with their own. Whenever possible, they also tried to dismiss Galileo's worldview either by claiming that it was not a coherent system at all or by saying that it was an illegitimate one. In particular, their tactics took the form of dismissals of the cognitive legitimacy of the mathematical method, criticisms of Galileo's definitions, and accusations of *petitio principii* and ad hoc-ness.

For instance, the Anonymous Academician responded to Galileo's attack on Aristotle's element-based explanation of movement:

> Although there are anomalies to the peripatetic doctrine, *its foundations are much sturdier and more reasonable than those of Galileo's views.* By relying on a great range of objections to Aristotle, on various experiences, and on new demonstrations, [Galileo's view] is able to pass at first as all rich and elegant, but, if we weigh and consider it carefully, we see that the objections melt and his experiences either vacillate or are recognized as *particular effects rather than actual causes of phenomena, showing that mathematical proofs and propositions cannot grasp the true cause of natural phenomena.*[152]

Di Grazia was even more categorical than the Anonymous Academician about the gap in cognitive status between philosophy and mathematics:

> *Before considering the demonstrations of Signor Galileo* we thought it necessary to show how far from the truth are those who *want to demonstrate natural phenomena through mathematical reasoning.* . . . In fact, I claim that all sciences and arts have their own principles and methods through which they study the specific accidents of their specific subject matter. Therefore, *it is not proper to use the principles of a science to demonstrate the accidents studied by another science.* Consequently, *those who want to demonstrate natural accidents through mathematical methods are delirious,* because these two sciences are most different. In fact, the natural philosopher [*scientifico naturale*] studies natural phenomena whose essence entails movement, while, instead, *the subject matter of mathematics does not comprehend movement.*[153]

A few pages later, di Grazia applied this methodological distinction and dismissed a demonstration of Galileo's precisely because "he wants to demonstrate natural things with mathematical reasons."[154] Later, he

152. *GO,* vol. 4, p. 165 (emphasis mine).
153. Ibid., p. 385 (emphasis mine).
154. Ibid., pp. 389, 423.

moved from an attack on the disciplinary trespassing implied by Galileo's theory of buoyancy to a sharp questioning of Galileo's qualifications as a philosopher: "I would like Signor Galileo to adopt some philosophical propriety, because, *although he adorns himself with that title,* he does not behave accordingly."[155]

Delle Colombe, too, stressed the cognitive gap between mathematics and philosophy. He had maintained that position also in the "Contro il moto della terra."[156] In his *Discorso apologetico,* he pressed the point again by saying that, in choosing between Archimedes and Aristotle, one should not have any doubt. The choice was not dictated by the authority of the Aristotelian corpus but by the cognitive superiority of philosophy over mathematics.[157] Coresio was not as direct as his three other colleagues, but he also dismissed a geometrical demonstration by Galileo because it was based on non-Aristotelian *physical* assumptions.[158]

The Aristotelians found it unacceptable that a mathematician could investigate physical nature, and saw such an investigation as a violation of the received disciplinary hierarchy. Therefore, they claimed that Galileo's assumptions and physical principles were either wrong or nonexistent. If they perceived them as wrong, they usually did so by indicating that they did not fit the conceptual categories of Aristotle.[159] Sometimes—as in the case of di Grazia[160]—they tried to see if Galileo had borrowed those principles from other philosophers such as Plato, probably because they perceived that borrowing from another legitimate *philosopher* was a lesser violation of disciplinary boundaries than a mathematician's developing his own physical principles. The concern for the maintenance of disciplinary boundaries was also at the base of the Aristotelians' inability (or unwillingness) to entertain the possibility of a professional role such as that of the mathematician-philosopher proposed by Galileo. The Anonymous Academician was unable to see Galileo as a unified "professional self": "Because the Author [Galileo] sometimes presents himself as a mathe-

155. Ibid., p. 391 (emphasis mine).

156. The first few pages of delle Colombe's work are a conglomerate of attacks on the cognitive status of mathematics. It is there that he compares epicycles to "girelle" (wooden tops). When he finally calms down, he claims: "And who does not know that it is more necessary to be a philosopher than a mathematician and to know more about the first than the second science to be able to judge correctly whether these interpretations [*teoriche*] and mathematical demonstrations can be appropriately applied to material entities, to place, and to motion. The judgment over all these three issues belongs to natural philosophy and not to mathematics, which abstracts from material qualities" (Ludovico delle Colombe, "Contro il moto della terra" [Florence 1611]; reprinted in *GO,* vol. 10, pp. 251–90. This quote is on p. 255.).

157. *GO,* vol. 4, p. 352.

158. Ibid., p. 233.

159. Ibid., pp. 217, 388.

160. Ibid., p. 386.

matician and some other times as a philosopher, he who finds himself alone in front of him has to be very careful in challenging him for he may find himself confronted not by one but two strong champions."[161]

More often, Galileo's physical principles were not even perceived or acknowledged as such by his adversaries. Also, many of his definitions, like those of specific and absolute weight (which did not fit the Aristotelians' taxonomy of categories), were routinely criticized by them as flawed.[162] This and the frequent accusations of *petitio principii* were the result of their professional inability to perceive (or to admit) Galileo's principles at all. Being unable to perceive the tautologies of their own discourse, the Aristotelians claimed that Galileo kept assuming what he was supposed to demonstrate.[163] In certain cases—like the geometrization of the small banks—their position was defensible even on Galileo's grounds, but most of the time their accusations of *petitio principii* or of ad hoc-ness were plain statements of Galileo's methodological and socioprofessional "otherness."[164]

Patronage and Lack of Closure

As one might expect, the dispute on buoyancy did not end with an explicit verdict. As we will see in the next chapter, Galileo and the philosophers adopted different tactics to convince Cosimo to dismiss the opposition's claims, but neither party obtained the grand duke's explicit endorsement. The dispute did not reach a closure; it simply died out.

Recently, sociologists and historians of modern science have focused on the intricate and quite fascinating processes through which disputes end, official histories are written up, results are canonized and included in textbooks, instruments are "blackboxed," and the winners are rewarded with professional recognition and influence.[165] In the patronage system,

---

161. Ibid., p. 171.

162. Ibid., pp. 187, 220, 354, 386.

163. Ibid., pp. 163, 233, 398.

164. As we will see in the next chapter when we discuss issues of bilingualism, Galileo had a similar and yet not identical attitude about the Aristotelians' categories. Although he usually dismissed them, he did not do so simply by saying that they were wrong because they did not fit his framework (as the Aristotelians did with his categories), but rather by making a gesture toward understanding them before refuting them.

165. See, for instance: Augustine Brannigan, *The Social Basis of Scientific Discoveries* (Cambridge: Cambridge University Press, 1981); H. Tristam Engelhardt, Jr., and Arthur L. Caplan, eds., *Scientific Controversies* (Cambridge: Cambridge University Press, 1987); Harry M. Collins, ed., *Knowledge and Controversy: Studies in Modern Natural Science,* a special issue of *Social Studies of Science* 11 (1981); idem, *Changing Order* (London: Sage, 1985); Simon Schaffer, "Scientific Discoveries and the End of Natural Philosophy," *Social Studies of Science*

CHAPTER THREE

however, we do not generally find the conditions for closure. Concerns for their status tended to prevent patrons from passing judgments while, in general, mathematical practitioners lacked the status to legitimize their own claims, especially when the claims implied trespassing in the jurisdiction of more established disciplines. Moreover, the dispute on buoyancy was not a debate within a scientific community but an argument between two parties reflecting two very different socioprofessional identities. Finally, the patronage system was not a social system whose survival depended on reaching closure. The historical actors we have encountered were not as *interdependent* as, say, the members of a modern scientific community. The only sense in which they were "connected" was by virtue of being dependent on a common patron.

When closure did take place—as in the debate over the reliability of Galileo's telescope and the reality of his astronomical discoveries—it was also because of the utter spectacularity of Galileo's achievements (as perceived through court culture) and the unusual patronage networks he had developed with and through the Medici. The telescope happened to be perceivable as a court marvel. Galileo could reach important European courts through the network of Medici diplomacy and patronage, have powerful princes and cardinals see through the telescope, and could represent his discoveries as dynastic emblems of the Medici. It was also this rare combination of factors, the unusual fit between Galileo's work and the taste and culture of his patrons, that allowed him to mobilize his patrons to a remarkable extent and achieve closure on at least some of his claims.

While setting issues of patronage and closure momentarily aside to return to them later, I would like to try to connect some of the considerations made in this chapter to the well-known philosophical problem of incommensurability. We have seen that Galileo and the Aristotelians were more concerned with articulating and defending the coherence of their positions than with engaging in dialogue with the other party. Also, we have seen how incompatible their claims and methodologies were. Not only could they not agree on what "in," "on," "to float," or "water surface" meant, but their claims about the cause of buoyancy (and about "cause" itself) were embedded in incompatible cosmologies.

To pursue the possible connection between the dispute on buoyancy and Kuhn and Feyerabend's notions of incommensurability, I will now turn to a more anthropological point of view and investigate the extent to which the socioprofessional identities contributed to the communication

16 (1986): 387–420; Martin Rudwick, *The Great Devonian Controversy* (Chicago: University of Chicago Press, 1985), esp. pp. 401–56; Peter Galison, *How Experiments End* (Chicago: University of Chicago Press, 1987); and Bruno Latour, *Science in Action* (Cambridge, Mass.: Harvard University Press, 1987), pp. 63–100.

deadlock analyzed above. Finally, we will see if by integrating the anthropological and conceptual analyses of the dispute we may come up with useful insights about the genealogy of incommensurability between scientific paradigms as well as about the pattern of socioprofessional speciation that characterized the scientific revolution.

FOUR

The Anthropology of
Incommensurability

Incommensurability and Sterility

SINCE IT ENTERED THE DISCOURSE of the history and philosophy of science with Feyerabend's "Explanation, Reduction, and
Empiricism" and Kuhn's *The Structure of Scientific Revolutions,* the
notion of incommensurability has figured prominently in the debate on the process of theory choice.[1] According to Kuhn, two scientific
paradigms competing for the explanation of a set of natural phenomena
may not share a global linguistic common denominator. As a result, the
very possibility of scientific communication and dialogue becomes problematic and the process of theory choice can no longer be reduced to the
simple picture presented, for example, by the logical empiricists. What I
want to argue here is that incommensurability is more than just an unfortunate problem of linguistic communication; that, quite to the contrary, it
plays an important role in the process of scientific change.

The type of communication breakdown we have encountered in the
dispute on buoyancy was not uncommon in the scientific revolution. In
canonical texts such as Descartes' *Le monde,* Galileo's various writings,

1. Paul K. Feyerabend, "Explanation, Reduction, and Empiricism," *Minnesota Studies in
the Philosophy of Science* 3 (1962): 28–97 (reprinted in idem, *Philosophical Papers* [Cambridge:
Cambridge University Press, 1981], pp. 44–96); and Thomas S. Kuhn, *The Structure of Scientific Revolutions* (Chicago: University of Chicago Press, 1962). Some of Feyerabend's later
views on incommensurability are in his "Consolations for the Specialists," in Imre Lakatos
and Alan Musgrave, eds., *Criticism and the Growth of Knowledge* (Cambridge: Cambridge
University Press, 1970), esp. pp. 219–29; idem, *Against Method* (London: Verso, 1975),
pp. 223–85; idem, *Science in a Free Society* (London: Verso, 1978), pp. 65–70; and idem,
Farewell to Reason (London: Verso, 1987), pp. 265–72. Although my analysis is indebted to
several of Feyerabend's specific ideas and insights, I refer more consistently to Kuhn's work
because his notion of paradigm can be related more easily to the framework I have sketched
out in the previous chapters. References to Kuhn's other works on the subject will be given
in the rest of the chapter.

211

Bacon's *Novum organum,* and Locke's *Essay,* we find the new philosophers claiming not to understand some of the Aristotelians' fundamental concepts and refusing to subject their own views to judgments based on the parameters of the older tradition. In *Le monde,* Descartes quoted Aristotle in Latin, claiming to do so because—being unable to understand the sense of Aristotle's definition of motion—he was unable to translate it into French.[2] Galileo's attacks on the Aristotelians for an alleged unwillingness to understand his views were all too common in his printed works and, in private, he also invited his supporters not to waste time in the hopeless attempt to dialogue with the philosophers.[3] The critiques by Bacon and Locke of the meaninglessness of the terms used by the Aristotelians are well-known aspects of their work.[4]

Some historians or philosophers of science would probably say that these frequent statements about the impossibility of communicating with competitors represented a mere rhetorical strategy. The new philosophers did not *want* to talk to the Aristotelians. Dialogue would have been possible, if only the two parties had been seriously committed to develop it. Strong relativists, on the other hand, would take the question of whether those statements of incommunicability were real or rhetorical to be beside the point.[5] All that matters is that an impossibility of communication was claimed by the members of a group. Whether or not they were right in their claims is not something that can be legitimately judged from any point of view external to that culture or group.

The analysis of the phenomenon of incommensurability does not

2. "To render it [the nature of motion] in some way intelligible, they have still not been able to explain it more clearly than in these terms: *motus est actus entis in potentia, prout in potentia est,* which terms are for me so obscure that I am constrained to leave them here in their language, because I cannot interpret them" (René Descartes, *Le monde, ou Traité de la lumière,* ed. and trans. Michael S. Mahoney [New York: Abaris Books, 1979], p. 63; the Latin quote is from Aristotle, *Physics III,* 1, 201a).

3. *GO,* vol. 10, no. 499, pp. 502–3; vol. 11, p. 47; vol. 5, p. 231.

4. For Bacon on the meaninglessness of Aristotelian definitions, see Francis Bacon, *Novum organum* (Indianapolis: Bobbs-Merril Company, 1960), Aphorisms, book one, nos. 60 and 63. In the *Essay,* Locke focuses on the unintelligibility of the same definition of Aristotle's *Physics* that had previously attracted Descartes' irony: "What more exquisite jargon could the wit of man invent, than this definition: *The act of a being in power, as far forth as in power* which would puzzle any rational man, to whom it was not already known by its famous absurdity, to guess what word it could ever be supposed to be the explication of" (John Locke, *An Essay Concerning Human Understanding,* ed. A. Fraser [New York: Dover, 1959], 2: 34–35). In the *Novum organum,* Aphorisms, book one, nos. 33 and 35, Bacon makes clear that his work cannot be judged according to the parameters of the old school of philosophy. His work is outside the jurisdiction of the Scholastics, whose authority he takes to be "on trial."

5. I am aware of the existence of a wide range of philosophical positions that have developed more mediated forms of relativism. The position sketched here could be one similar to Bloor's "strong program" for the sociology of science.

need to be caught between these opposite positions. Without following either the rationalists in ruling out the very existence of incommensurability, or the relativists in taking it as a datum, we may develop a third way and analyze the *emergence* of incommensurability diachronically in relation to the socioprofessional identity and relative power and status of those involved in the nondialogue.

The constructive role of incommensurability in the process of scientific change can be glimpsed through what I would call a "Darwinian metaphor" derived from an analogy between Kuhn's concept of paradigm and Darwin's notion of species.[6] Both Kuhn's paradigm and Darwin's species refer to *populations* of individuals who interbreed either sexually (in Darwin's case) or intellectually (in Kuhn's case).[7] Consequently, the barrier of sterility among species observed by Darwin could be metaphorically compared to the incommensurability Kuhn has perceived among competing paradigms. Like sterility, which acts as an antiswamping device preventing the characters of the new species from being absorbed back into the old one, incommensurability brings about intellectual sterility, that is, the impossibility of breeding intellectually. While admitting that incommensurability results in a communication breakdown, this

---

6. Kuhn himself hinted at the evolutionary dimensions of his model in *Structure*. An extensive and quite systematic attempt to present conceptual change as an evolutionary process has been presented by Stephen Toulmin in his *Human Understanding* (Princeton: Princeton University Press, 1972). A more recent (and more challenging) empirically based evolutionary view of science is provided by David Hull, *Science as a Process* (Chicago: University of Chicago Press, 1988). The book's main argument is summarized in idem, "A Mechanism and Its Metaphysics: An Evolutionary Account of the Social and Conceptual Development of Science," *Biology and Philosophy* 3 (1988): 123–55. See also Gerard Lemaine, "Social Differentiation and Social Originality," *European Journal of Social Psychology* 4 (1974): 17–52. A species-based view of medieval and Renaissance Aristotelianism has been put forward by Edward Grant, "Ways to Interpret the Terms 'Aristotelian' and 'Aristotelianism' in Medieval and Renaissance Natural Philosophy," *History of Science* 25 (1987): 336–58. A quite different evolutionary view of conceptual change is provided by "evolutionary epistemology," an approach (initially proposed by Donald T. Campbell) based on a Popperian interpretation of Darwin. "Evolutionary epistemology" draws a close analogy between Darwinian natural selection and Popperian falsification and concludes that natural selection is a "rational" process of "error elimination." As shown by the remainder of this chapter, I do not share this Popperian reading of Darwin and, in particular, do not see "natural selection" as a "rational" process, nor do I use categories such as progress or "directed evolution." On evolutionary epistemology, see Donald T. Campbell, "Evolutionary Epistemology," in Paul A. Schilpp, ed., *The Philosophy of Karl Popper* (La Salle: Open Court, 1974), 1: 413–63; and Kai Hahlweg and C. A. Hooker, eds., *Issues in Evolutionary Epistemology* (Albany: State University of New York Press, 1989).

7. The analogy between species and paradigm can be taken a bit further. In fact, Darwin's notion of species as an interbreeding population differs from previous views of species as defined by a set of morphological attributes. This reference is reminiscent of the one between Kuhn's notion of paradigm, as referred to a community of scientists, and older views of scientific theories as uninterpreted logical systems, as proposed by the logical empiricists.

CHAPTER FOUR

Darwinian metaphor suggests that it may have a productive role in the conceptual speciation of a new paradigm. In a sense, this metaphor presents incommensurability more as a "bet" than a plain "cost."

The metaphor also suggests that the interaction between scientific paradigms is a process mediated by something comparable to natural selection, something we may provisionally call the reward system of science. To survive, competing paradigms do not need to engage in a fully constructive dialogue during a process of theory choice. Similarly, a theory does not need to be falsified or a research program superseded by a new one in order to be dropped. Like species that die off not necessarily because they cannot compete with others but because they do not fit their environment any longer, paradigms can come to an end, not because they are refuted but because their worth is no longer valued within their reward system. Just as species are not necessarily confined to a limited area of the environment that would force them to compete directly with other species but can migrate (or find themselves) in a fairly safe ecological niche, scientific paradigms can also develop relatively undisturbed by competition if they manage to get into an isolated area of the reward system.[8] In short, the very idea of theory *choice* is problematized by this interpretation of incommensurability.

Consequently, while incommensurability may appear as a problem to those who view it synchronically, that is, as a *result* of the linguistic structures of already existing theories, its diachronical analysis suggests important clues about the *process* through which paradigms and socioprofessional identities develop out of previous ones.

Socioprofessional Identities and Communication Breakdowns

Although the Darwinian metaphor provides a suggestive starting point by providing heuristic hints about the possible role of incommensurability in the process of scientific change, a closer analogy between biological and cognitive speciation would not match the available evidence. Whereas in a process of biological change the emergence of sterility is symmetrical, in the case of scientific change one can encounter sharp asymmetries in claims of incommunicability.[9] Sometimes one group claims to be able to

8. The analogy between paradigms and species suggests that groups are the objects being discussed here. However, that is not the case. As indicated by this volume, the category I find most appropriate to discuss scientific change is "socioprofessional identity." If there is an apparent discontinuity between the terminology adopted here and that in other chapters, its roots are pragmatic and conventional rather than methodological. Because I prefer to fit this case study into the more common framework of received studies of incommensurability as something emerging from activities of groups or "tribes," I have chosen to use terms such as "groups" or "paradigms."

9. However, this is not Kuhn's opinion. In his case, incommensurability is a symmetrical phenomenon.

understand the other, but denies that the reverse is true. As we have seen, Galileo claimed that his Aristotelian interlocutors did not understand his treatment of buoyancy because they were mathematically illiterate. In contrast, he boasted of a perfect grasp of Aristotelian philosophy.[10] Another important difference between biological and scientific change is that the possible emergence of incommensurability during the process of scientific change does not seem to be determined by the paradigm's "genotype" but depends on the context in which "scientific speciation" takes place. In the nondialogue between Galileo and the Aristotelians on buoyancy, the inability to communicate was moralistically blamed by Galileo on the interlocutors' unwillingness to engage in a constructive dialogue. In other cases, we find that one side states authoritarianly that it did not need or want to talk to the other. Claims of ability or will to communicate seem to reflect one's conscious or unconscious strategies and not just the linguistic dimensions of one's paradigm.

Additionally, the Darwinian metaphor does not allow us to differentiate incommensurability from incommunicability. As we will see, these two phenomena are connected, but their relationship is more complex and harder to disentangle than a Darwinian metaphor would suggest. Historical cases of scientific change indicate that the breakdown of communication does not need to be directly caused by the different linguistic structures of the competing paradigms. Rather, it is often associated with instances of *trespassing professional or disciplinary boundaries and violating socioprofessional hierarchies.* The breakdown of communication between Galileo and the Aristotelian philosophers during the dispute on buoyancy was also precipitated by the subordination of mathematics to philosophy required by the hierarchy of disciplines that provided the framework for the debate. As expressed by di Grazia, Galileo's arguments could be dismissed a priori on grounds that Galileo, a mathematician, could not put forward interpretations about phenomena that fell in the philosophers' domain. On the other hand, we find instances in which communication was maintained across radically different positions when the practitioners shared comparable socioprofessional identities. Kepler (Copernican), Magini (Ptolemaic), and Tycho (Tychonic)—all technical astronomers— were able to sustain a long dialogue although their work reflected radically different cosmologies.[11]

A similar pattern of communication despite theoretical divergence can be found among the Aristotelians. Although recent studies have

10. *GO*, vol. 4, pp. 31–32, 50, 124–25.

11. See the correspondence between Magini, Tycho, and Kepler in Antonio Favaro, ed., *Carteggio inedito di Ticone Brahe, Giovanni Keplero e di altri astronomi e matematici dei secoli XVI e XVII con Giovanni Antonio Magini* (Bologna: Zanichelli, 1886). See also Robert S. Westman, "The Melanchthon Circle, Rheticus, and the Wittenberg Interpretation of the Copernican Theory," *Isis* 66 (1975): 165–93.

shown that Aristotelianism, far from constituting a homogeneous philosophy, included a varied array of quite independent trends, it is evident that when they came to oppose Galileo and the Copernicans, the Italian Aristotelians did not get caught up in problems of conceptual incompatibility that may have existed among them. Beyond the many different and often contrasting views, they tended to share a unified socioprofessional identity: usually, Aristotelians were university professors of philosophy.[12]

On the other hand, we encounter cases of irreducible methodological differences within one discipline that can be traced back to the radically different social status and drive to upward mobility of the upholders of those two methodological styles. Elsewhere, I have indicated that the lack of dialogue between the sixteenth-century mathematicians of the Urbino and northern Italian schools of mechanics is an example of this phenomenon.[13] In that case, the difference in social identity proved more influential than a common disciplinary background in determining the possibility of dialogue.

Further evidence for the importance of socioprofessional identities in regulating communication between scientific practitioners comes from an analysis of the rhetorical strategies of nondialogue adopted by the opposing parties in cases of cross-disciplinary disputes. The epideiectic rhetoric utilized by Galileo in his *Dialogue on the Two Chief World Systems* is a good example of one such strategy.[14] Despite his declared purpose of convinc-

12. Grant, "Ways to Interpret the Terms 'Aristotelian' and 'Aristotelianism' "; Charles B. Schmitt, *Aristotle and the Renaissance* (Cambridge, Mass.: Harvard University Press, 1983), esp. pp. 10–33; idem, *The Aristotelian Tradition and Renaissance Universities* (London: Variorum, 1984); idem, *Studies in Renaissance Philosophy and Science* (London: Variorum, 1981); and *Les études philosophiques* 3 (1986), special issue on "L'Aristotélisme au XVIᵉ siecle."

13. Mario Biagioli, "The Social Status of Italian Mathematicians, 1450–1600," *History of Science* 27 (1989): 56–67.

14. This is the case with so-called epideiectic rhetoric, also known as ceremonial rhetoric. It is a kind of rhetoric that presupposes a basic agreement between the values of the speaker and those of his/her audience. Examples of this rhetoric could be the éloges of a late academician. The rhetoretician narrates the late academician's life as an emblem of the academy's corporate values in such a way that, when the speech is received by a sympathetic audience, it helps to reinforce the corporate values of that group. See Brian Vickers, "Epideiectic Rhetoric in Galileo's *Dialogo*," *Annali dell'Istituto e Museo di Storia della Scienza di Firenze* 8 (1983): 69–101; and J. W. O'Malley, *Praise and Blame in Renaissance Rome* (Durham: Duke University Press, 1979). Although Dorinda Outram does not use epideiectic rhetoric as one of the interpretative categories of her "The Language of Natural Power: The Eloges of Georges Cuvier and the Public Language of Nineteenth-Century Science," *History of Science* 16 (1978): 153–78, her analysis is of relevance to this point. On page 159, she claims that through the éloges, Cuvier described "his audience to themselves." This is a description of the workings of epideiectic rhetoric. The epideiectic dimensions of the *Eloges* of the members of the Paris Academy of Sciences are also discussed in Charles B. Paul, *Science and Immortality* (Berkeley: University of California Press, 1980), pp. 1–12.

ing the Aristotelians of the truth of his views, Galileo did exactly the opposite. He assumed a sympathetic audience—one of courtiers and "freethinkers" rather than of Scholastics—and made fun of his competitors by representing them through an unrealistically simpleminded and dogmatic straw man (Simplicio) who was systematically ridiculed by the Galilean champions (Sagredo and Salviati).[15] Simplicio was made to represent the stereotypical pedant—a figure that may have attracted the courtiers' laughter but certainly not their sympathy. The *Dialogue* turned out to be a sort of insiders' joke at the expense of the Aristotelians. Its function was to make readers who were already sympathetic to Galileo (or to the culture reflected in his argumentative style) and who identified themselves with Sagredo or Salviati laugh with them at the Aristotelians. Despite its title, the *Dialogue* was not meant to be a dialogue.[16] It was not meant to convince the "other," but rather to confirm and preserve the identity of the "one."

Galileo was not alone in adopting rhetorical tactics aimed at creating a cohesive spirit among his potential supporters. In looking at the four texts produced in response to Galileo's *Discourse On Bodies in Water* one is puzzled by the philosophers' quantitatively overwhelming and amazingly repetitive refutations of Galileo's theses. These texts seem to represent more of a hysterical reaction to the "other" than a constructive assessment of the competitor's claims. Castelli—whom Galileo put in charge of responding to his critics—noticed quite perceptively that the function of those prolix texts was to reassure the supporters of the Aristotelians that Galileo had been attended to and that the invasion of their disciplinary domain and the violation of the proper disciplinary hierarchy had been controlled. As Castelli put it, the sight of so many printed characters must have quenched their anxieties.[17] As with Galileo's *Dialogue,* the responses of the League were not attempts to convince the adversary. Instead, they were forms of nondialogue whose function was to maintain the cohesion— this time by reducing the anxieties—of the group to which the writer belonged.

These considerations are not intended to suggest that claims of in-

---

15. After reading the *Dialogue,* Campanella wrote Galileo that "Simplicius is the joke of this philosophical comedy. He displays at once the stupidity of his sect, the way they speak, their incoherence, their obstinacy, and everything else" (*GO,* vol. 14, no. 2283, p. 366).

16. This was quite typical of sixteenth-century courtly dialogues, which (like Castiglione's *Book of the Courtier*) were both *about* and *for* the court. They were aimed at celebrating the identity and values of those who read (and were represented by) them (Nuccio Ordine et al., *Il dialogo filosofico nel '500 europeo* [Milan: Angeli, 1990], p. 20).

17. ". . . once piled up, these [propositions] satisfy the expectation of the plebs who, because they do not understand the sense of the text, calm down just by seeing the printed characters and by being able to say that [the challenge] has been attended to" (*GO,* vol. 4, p. 462 [emphasis mine]).

communicability, rhetorical strategies of nondialogue, and linguistic in-commensurability amount to the same thing. Similarly, I do not claim that difference in socioprofessional identity determines the possibility of com-munication or the emergence of incommensurability. Rather, such claims, strategies, and linguistic phenomena are related in that they all play a cru-cial role in the formation and preservation of a group's cohesion and socioprofessional identity. Although incommensurability is a very specific (and quite rare) phenomenon tied to the linguistic dimensions of compet-ing theories, the *development* of incommensurability depends also on the various processes through which socioprofessional identities are formed around theories, and on the way the formation of those identities allows, in turn, for the further articulation of the theories.

The work of some philosophers, cultural anthropologists, and so-ciologists of scientific knowledge has provided tools for linking the emergence of incommensurability to the process through which so-cioprofessional identities are shaped and maintained. Imre Lakatos ana-lyzed the responses of a group of mathematicians to anomalies to their paradigm that were discovered by mathematicians belonging to a compet-ing group.[18] He classified some of those responses as "monster-barring" strategies. Lakatos's remarkable anthropological insight suggests that anomalies or novelties can be perceived as expressions of the "other." Ex-panding on Lakatos, David Bloor has interpreted the pattern of mathe-maticians' responses to conceptual novelties in terms of Mary Douglas's "grid and group" model.[19] In doing so, Bloor has extended Lakatos's "conceptual other" into a "social other" and related a given community response to the "other" to the internal structure and external boundaries of that community. My interpretation of the emergence and role of in-commensurability in the process of scientific change evolves from these studies. But, differently from Bloor, I extend the analysis of a group's re-sponse to the "other" so as to cover also the phenomenon of incommen-surability.

Philosophers and Mathematicians

Although hierarchies among disciplines are still present—if not always explicitly acknowledged—in today's science, they are less rigid than those

18. Imre Lakatos, *Proofs and Refutations* (Cambridge: Cambridge University Press, 1976).

19. David Bloor, "Polyhedra and the Abominations of Leviticus: Cognitive Styles in Mathematics," *British Journal of the History of Science* 11 (1978): 245–272, reprinted in Mary Douglas, ed., *Essays in the Sociology of Perception* (London: Routledge and Kegan Paul, 1982), pp. 191–218; and idem, *Wittgenstein: A Social Theory of Knowledge* (New York: Columbia University Press, 1983), especially "Strangers and Anomalies," pp. 138–59. For Mary Douglas's early views on the relationship between the response to the "other" and the social taxonomy of the respondent group, see her *Purity and Danger* (London: Routledge, 1966). For the

found during the scientific revolution. More specifically, while today's hierarchies tend to reflect status differences *within* the scientific community, Galileo's example shows that no clear distinction could then be found between social and professional hierarchies or between scientific credibility and social status and honor. Before the institutionalization of science, beliefs in theories, paradigms, worldviews, or the adoption of specific scientific practices did not represent simply what we may now call a professional culture. As we have seen, what was often at stake in disputes was not just a philosopher's view of a specific phenomenon and that person's ranking within a professional community, but the entire social status and identity of the person.

Consequently, it is not surprising to see that mathematicians such as Rheticus, Copernicus, or Galileo did not perceive disputes about the new astronomy and natural philosophy merely as scientific controversies but rather as something in-between disciplinary crusades and legal suits about disciplinary boundaries, domains, and hierarchies.[20] In the *Narratio prima,* Rheticus wrote that the worth of Copernicus' hypothesis was to be determined "by geometers and philosophers (who are mathematically equipped). For the trial and decision of such controversies, a verdict must be reached in accordance with not plausible opinions but mathematical laws (the court in which this case is heard). The former manner of rule has been set aside, the latter adopted."[21]

Therefore, the legitimation of a new worldview required revolutions in the social hierarchy of disciplines and the emergence of new socioprofessional identities. The Copernican revolution is a good example of this process. As perceptively noticed by the Protestant theologian Osiander in his preface to *De revolutionibus,* Copernicus's work—by claiming that astronomers dealt with the true structure of the cosmos and that, consequently, they were philosophers—could (and eventually did) trigger such a *double* revolution:

> There have already been widespread reports about the novel hypotheses of this work, which declares that the earth moves whereas the sun is at rest in the center of the universe. Hence, certain scholars, I have no doubt, are deeply offended and believe that the liberal arts,

"grid-and-group" model, see idem, *Natural Symbols* (New York: Pantheon, 1970). Further reflections on the model are found in idem, *Cultural Bias* (London: Royal Anthropological Institute, 1978).

20. Galileo's forensic tropes are especially evident in his 1615 "Letter to the Grand Duchess," in Maurice Finocchiaro, *The Galileo Affair* (Berkeley: University of California Press, 1989), pp. 87–118. There he argues about the cognitive privileges, methods, and disciplinary boundaries of astronomy and theology. Galileo's statements of the form "mathematics is for mathematicians" are scattered throughout his writings.

21. Edward Rosen, ed., *Three Copernican Treatises* (New York: Dover, 1939), p. 139.

which were established long ago on a sound basis, should not be thrown into confusion.[22]

Galileo must have been very aware of these disciplinary implications of Copernican astronomy if in his *Dialogue* he had Simplicio comment on Salviati's claims by saying that "This way of philosophizing tends to subvert all natural philosophy, and to disorder and set in confusion heaven and earth and the whole universe."[23]

By indicating that scientific change and social change are closely related, the view of the emergence of incommensurability outlined above fits well with the scenario described by these quotes. Such a view also suggests that incommensurability may emerge most clearly and the breakdown in communication may be most severe when the legitimation of a new paradigm requires not only the acceptance of a radically new worldview but also a revolution in the received hierarchy of disciplines. Consequently, the breakdown of communication between Galileo and the Aristotelians on methodological and cosmological issues during the dispute on buoyancy should also be interpreted in the context of the disciplinary hierarchy characteristic of a scientific revolution at whose beginning mathematics was subordinated to philosophy.

To Aristotle and his followers, mathematical demonstrations were necessary demonstrations only when they were *not* applied to material entities. As di Grazia and delle Colombe reminded Galileo, the proper subject matter of mathematics was abstract entities.[24] The truth of a theorem could not be transferred from the domain of mathematics to that of physics, that is, from immaterial to material entities. Similarly, the mathematicians were expected to remain within the boundaries of statical and kinematical analyses of natural phenomena. In fact, mathematics (being an abstract, that is, nonphysical discipline) could not explain the causes of change and, more specifically, of motion. That required adequate *physical* principles which were not the jurisdiction of mathematics but of philosophy.[25] As we have seen, one of the points the Aristotelians criticized Galileo for was his inability to produce tenable physical principles on which to ground his claims and demonstrations. As his adversaries reminded him, mathematics—being alien to the "real" principles of the

22. Nicolaus Copernicus, *On the Revolutions*, in *Complete Works*, trans. Edward Rosen, ed. Jerzy Dobrzycki (Warsaw-Cracow: Polish Scientific Publishers, 1978), 2: xvi. On the disciplinary agenda of Osiander's preface, see Robert S. Westman, "The Astronomer's Role in the Sixteenth Century: A Preliminary Study," *History of Science* 18 (1980): 105–47.

23. Galileo Galilei, *Dialogue Concerning the Two Chief World Systems*, trans. Stillman Drake (Berkeley: University of California Press, 1967), p. 37.

24. *GO*, vol. 4, p. 385. Analogous statements can be read in delle Colombe's "Contro il moto della terra," reprinted in *GO*, vol. 3, p. 255.

25. *GO*, vol. 4, p. 423.

physical world—could only aspire to measure the quantities, that is, the *accidental* aspects of phenomena.[26] Moreover, the different cognitive status of philosophy and mathematics reflected also the different social status of philosophers, mathematicians, and technical astronomers in the Renaissance.[27]

According to the methodological boundaries entailed by this hierarchy, the philosophers developed qualitative cosmologies usually based on Aristotle's theory of the homocentric spheres. Technical astronomers, instead, were expected to produce quantitative predictions of planetary motions by means of various geometrical devices. The superior cognitive and social status of philosophy over technical astronomy was reflected in the philosophers' dismissal of the cognitive legitimacy of the mathematicians' method when this was applied to the explanation of physical processes. The mathematicians' geometrical constructions were perceived by the philosophers not as true representations of the cosmos but as mere computational devices or, worse, as tricks. In a statement that epitomizes the philosophers' "professional infantilizing" of the mathematicians, delle Colombe claimed that geometry was children's knowledge and that epicycles were the mathematicians' wooden tops.[28] This trope must have been a common one; even Galileo referred to it:

> Here I expect a terrible rebuff from some of the adversaries. I already
> seem to hear somebody shouting in my ears that it is one thing to

26. Ibid., pp. 389, 423. For the sixteenth-century Italian debates on the cognitive status of mathematics, see Paolo Galluzzi, "Il Platonismo del tardo Cinquecento e la filosofia di Galileo," in Paola Zambelli, ed., *Richerche sulla cultura dell'Italia moderna* (Bari: Laterza, 1973), pp. 39–79; Giovanni Crapulli, *Mathesis universalis* (Rome: Edizioni dell'Ateneo, 1969), pp. 33–62; Alistair C. Crombie, "Mathematics and Platonism in the Sixteenth-Century Italian Universities and in Jesuit Educational Policy," in Y. Maeyama and W. G. Saltzer, eds., *Prismata* (Wiesbaden: Steiner Verlag 1977); Peter Dear, "Jesuit Mathematical Science and the Reconstitution of Experience in the Early Seventeenth Century," *Studies in History and Philosophy of Science* 18 (1987): 133–75; G. C. Giacobbe, "Il *Commentarium de certitudine mathematicarum disciplinarum* di Alessandro Piccolomini," *Physis* 14 (1972): 162–93; idem, "Francesco Barozzi e la *Quaestio de certitudine mathematicarum*," *Physis* 14 (1972): 357–74; idem, "La riflessione metamatematica di Pietro Catena," *Physis* 15 (1973): 178–96; and idem, "Epigoni del Seicento della *Quaestio de certitudine mathematicarum*: Giuseppe Biancani," *Physis* 18 (1976): 5–40. On the debate on the cognitive status of mathematics in astronomy, see note 27 below. On the classification of disciplines, see James A. Weisheipl, "The Nature, Scope, and Classification of the Sciences," in David C. Lindberg, ed., *Science in the Middle Ages* (Chicago: University of Chicago Press, 1978), pp. 461–82; and Steven J. Livesey, "William of Ockam, the Subalternate Sciences and Aristotle's Theory of 'Metabasis,'" *British Journal for the History of Science* 19 (1985): 127–45. Also of interest is Peter Machamer, "Galileo and the Causes," in Robert E. Butts and Joseph C. Pitt, eds., *New Perspectives on Galileo* (Dordrecht: Reidel, 1978), pp. 161–80.

27. For the Italian case, see Biagioli, "Social Status of Italian Mathematicians, 1450–1600."

28. *GO*, vol. 3, pp. 253–54.

CHAPTER FOUR

treat things physically and another to treat them mathematically, and that the geometers should remain among their spinning tops without bothering with philosophical matters, whose truths are different from mathematical truths—as if truth could be more than one.[29]

However, the philosophers' disciplinary pride did not match their performance. Although only the philosophers could claim access to the true causes of planetary motions, that is not the same thing as actually finding the causes. As the fictitious Alimberto Mauri (probably a pseudonym of Galileo) wrote in response to delle Colombe's work on the new star of 1604:

Philosophers want uniformity [of motion] in the stars, and not imaginary or feigned uniformity, but true and real. . . . So they run to the astronomers for help (since the Philosophers cannot manage this matter for themselves), in order that they may bring forward the reasons for such appearances and thus maintain in men's minds as truths these ideas of Philosophers about uniform and regular skies. So the astronomers, as faithful friends of theirs, have thought day and night about epicycles, eccentrics, and equants. . . . But now behold how those instruments, formerly not in their possession, are by Philosophers harmfully *vilified out of contempt for the donors,* or else are *abused through ignorance in such matters.*[30]

However, even though the philosophers were unable to live up to their disciplinary duties to provide the mathematicians with physical principles upon which to develop physically true accounts of planetary motions, they could still rely on the received hierarchy of disciplines and blame their philosophical failures on the mathematicians. I would say, somewhat anachronistically, that as a result of the philosophers' disciplinary power the mathematicians were forced into some sort of *nominalist* methodological position on cosmological matters while the philosophers were entitled to take a *realist* stance on those issues.[31] This a priori dis-

29. *GO,* vol. 4, p. 49.

30. Alimberto Mauri, *Considerations of Alimberto Mauri on Some Places in the Discourse of Lodovico delle Colombe about the Star Which Appeared in 1604,* in Stillman Drake, trans., *Galileo Against the Philosophers* (Los Angeles: Zeitlin and Ver-Brugge, 1976), p. 102.

31. The debate on nominalism and realism in astronomy has often been structured around the critique of Pierre Duhem's overly schematic categories of "instrumentalism" and "realism" as presented in his *To Save the Phenomena* (Chicago: University of Chicago Press, 1969). G. E. R. Lloyd has criticized a number of Duhem's translations and interpretations of original texts in his "Saving the Appearances," *Classical Quarterly* 28 (1978): 202–22. Nicholas Jardine has presented the debate on instrumentalism and realism within the context of the astronomers' reaction against the skeptics' critique of the cognitive claims of astronomy in his "The Forging of Modern Realism: Clavius and Kepler Against the Scep-

missal of the mathematicians' claims to the physical reality of their mathematical hypotheses was also related to the breakdown of constructive dialogue between two disciplines: the philosophers *did not need to listen* to mathematicians. Given the hierarchical setting in which they operated, they were not obliged either to learn the mathematicians' language or to take their physical principles seriously.

Copernicus rejected the mathematical nominalism imposed upon astronomers by the philosophers and upheld the cognitive legitimacy of mathematical realism. By mathematical realism I do not mean to say that the Copernicans necessarily thought that spheres, eccentrics, and epicycles were physically real. Nor did Galileo think so.[32] Rather, they were realists in thinking that mathematics was the key to finding the real physical structure of the cosmos.[33] Although the details of the Copernican sys-

tics," *Studies in History and Philosophy of Science* 10 (1979): 141–73. Jardine subsequently expanded his analysis in *The Birth of History and Philosophy of Science* (Cambridge: Cambridge University Press, 1984), esp. pp. 225–57. See also idem, "The Significance of the Copernican Orbs," *Journal for History of Astronomy* 13 (1982): 168–94; and idem, "Epistemology of the Sciences," in Charles B. Schmitt and Quentin Skinner, eds., *The Cambridge History of Renaissance Philosophy* (Cambridge: Cambridge University Press, 1988), pp. 685–711. Robert S. Westman has approached the problem from a more sociological point of view in his "Astronomer's Role in the Sixteenth Century." See also idem, "Kepler's Theory of Hypothesis and the 'Realist Dilemma,'" *Studies in History and Philosophy of Science* 3 (1972): 233–64. Although my thesis moves from Duhem's taxonomy, it is not at all weakened by these critiques of Duhem's schematic (and sometimes philologically untenable) picture. This is because I do not take a "realist" view of Duhem's taxonomy. I do not intend to defend Duhem's distinction between realists and instrumentalists as actually representing two different *intellectual* traditions stretching from the Greeks to the Jesuits. Rather, I think that Duhem's observations can be seen as indicating a pattern of anthropologically understandable interactions (or struggles) between socioprofessional groups (or roles) with different statuses (both social and cognitive). These patterns of interactions are the focus of my study. Therefore, I do not consider Duhem's *To Save the Phenomena* as an "authority" that legitimizes my thesis, but rather as a pretext that triggered it.

32. "Lettera a Monsignor Pietro Dini," in *GO*, vol. 5, pp. 297–99, esp. p. 299. See Westman's very interesting contextualization of Copernicus's complex attitude about the physical reality of the spheres ("Astronomer's Role in the Sixteenth Century," 112–16). On the same subject and for a review of the debate on the topic, see Jardine, "Significance of the Copernican Orbs."

33. The "philosophical astronomers" were in a bind. To claim the physical reality of geometrical devices would have been self-defeating. If different configurations of devices could account for the same planetary motion, then it followed that those devices were not physically real (or at least that one could not decide which was the "real" combination). At the same time, by saying that those devices were not real, the "philosophical astronomers" would have made themselves vulnerable to questions from traditional philosophers regarding the causes of planetary motion (causes that, as "philosophers," they were supposed to uncover). This predicament is epitomized by Galileo's telling Dini in 1615 that epicycles, eccentrics, and spheres are evidently not real in a physical sense but that, at the same time, it is absolutely true that planets move around in epicycles (that is, *circular trajectories* that do not enclose the earth—as, after his telescopic discoveries, with Venus or the satellites of Jupiter) or in eccentrics (that is, in circles that are not centered on the earth, as in the case of Mars). In

tem may have not been fully satisfactory, the internal mathematical coherence of the system and its ability to account directly from its physical assumptions for a number of celestial phenomena turned it into a strong candidate for offering a true physical description of the cosmos. As a result of Copernicus's views, some mathematicians became self-appointed "new philosophers." Galileo was among them.[34] As he wrote in the first letter on sunspots,

> [Scheiner] continues to adhere to eccentrics, deferents, equants, epicycles, and the like as if they were real, actual, and distinct things. These, however, are merely assumed by *pure astronomers* in order to facilitate their calculations. However, they should not be retained by *astronomers-philosophers* who, going beyond the demand that they somehow save the appearances, seek to investigate the true constitution of the universe—the most important and most admirable problem that there is. For such constitution exists; it is unique, true, real, and could not possibly be otherwise.[35]

Not only did Copernicus and other "philosophical astronomers" embrace mathematical realism, they also began to dismiss the language and method of the philosophers by claiming that mathematics was the only language in which astronomical matters *should* be discussed and judged.[36] This form of mathematical elitism—epitomized by Copernicus's statement that "mathematics is for mathematicians"—signified an attempted inversion of the previously accepted rules of the game. Copernicus declared himself unwilling to listen to the philosophers' arguments in the same way the philosophers had previously dismissed the mathematicians'. Rheticus followed suit by claiming that astronomy should be judged in mathematical courts only, while Galileo stated that philosophers should not criticize his arguments unless they understood mathematics.

a sense, Galileo tries to argue that epicycles and eccentrics *do* and *do not* exist (and he argues that they exist not as components of complex planetary models by pointing to very simple, qualitative cases). This puzzling tension is the mirror image of Galileo's socioprofessional predicament. He wants to be a philosopher: that is, he wants to uncover causes, but cannot find them yet. At the same time, he cannot argue that the astronomers' devices are physically real because that would disqualify him as a "philosophical astronomer" (*GO*, vol. 5, p. 299). Eventually, he asks for God's help: ". . . as God does not lack the means to make the stars move in the immense celestial spaces, within well-defined paths, but without having them chained and forced" (ibid., as translated in Finocchiaro, *Galileo Affair*, pp. 61–62). The problem of the reality of geometrical devices in Galileo indicates that perhaps his apparent lack of interest in technical planetary theory is also a result of his desire or need not to confront the status of geometrical devices head on.

34. See Galileo's statement in the *Dialogue Concerning the Two Chief World Systems*, 341.

35. *GO*, vol. 5, p. 102; the translation is adapted from Stillman Drake, ed., *Discoveries and opinions of Galileo* (Garden City, N.J.: Doubleday, 1957), pp. 96–97 (emphasis mine).

36. In the dedication of *De revolutionibus* to Pope Paul III, Copernicus claimed that "astronomy is written for astronomers" (Copernicus, *On the Revolutions*, 2: 5).

Disciplinary hierarchies and strategies of emancipation did not characterize the interaction between Copernicans and philosophers only, but
they framed the dialogue between philosophers and practitioners of mixed
mathematics in general. As I have discussed elsewhere, the sixteenth-
century methodological debate on the *certitudo mathematicarum* did not
focus on astronomy but dealt with the cognitive status of mathematics in
general vis-à-vis that of philosophy.[37] Similarly, during the dispute on
buoyancy, Delle Colombe attacked Galileo's mathematization of buoyancy with methodological arguments analogous to those he had previously employed in his critique of the Copernican implications of the
discoveries of 1610.[38]

Both the Copernican hypothesis and Galileo's treatment of buoyancy
represented instances of the mathematicians' trespassing in the philosophers' domain and of the attempt to upset received disciplinary hierarchies. As such, the debates they started were characterized by comparable
types of noncommunicative behaviors, a priori dismissals of the other's
positions, disciplinary name-calling, and attempts to enforce or change
the received rules of the game rather than to engage in a constructive dialogue.

It is important to realize that these invasions and the communication
breakdowns that ensued were not just the result of power (or survival)
struggles among different parties or disciplines. The mathematicians did
not have the disciplinary status and power to attack successfully the philosophers on just any pretext. They needed very good resources to do so.
The Copernican astronomy and the Archimedean theory of buoyancy
were two of them. Copernicus's work was particularly important in this
regard because it allowed mathematicians such as Galileo to present themselves as philosophers. Different from the many (and not always coherent)
mathematical hypotheses that had populated Ptolemaic astronomy during its long career, Copernicus's theory provided a coherent and *profession-
ally unifying* worldview. In a sense, the chaotic state of traditional
planetary astronomy described by Copernicus in his preface to *De revolutionibus* mirrored the fragmentation of the professional identity of the
astronomers themselves. If, as Copernicus put it, early sixteenth-century
astronomy was a "monster"—a methodologically incoherent assemblage
of disjunct descriptions of the motions of specific planets—then to be an
astronomer meant to be somebody specializing in piecemeal astronomical
repair jobs. Copernicus's heliostatic hypothesis promised to change all
this by providing the mathematicians with a "dogma" around which *both*
a coherent astronomy *and* a stronger and more unified socioprofessional
identity could be developed. It could bring coherence to the cosmos

37. Biagioli, "Social Status of Italian Mathematicians," 52–54.
38. *GO*, vol. 3, pp. 254–55; vol. 4, p. 352.

CHAPTER FOUR

and professional cohesion among astronomers. Copernicans could think of themselves as philosophers and have a chance to be taken seriously. Ptolemaics could not.

These considerations may help us to restate the issue of whether Galileo's commitment to Copernicanism was a cause or an effect of his move to court. As the reader may have noticed, I have avoided the question as presented in these terms. Rather, I have argued that Galileo's Copernicanism and courtly aspirations went hand in hand. This claim can be now made more specific. Copernicus provided Galileo with the resources he needed to represent himself not as a mathematician but as a philosopher (and a nonpedantic one) while the court allowed him actually to obtain that title. In a sense, Copernicanism was the "natural" choice for someone such as Galileo who aspired to a higher socioprofessional status, while the court was the social space that could best legitimize such an unusual socioprofessional identity.[39] Obviously, I am not suggesting that Galileo decided to become a Copernican in order to move up in the social scale. As shown by Pierre Bourdieu, while certain cultural tastes appear as "natural" to the persons who hold them, such tastes usually happen to reflect the persons' actual or desired social status.[40] That Galileo's Copernicanism became more explicit as he became an increasingly established court philosopher reflects, I think, not only the increasing (but still not decisive) evidence he had in favor of heliocentrism but his new courtly and philosophical "habitus" as well. Moreover, Galileo's development of further evidence in favor of Copernicanism was also the result of his need to live up to his new socioprofessional identity.

Consequently, whether Galileo became a full-fledged Copernican as a result of going to court or whether he wanted to become the grand duke's philosopher in order to come out of the Copernican closet is, I think, a chicken-or-the-egg type of question. Instead of looking for one *cause,* we may investigate the *process* of mutual reinforcement between Galileo's new socioprofessional identity and his commitment to Copernicanism. A related and equally circular question is whether "society" or "nature" is more important in triggering scientific change. For Copernicanism to become a powerful resource for socioprofessional mobility, the Copernican hypothesis needed to provide an interpretation of the cosmos that was more satisfying than the one provided by Ptolemaic astronomy, at least to a given group of astronomers. By "satisfying" I mean that it was both a

39. The Tychonic system did not offer the same philosophical self-fashioning opportunities as that of Copernicus. Because Tycho's model was mathematical (its physical interpretation would have been quite nightmarish), its endorsement would not have provided a powerful resource to present oneself as a "philosopher."

40. Pierre Bourdieu, *Distinction* (Cambridge, Mass.: Harvard University Press, 1984), pp. 11–96.

social and epistemological resource: it seemed to provide a better representation of the cosmos *and* allowed for socioprofessional legitimation. I would argue that these two dimensions are not separable. It was only by being *both* a social and cognitive resource that it attracted a few astronomers and mobilized them to articulate it further. It was as a result of these further articulations (think of Kepler's and Galileo's contributions) that Copernicanism became an increasingly convincing option *for a larger audience*—something that it was not in 1543. Through becoming a hypothesis acceptable to people outside the handful of astronomers who first adopted it, Copernicanism helped those astronomers' establishment of their credibility and a new socioprofessional identity for themselves—that of the mathematics-oriented natural philosopher. Scientific and social change occurred together through an ongoing exchange of resources.

Although the Archimedean theory of buoyancy may not have had the same "emancipatory power" as Copernicus's astronomy, Galileo's *dynamical* reinterpretation of Archimedes' hydro*statics* also allowed him to move from the traditional domain of mathematics into that of philosophy. As his adversaries were quick to notice, Galileo's theory of buoyancy was a bit of a Trojan horse through which he was trying to invade their domain. If Copernicus offered the mathematicians a chance to displace the philosophers from the superlunary sphere, Galileo's theory of buoyancy could have started the mathematicians' invasion of the sublunary realm. Similar armies, weapons, and tactics were facing each other on different fields.

Philosophers *Inter Pares*

Although the title of philosopher was a crucial resource for Galileo, it did not automatically empower him to dismiss the claims of the traditional philosophers or to impose his worldview on them. What the Medici gave him was the right to argue with the philosophers on equal footing.[41] As a result, the philosophers could no longer assume that their a priori dismissal of the cognitive validity of the mathematical method would be taken seriously.

Given the multilevel deadlock in the dispute on buoyancy, it is not surprising that the only strategy left to each side was to attempt to gain enough power from the reward system so as to *dismiss* the adversary. Such a strategy took a range of expressions. In general, both Galileo and the Aristotelians tried to discredit each other's credibility. Galileo claimed that his adversaries were completely ignorant of mathematics and unable

---

41. Actually, Galileo presented his performance in philosophical debates as the "test" of his philosophical knowledge and abilities (*GO*, vol. 10, no. 307, p. 353).

to understand his arguments, while the philosophers questioned Galileo's competence in interpreting Aristotle.[42] However, the philosophers' claims were less convincing. They could not really claim that Galileo was philosophically ignorant, for he had shown—both in his refutation of Bonamico and in his unorthodox exegesis of *On the Heavens*—that he was quite comfortable with Aristotle.[43] In contrast, none of the philosophers took up Archimedes' or Galileo's geometry. Therefore, all the philosophers could claim was that Galileo was heretical in his interpretation of Aristotle. Di Grazia was by far the most aggressive in questioning Galileo's competence in Aristotelian philosophy, by bringing in the Greek text and arguing on philological grounds that Galileo's interpretations were incorrect.[44] However, precisely because of their extreme detail, di Grazia's attacks showed that Galileo's competence in Aristotelian philosophy could be argued but not dismissed.

The philosophers responded very mildly to Galileo's attack on their mathematical illiteracy. It seems they did not want to recognize it as an issue by answering it.[45] Confirming the philosophers' attempt to dismiss mathematics rather than acknowledge their incompetence in it, Coresio claimed that the mathematics of buoyancy was so simple that one did not need to be a mathematician to grasp (and therefore dismiss) Galileo's argument.[46] If the Aristotelians were not so sweeping in their attacks on Galileo's professional qualifications it was because they could not deny him the title of philosopher (as he was denying them that of mathematicians), for it had been given him by their common patron, Cosimo II. All they could do was to undermine him obliquely by presenting him as an "improper" philosopher.[47]

Besides attempting to undermine each other's professional qualifications, Galileo and the Aristotelians tried to end the deadlock by having Cosimo II give them enough power to dismiss the adversary. We can get a sense of Galileo's thoughts about capturing the support of Cosimo from

42. *GO*, vol. 4, pp. 50, 158, 467.

43. Ibid., pp. 31, 36, 42–43, 97–98, 124–25.

44. Ibid., pp. 420, 426.

45. Coresio, who did not defend his own or his colleagues' mathematical competence but rather that of Aristotle, claimed: ". . . it is known that, at that time, students of philosophy devoted much more [time] to the mathematical sciences than today, and nobody ever studied logic without having first worked at them. This was particularly true of Plato's students. Therefore, how can it be believed that his best student Aristotle could have entered [Plato's school] without knowledge of mathematics?" (ibid., p. 240).

46. "Besides, the proposition in our text was not so difficult that one would not have been able to understand or use it without a very precise understanding of mathematics" (ibid.).

47. Ibid., p. 391.

the preliminary sketches of the *Discourse,* where he presented himself as Cosimo's scientific paladin in need of his king's support:

> Having been chosen by Your Highness as your personal mathematician and philosopher, I should not tolerate that anybody's malignity, envy, or ignorance (and perhaps all three) may stupidly insult your prudence; for that would be to abuse your incomparable benignity. On the contrary, I shall always put down (and with very little trouble) their every impudence.[48]

However, Galileo suggested to the grand duke that even the strength of a paladin had its limits and reminded him that, having given him the title of philosopher, it was now time for the prince to stand by his knight:

> Most Serene Lord, I have taken the trouble (as your Lordship has seen) to keep alive my true proposition, and along with it many others that follow therefrom, preserving it from the voracity and falsehood overthrown and slain by me. I know not whether the adversaries will give me credit for the work thus accomplished, or whether they, *finding themselves under a strict oath obliged to sustain religiously every decree of Aristotle* (perhaps fearing that if disdained he might invoke to their destruction a great company of his most invincible heroes), have resolved to choke me off and exterminate me as a profaner of his sacred laws. In this they would imitate the inhabitants of the Isle of Pianto when, angered against Roland, in recompense for his having liberated so many innocent virgins from the horrible holocaust of the monster, they moved against him, *lamenting their strange religion and vainly fearing the wrath of Proteus,* terrified of submersion in the vast ocean. And indeed they would have succeeded had not he, impenetrable though naked to their arrows, behaved as does the bear toward small dogs that deafen him with vain and noisy barking. *Now I, who am no Roland, possess nothing impenetrable but the shield of truth; for the rest naked and unarmed, I take refuge in the protection of Your Highness,* at whose mere glance must fall anybody who—out of his mind—imperiously attempts to mount assaults against reason.[49]

In this remarkable metaphor (lifted from Ludovico Ariosto's *Orlando furioso*) for the dispute and for Cosimo's power in controlling it, Galileo indicated explicitly that the only possible way out of a situation rendered

48. Ibid., p. 31. The translation is adapted from Stillman Drake, *Galileo at Work* (Chicago: University of Chicago Press, 1978), p. 172.

49. *GO,* vol. 4, p. 51. The translation is adapted from Drake, *Galileo At Work,* pp. 173–74 (emphasis mine).

irrational by "those fanatics" was to be given the "power of impen-
etrability" by Cosimo, so that—like the bear that ignores the puppies
barking around it—he could dismiss his enemies and walk away.

The Aristotelians, too, understood that the favor of the Medici could
decide the dispute. They dedicated all their works to members of the
Medici family, from Cosimo's wife and brothers to Galileo's longtime
enemy—Giovanni de' Medici. But, while Galileo stressed his *personal* link
to Cosimo II (as his paladin), the philosophers stressed the *institutional*
link between the Studio Pisano and the house of Medici. In presenting the
Considerazioni of the Anonymous Academician to Maria Maddalena, the
provveditore (i.e., chancellor) of the University of Pisa endorsed this critique of Galileo in the name of the Pisan academic community "because it
is the duty of the provveditore of the University of Pisa to publish the defenses of the doctrine here professed—a doctrine that is taught here by
most excellent philosophers hired and paid by this institution for this purpose."[50] The provveditore, Count Pannocchieschi d'Elci, claimed that
Aristotle—the greatest of the philosophers—was protected by the greatest of the ancient kings (Alexander the Great) and hoped that the Medici,
as *nuovi Alessandri,* would continue to protect him. If, instead,
the Medici supported Galileo,

> [the glory of Aristotle] would either decline or fall altogether be-
> cause most students—full of youthful exuberance, anxious to find
> some doctrine to follow, or bored by the received philosophy—
> would orient themselves toward a doctrine that proposes new
> ideas—though less reliable ones—especially if these were perceived
> as accepted by the sovereigns.[51]

In a similar vein, the Anonymous Academician warned the Medici:

> Could it be that many bright youths, curious to know many things
> and captured by the novelty of this doctrine, would abandon the
> straight and safe road of the peripatetic doctrine to adopt a different
> one that—full of curves—presents different interpretations of all
> the phenomena of the universe? *If this were to happen, the universities
> and the public schools would lose too many students, and the great teachers
> who have taken Aristotle as their guide and first master would be barely
> listened to.*[52]

This dispute on buoyancy did not end with a clear verdict. With his
usual diplomatic attitude, Cosimo avoided taking a stand. And he could
not have made any clear-cut decision. Galileo's international prominence

50. Ibid., p. 147.
51. Ibid.
52. Ibid., pp. 177–79 (emphasis mine).

had increased during the dispute on buoyancy because of the discovery and debate on sunspots, and Cosimo would have probably damaged his own image had he let him down. While Galileo's adversaries were publishing their responses to the *Discourse,* his (and his audience's) attention was focused on the more spectacular issue of sunspots. Judging from Galileo's correspondence, his *Discourse* did not seem to get much attention and its publication was overshadowed by the enthusiasm aroused by the exchange between him and Apelles on sunspots. In many ways, the debate on buoyancy remained limited to Florence.

The Grand Duke could not drop Galileo; yet he could not rule against the Aristotelians either, for that would have undermined the credibility of Aristotle's doctrine and, consequently, of the curriculum of the Studio Pisano. Cosimo was caught between an important personal patronage relationship with Galileo and an institutional link between his house and the university. As a result, although nobody won, nobody was declared defeated, for that would have reduced the prestige of Cosimo either as a patron or as the prince of the university. We can infer that Cosimo implicitly supported Galileo, since Galileo did not feel he had to answer the philosophers' critiques but—quite insultingly—passed the task on to his disciple and client Castelli. And Castelli did not rush. Before his response came out, both authors who might have hidden behind the "Anonymous Academician" (Pannocchieschi d'Elci and Papazzoni) died, and Coresio (a native Greek and a follower of the Greek Orthodox Church) had run into trouble with the Florentine Inquisition.[53] Consequently, Castelli responded only to delle Colombe and di Grazia. His fat tome (widely emended by Galileo) came out in 1615, when Castelli was already a professor of mathematics at Pisa.[54]

However, Galileo's "victory" was short-lived. The League continued to be active against him and seemed to gain the support of Florence's Archbishop Marzimedici. There is evidence that the Dominican friar Tommaso Caccini, who publicly accused Galileo and his followers of religious heterodoxy—accusations that would lead the Inquisition to take an interest in Galileo's Copernican beliefs—was connected to the League.[55] In the end, the philosophers may have found in the Roman

53. Drake, *Galileo at Work,* p. 446.

54. Benedetto Castelli, *Risposta alle opposizioni del S. Lodovico delle Colombe e del S. Vincenzio di Grazia contro al trattato del Sig. Galileo Galilei* . . . (Florence: Giunti, 1615); reprinted in *GO,* vol. 4, pp. 448–691.

55. Caccini's brother Matteo wrote to him from Rome in January 1615: "I hear a most extraordinary thing about Your Reverence that surprises and disgusts me. You should know that rumors of what has happened have reached here and that you will receive such a rebuke that you will regret having learned to read. But what kind of silly thing to let yourself be set up like a pigeon or a fool [literally 'testicle'] by certain pigeons [a pun on delle Colombe's name]. . . . I beg you to stop, not to preach [on these matters] anymore" (reproduced in

Curia a more powerful reward system, one in which their arguments found a better reception than they had with the Medici.

Bilingualism in Context

Although Thomas Kuhn has addressed the importance of sociological factors in the process of theory choice, such as the scientist's age and level of professional initiation, his treatment of incommensurability has been characterized by a more specifically linguistic approach. Kuhn has presented his notion of paradigm as integrating both the conceptual and sociological dimensions of scientific activity, but actually he privileged its linguistic and conceptual dimensions in his interpretation of incommensurability. In this instance, the notion of paradigm did not carry with it the sociological dimensions of the scientific community but seemed to be reduced to what—in his *Copernican Revolution*—Kuhn called a "conceptual scheme." Instead, not all the forms of noncommunicative behavior which emerged during the dispute on buoyancy were rooted in the linguistic dimensions of the competing paradigms; they depended also on the participants' attempts to establish or preserve their socioprofessional identity which, in turn, contributed to the emergence of incommensurability.

Although in more recent analyses of incommensurability Kuhn has dropped "paradigm" and introduced "lexical structure," he has maintained a primarily linguistic approach to the issue.[56] Developing themes already present in *The Structure of Scientific Revolutions,* Kuhn has recently focused on the crucial difference between learning a language and translating from it. While the existence of incommensurability between two paradigms means that global translation between them is impossible, that does not preclude members of one paradigm from understanding the opponents' worldview by learning their language. This is not achieved by learning a set of rules in order to attach the categories of the new language

---

Antonio Ricci-Riccardi, *Galileo Galilei e Fra Tommaso Caccini* [Florence: Le Monnier, 1902], pp. 69–70; another reference to delle Colombe is in a letter reproduced at p. 80).

56. Thomas S. Kuhn, "Commensurability, Comparability, Communicability," in Peter D. Asquith and Thomas Nickles, eds., *PSA 1982,* (East Lansing, Philosophy of Science Association, 1983), 2: 669–88; idem, "Scientific Development and Lexical Change," (Paper delivered for the Thalheimer Lectures, Johns Hopkins University, 12–19 November 1984); and idem, "The Presence of Past Science" (Paper delivered for the Shearman Memorial Lectures, University College, London, 23–25 November 1987). See also Kuhn, "What Are Scientific Revolutions?" in Lorenz Krüger, Lorraine J. Daston, and Michael Heidelberger, eds., *The Probabilistic Revolution* (Cambridge: MIT Press, 1987), 1: 7–22; idem, "Possible Worlds in History of Science," in Sture Allen, ed., *Possible Worlds in Humanities, Arts, and Sciences,* Proceedings of Nobel Symposium 65 (Berlin: Walter de Gruyter, 1989), pp. 9–32; idem, "The Road Since *Structure,*" in A. Fine, M. Forbes, and L. Wessels, eds., *PSA 1990* (East Lansing: Philosophy of Science Association, 1991), 2: 3–13.

to their referents, but rather through a series of *ostensions,* that is by pointing to the objects to be connected to those terms. Consequently, to perceive (as Quine did) the linguistic understanding of a different culture only in terms of translation is both limiting and misleading. To Kuhn, Quine's "radical translator"—somebody who is trying to translate "a language of hitherto untouched people"—is not a *translator* at all but somebody who is *learning* a new language and its taxonomy of the world.[57]

Kuhn's shift of focus from translation to language acquisition reflects his views of how linguistic categories are developed and renegotiated. Siding with the structural linguists, he argues that a given linguistic category is shaped by the differences between it and those that surround it.[58] What we mean by "swan" depends also on what we mean by "duck" and on how "swan" differs from "duck." Once it is admitted that the relationship between a term and the object is constructed as a result of an array of *differences* between that term and object and the other terms and objects around them, it follows that the referent of a term cannot be established locally. Piecemeal translation will not do. To understand the meaning of a specific term one has to reconstruct the *entire* linguistic grid peculiar to that language.[59] And it may turn out that such a grid and the worldview associated with it would not be fully homologous with that associated with the interpreter's native language. In that case, the interpreter would face linguistic incommensurability and complete translation would not be possible. For instance, how can we translate our "swan" (which we define also in terms of "duck") into a language of a culture whose world does not contain ducks?

To sum up Kuhn's argument, incommensurability is the result of non-homology between linguistic grids, which, in turn, reflects the differences between two cultures and their environments as those two cultures have experienced them. Although incommensurability precludes full translation of one language into the other, access to two incommensurable linguistic grids is still possible by learning the other language together with the world taxonomy associated with it. However, if bilingualism is a way *around* incommensurability, it cannot resolve it. To be *bi*lingual does not

57. Willard V. O. Quine, *Word and Object* (Cambridge, Mass.: MIT Press, 1960), p. 28. On Quine's analysis of the problem of translation, see esp. pp. 26–79. Kuhn's critique of Quine is in his "Commensurability, Comparability, Communicability," pp. 673–75, 680–82.

58. Ferdinand de Saussure, *Cours de linguistique générale* (Paris: Payot, 1986), pp. 155–62; Claude Lévi-Strauss, *The Savage Mind* (Chicago: University of Chicago Press, 1966), p. 115; and Kuhn, "Commensurability, Comparability, Communicability," pp. 680–82. Kuhn presented similar views in his "Second thoughts on Paradigms," *The Essential Tension* (Chicago: University of Chicago Press, 1977), pp. 293–319. Kuhn's notion of grid is not unlike the network model of universals presented by Mary Hesse in her *Structure of Scientific Inference* (Berkeley: University of California Press, 1974), pp. 45–73.

59. Kuhn, "Commensurability, Comparability, Communicability," pp. 673–75, 680–82.

mean to be *meta*lingual. Bilingualism may make one aware of incommensurability, but does not solve it. Full translation remains impossible.

Although I share Kuhn's views about the linguistic dimensions of incommensurability and about the impossibility of complete translation between incommensurable linguistic grids, I think that his linguistic focus may have prevented him from realizing how access to bilingualism is something deeply regulated by social dynamics. I want to expand on this point because it helps to highlight some of the differences between Kuhn's synchronic view of incommensurability and the more diachronic view presented here.

Although Kuhn's approach is essentially historical and (more recently) ethnographic, he does not apply these interpretive tools to studying the process through which people *become motivated* to develop a certain theory or *become committed* to a certain worldview. Kuhn's apparent lack of interest in this problem suggests that he takes the "engine" of scientific change to be unproblematically given. What seems to compel people to develop their lexicons is the puzzle-solving mentality that Kuhn has placed at the center of the model of scientific change in his *Structure of Scientific Revolutions*—an ethos that is particularly conspicuous in what he has called normal science. Kuhn's reliance on puzzle-solving as the engine of scientific change may be a result of his own training as a physicist, or of the belief that puzzle-solving is ultimately connected to some survival-oriented feature of the human mind. Be that as it may, I believe that—without denying the survival-rooted dimensions of human knowledge—we do not need to follow Kuhn in naturalizing to this extent the puzzle-solving drive. Instead, we can identify specific cultural dynamics shaping that drive—dynamics that are also closely connected to the emergence of incommensurability.

I want to argue that by not pushing his historico-anthropological analysis far enough to include the social and cultural construction of the knowing subject and his or her motivations, Kuhn ends up naturalizing the engine of scientific change and missing an important genealogical link between this engine and the emergence of incommensurability. A study of the social constraints on bilingualism provides a key to uncovering this common genealogy.

The case study presented earlier indicates how the adoption and articulation of different worldviews are linked to the development and maintenance of socioprofessional identities and that, therefore, the *choice* of puzzles to be addressed depends also on the practitioners' identity, desires, and opportunities for socioprofessional mobility. For instance, most competent astronomers remained Ptolemaic well after 1543, others adopted the Tychonic system when it became available, and only a minority of them (who tended to be located in princely courts) endorsed

Copernicus's theory as a *physical* description of the cosmos. In short, mathematicians could either keep solving puzzles that would keep them within the traditional boundaries of mathematics or, as in Galileo's case, pursue puzzles leading them into the philosophers' domain. As I hope I have shown, Galileo's gradual commitment to Copernicanism was also informed by his peculiar social position and background and the perception of social mobility and possible identities that went with it. One may try to apply similar considerations to Copernicus himself.[60]

I want to argue that the dynamics of identity development and maintenance that direct one's puzzle-solving in a certain direction are also responsible for one's deciding to become bilingual. In particular, once we integrate the view of language as a taxonomical grid with considerations of the role language plays in maintaining the group's identity and cohesion, we begin to realize that to learn the language of the "other" entails losing one's socioprofessional identity. Given the hierarchical disciplinary setting in which Galileo and the philosophers were debating, learning the other's language was not a viable option. I want to expand on this by analyzing the implications of Galileo's claim that the deadlock in the debate on buoyancy was caused by the Aristotelians not understanding him because of their ignorance of mathematics, while he could understand Aristotle perfectly.[61]

Galileo's claim about the stubborn ignorance of the philosophers in mathematical matters reflects a specific socioprofessional ethos. In a fashion that is reminiscent of his earlier attempts to define the rules of the game by setting its experimental and methodological standards, Galileo acted as if the other party should share his ethos, one geared toward the production of *new* knowledge. As a result, he assumed a moralistic voice and characterized the Aristotelians' ignorance of mathematics as unethical, claiming that they were viciously stubborn not to see the truth that stood in front of their eyes. However, Galileo's commitment to the production of new knowledge was by no means "natural." His identity as a producer of controversial novelties and as a thinker not bound to a philosophical system reflected also a courtly identity. And he used these local cultural codes to delegitimize his adversaries by representing them as pedants, bores, and Aristotle's philosophical *slaves* (as opposed to Galileo himself and his gentlemanly audience of *free* thinkers).[62]

But Galileo was not alone in using a moralistic rhetoric. The Aristo-

60. It seems to me that Robert S. Westman, "Proofs, Poetics, and Patronage: Copernicus's Preface to *De revolutionibus*," in David C. Lindberg and Robert S. Westman, eds., *Reappraisals of the Scientific Revolution* (Cambridge: Cambridge University Press, 1990), suggests this possibility.

61. *GO*, vol. 4, pp. 31–32. 65.

62. In addition to the references given in chapter 2, see also *GO*, vol. 5, pp. 102, 136.

telians, too, used their own corporate ethos, as commentators rather than original authors, for a "natural" standard of reference and attacked Galileo for his intellectual narcissism, his "lust for novelty," and his attempted subversion of traditional disciplinary hierarchies.[63] These reciprocal moralistic accusations do not seem to have much to do with buoyancy. Rather, they reflect quite accurately the irreducible differences in the ethos and socioprofessional identities of those who uttered them. The pervasive use of moralistic (rather than "rational") arguments on both sides also indicates the lack of (or disinterest in) dialogical alternatives. Moralism is there to defend received dogmas and identities, not to negotiate them. Moralism is a sign of the possible emergence of incommensurability.

Therefore, the differences between Galileo and the philosophers were not limited to their views of the physical world, but extended well into their own professional ethos. Although sixteenth-century Aristotelianism did not constitute a homogeneous philosophy, its practitioners shared a fairly consistent socioprofessional identity. As they tended to be university professors, they went through a quite homogeneous and long professional training and had a strong corporate identity both as members of an internally structured institution such as the university and as keepers of a set of canonical texts. Galileo saw them as members of a peculiar religious order.[64] The discovery of novelties was not one of the corporate duties for which they were trained and paid.[65] To use Mary Douglas's terminology, they were a high-grid/high-group culture.[66] Their corporate identity was explicitly portrayed in the Anonymous Academician's "call to arms" to prevent Galileo's doctrine from entering the university:

> My fellow Peripatetics, it is no longer time for jokes. The honor and status of your prince is now threatened. With waving flags, the Author [Galileo] is boldly moving against the previously undefeated fortress of the Peripatetic doctrine. Although these types of arguments have been deployed against it on other occasions but were eventually refuted and vanquished, nevertheless it is a much lauded military rule to control the enemies continuously to prevent them

63. *GO*, vol. 4, pp. 147, 156, 177–78, 335. It is interesting that in those quotes, Galileo's intellectual abilities are never questioned. What are stressed are his conceptual aggressiveness and the dangerous effects of the "spell" of his ideas on excitable youths.

64. Ibid., p. 51.

65. Cremonini, in one of his answers to the Inquisition, gives a clear (and arrogant) statement of his corporate identity: "I cannot and do not wish to retract my exposition of Aristotle because this is how I understand him, and *I am paid to present him as I understand him, and, were I not to do so, I would be obliged to give back my pay*" (quoted in Maria Assunta del Torre, *Studi su Cesare Cremonini* [Padua: Antenore, 1968], p. 60). On Cremonini, see also Charles Schmitt, "Cesare Cremonini, un aristotelico al tempo di Galilei," *Aristotelian Tradition and Renaissance Universities*.

66. Douglas, *Natural Symbols*, pp. 103–6.

from increasing their confidence and power, especially when they are ingenious, ambitious, and subtle.[67]

Also,

> I believe that in order to preserve the jurisdiction of this lady [philosophy] it would be enough that her confederates and followers—honoring their corporate duties—would help her destroy the war machine of the enemies and withstand this dangerous siege. Through a simple defense strategy and without attacking the air back [a reference to Galileo's "small banks" discussed above], they will be able to preserve philosophy in her jurisdiction. Eventually, the air—lacking any firm stand and relying for its force only on foreign powers—will be forced to withdraw to its own region.[68]

In contrast to the Aristotelians, Galileo was not a "confederate." He had obtained his title of philosopher through court patronage rather than through a regular training in philosophy. He did not undergo a standard professional initiation. As the Anonymous Academician put it, Galileo was an "alien" coming from "foreign regions" to raid the possessions of philosophy. He too had a corporate identity but only in the peculiar sense that courtiers had one. As we can see from the negotiations with the Medici for the position of philosopher and mathematician at court, Galileo represented himself as a producer of novelties. Anthropologists would probably classify Galileo as a "Big Man." Weber would have probably termed him a "charismatic personality."[69]

Two radically different socioprofessional cultures (associated with different cosmologies as well as social institutions) were confronting each other behind the nonmomentous issue of buoyancy. This helps us to understand the implications of Galileo's insistence that the philosophers learn mathematics. For an Aristotelian to learn mathematics, or rather to accept it as a method for the physical explanation of nature, would have meant learning the language of a previously subordinate "other" now turned alien invader. Given the institutional and power dimensions entailed by this decision, Galileo was inviting the Aristotelians to commit suicide.

These dynamics of identity maintenance may help reframe our under-

---

67. *GO*, vol. 4, p. 177.

68. Ibid., p. 156.

69. Douglas, *Natural Symbols*, pp. 128–29; and Max Weber, "The Sociology of Charismatic Authority," in H. H. Gerth and C. Wright Mills, eds., *From Max Weber* (New York: Oxford University Press, 1946), pp. 245–52. The analogy between the identities shaped by the system of early modern patronage and the so-called "Big Man" has been made in Werner L. Gundersheimer, "Patronage in the Renaissance: An Exploratory Approach," in Guy Fitch Lytle and Stephen Orgel, eds., *Patronage in the Renaissance* (Princeton: Princeton University Press, 1981), pp. 3–23, esp. pp. 12–13.

standing of the well-known refusal by the philosopher Cremonini (a good social friend of Galileo's) to look through the telescope. Apparently, he excused himself by saying that it would have given him a headache.[70] While Cremonini's conspicuous refusal has usually been presented as a handy epitome of the philosophers' silliness, it may be a sign of something more relevant. In principle, the use of instruments and the belief in the evidence they could provide were as alien to Aristotelian philosophy as the use of mathematics to explain physical phenomena. Not only could Aristotle's theory of vision hardly have made sense of the telescope's operation, but the very idea of an evidence-making machine was incompatible with Aristotelian philosophy. Cremonini's refusal to look through the telescope or to mention Galileo's discoveries in his 1613 *Disputatio de coelo* (or the Pisan Aristotelians' failure to recognize Galileo's physical principles as such) are all symptoms of the same identity-preserving dynamics which prevented the Aristotelians from taking mathematics seriously. What Galileo was proposing (and what threatened them) was not just a theory of buoyancy and a telescope but a new philosophical "form of life."

For quite similar reasons, the Aristotelians resisted Galileo's *momento.* In the *Discourse,* Galileo claimed to borrow the meaning of *momento* from the science of mechanics.[71] But, dismissing Galileo's quite explicit "ostension," the Anonymous Academician realized that this was a crucial principle for Galileo and criticized him for not defining it. He claimed that *momentum* was a Latin term but that Galileo did not use it according to its traditional meaning. The Academician then looked it up as a vernacular term in the newly printed canon of the Florentine vernacular—the dictionary of the Accademia della Crusca—but, to his surprise, *momento* was not listed.[72] Although the Anonymous Academician knew well where the

70. Gualdo wrote Galileo in July 1611 that he had spoken to Cremonini, who justified his not dealing with Galileo's discoveries in his forthcoming book because "I do not wish to approve of claims about which I do not have any knowledge, and about things which I have not seen . . . and then to observe through those glasses gives me a headache. Enough! I do not want to hear anything more about this" (*GO,* vol. 11, no. 564, p. 165).

71. In the first edition of the *Discourse,* Galileo wrote that he was borrowing the category of "momento" from the "scienze meccaniche." In the second edition, which appeared before the end of 1612, he added a longer description of "momento," probably in response to the critique of the Anonymous Academician. However, the vagueness of his later addendum confirms the crucial yet problematic status of "momento" in Galileo's dynamics. It is still a quite vague, "metaphorical" notion: "Among mechanicians, *momento* signifies that virtue, that force, that inclination, with which the mover moves and the moved resists." After listing some of the common meanings of *momento,* he concludes by saying that these are "metaphors, I would say, taken from mechanics" (*GO,* vol. 4, p. 68). On the various meanings attributed by Galileo to *momento,* see Paolo Galluzzi, *Momento* (Rome: Edizioni dell'Ateneo, 1979), pp. 227–46; and Maria Luisa Altieri-Biagi, *Galileo e la terminologia tecnico-scientifica* (Florence: Olschki, 1965), pp. 44–55.

72. "This term *momento* is Latin and Ptolemaic [Ptolemy was traditionally credited with having written a treatise *De momentis*], but is not used with this meaning in our modern

meaning of *momento* was to be found, for Galileo had indicated that he was borrowing it from the science of mechanics, he *could not* accept a notion that came from a lower discipline like mechanics and that was threatening for his entire conceptual edifice and professional identity.[73] Interestingly enough, in 1623 the second edition of the dictionary of the Crusca (an academy to which Galileo was elected in 1605) did contain a Galilean definition of *momento*.[74]

Therefore, to treat bilingualism simply as a linguistic notion is to overlook the dynamics of identity formation and maintenance at the basis of one's intellectual and professional choices. Moreover, although the implications of learning the "other's" language are not always as drastic as those faced by the Aristotelians, nevertheless, when learning the language of the "other" implies entertaining another socioprofessional identity, then to be bilingual means, in a sense, to be "schizophrenic." Or, to use a different metaphor, by looking at the same thing from two different points of view simultaneously one does not become objective but simply cross-eyed.

But if Galileo was rhetorical in blaming the Aristotelians for not learning mathematics or taking it seriously, he was right in saying that he could understand Aristotle. And yet, I believe, Galileo was not "schizophrenic" either. I have compared bilingualism to "schizophrenia" in those cases in which adopting a language implies adopting a different socioprofessional identity. This was not the case with Galileo. Aristotelianism was the language of his *past,* when he was a medical student at Pisa.[75] His continuing interest in selected aspects of Aristotelian philosophy (especially

---

vulgar and even less in the old one. In fact, there is no mention of it in the most extensive and exquisite dictionary of the Crusca. I am saying this not to make a point of linguistic purity and propriety, but because this issue is very important for the true intelligibility and definition of the present matter" (*GO*, vol. 4, p. 158). Di Grazia adopted the same tactic and indicated that Galileo's notion of ice (as expanded water) went against the Crusca's definition (ibid., p. 403).

73. For an analogous pattern, see Biagioli, "Social Status of Italian Mathematicians," 63–65.

74. Paola Manni, "Galileo accademico della Crusca," *La Crusca nella tradizione letteraria e linguistica italiana* (Florence: Accademia della Crusca, 1985), pp. 128–29. The politics of terminology are telling here. The Crusca dictionary had a quasi-official status in Florence because of the Medici's strong investment in the linguistic identity of Tuscany, and of their direct support of the Crusca. In a sense, the Crusca was the "Medici dictionary." Therefore, the Aristotelian critique that Galileo's definitions were not compatible with the Crusca dictionary was aimed at presenting him outside the canon of both philosophy and Florentine courtly language. Galileo's response was to use his connections to legitimize his paradigm by having his definitions canonized by the Crusca.

75. Galileo's familiarity with, and yet distance from, Aristotelian discourse is quite clear in a passage in the *Assayer:* ". . . at this point I get a most overwhelming nausea from those altercations that, during my childhood, when I was still studying under the pedant, used to give me great delight" (*GO*, vol. 6, p. 245).

Aristotle's logic and theory of demonstration) ensured Galileo's competence in that philosophy without tying his socioprofessional identity to it.[76]

More generally, members of the emerging (or invading) paradigm can be bilingual if they were previously trained in the old one and dropped it early in their careers, as Kuhn has often noted in the case of the members of the new paradigm. Like traders who speak different languages to different people they visit, without sharing in their cultural identity, the "invaders" can use the language of the adversary without adopting the socioprofessional identity attached to it. Unlike the members of the older paradigm, the supporters of the new paradigm can be bilingual without being "schizophrenic". Similarly, an anthropologist who *visits* an alien culture can reconstruct and appreciate its cosmology without necessarily ending up questioning his or her own culture. The historian's own predicament is an analogous one.[77]

Moreover, while members of established socioprofessional groups have much to lose from becoming bilingual, bilingualism is strategically important to the "invaders" (or the traders). If they want to invade a disciplinary domain (or a hostile market) they must know or learn something about it. To put it differently, because the emerging group is likely to have little or no power and to hold a still unarticulated paradigm, it cannot simply dismiss the powerful and established adversary; that would require a power it does not yet have. In the beginning (before they have established themselves socioprofessionally) the members of the emerging paradigm need to be bilingual in order to erode the authority of the old one. This is particularly evident in Galileo's *Dialogue,* where he manages to gain credibility precisely by including Simplicio (the voice of Galileo's bilingualism), who provides the alternative position (and a somewhat rigged term of comparison) for Galileo's claims.[78] Once some power is achieved and a new socioprofessional identity is forged, bilingualism loses importance as a resource.

76. Among the various texts addressing this issue, see William A. Wallace, *Galileo and His Sources* (Princeton: Princeton University Press, 1984); Alistair C. Crombie, "Sources of Galileo's Early Natural Philosophy," in Maria Luisa Righini-Bonelli and William R. Shea, eds., *Reason, Experiment, and Mysticism* (New York: Science History Publications, 1975), pp. 157–75; and Adriano Carugo and Alistair C. Crombie, "The Jesuits' and Galileo's Ideas of Science and of Nature," *Annali dell'Istituto e Museo di Storia della Scienze di Firenze* 8 (1983): 3–67.

77. On this issue, see Mario Biagioli, "From Relativism to Contingentism," in Peter Galison and David Stump, eds., *Disunity and Contextualism* (Stanford: Stanford University Press, forthcoming).

78. Nuccio Ordine et al., *Il dialogo filosofico nel '500 europeo* (Milan: Angeli, 1990); and Maria Luisa Altieri-Biagi, "Il dialogo come genere letterario nella produzione scientifica," *Giornate lincee indette in occasione del 350 anniversario della pubblicazione del "Dialogo sopra i massimi sistemi" di Galileo Galilei* (Rome: Accademia dei Lincei, 1983), pp. 143–66.

However, one's becoming bilingual is not just a strategic matter. One needs also to be in a specific cultural predicament to become bilingual. As we have seen, those who can have access to bilingualism (Galileo, the traders, the anthropologists) are those whose own culture and identity have some degree of fluidity. While in the case of the anthropologist that fluidity may result from a specific training, Galileo's ability to become bilingual was related to his having an "unfinished" socioprofessional identity. He did not have a new, comprehensive philosophical system yet, and his very identity as a "mathematical philosopher" was far from being established or canonical. This further explains why bilingualism was an option open to the emerging rather than to the established paradigm.

Bilingualism does not need to produce dialogue across incommensurable lexical structures, but it can help reinforce the confidence of the invaders and provide them with a tool for convincing potential supporters (but not necessarily their own adversaries). Also, by sharing the lexicon but not the socioprofessional identity of the Aristotelians, Galileo may not have fully understood how and in what ways they felt threatened by his demand that they learn mathematics. His apparent inability to understand what he perceived as their "stubbornness" (and therefore unethicalness) gave him further arguments to dismiss them.

Therefore, bilingualism ended up reinforcing (rather than questioning) Galileo's own socioprofessional identity in the same way the refusal of bilingualism was crucial to the Aristotelians' maintainance of their identity.[79] In turn, Galileo's establishment of a new identity was a crucial component of his drive to articulate Copernican astronomy and mathematical physics. Scientific change was also tied to the desire to establish a new socioprofessional identity, and this influenced Galileo's attitude toward bilingualism. Therefore, access to bilingualism (and the asymmetric pattern displayed by Galileo and the Aristotelians) is not a contextual accident but an essential ingredient of scientific change. Identity dynamics (and access to bilingualism) are not only behind the engine of scientific change but they also set up the conditions in which that engine can actually produce paradigm speciation and, sometimes, incommensurability.

If a given worldview can be developed only by people sharing and articulating a certain language game, then those people must maintain cohesion in order to make cognitive activity possible. This is the sense in which noncommunicative behaviors play a role in cognitive and socioprofessional speciation. The situation that results from everyone being

79. This pattern bears striking similarities with Bruno Latour's discussion of the "great divide" in his *Science in Action* (Cambridge, Mass.: Harvard University Press, 1987), pp. 210–13. In particular, Latour's "rational westerners" are those who, like traders, are bilingual or multilingual. Also interesting on the role of the "trader" in science is Peter Galison, "The Trading Zone: Coordinating Action and Belief," paper presented at UCLA in November 1989.

willing to learn the "other's" worldview would not be characterized by a perfectly ecumenical and consequently totally rational science, but by the absence of *different* groups, disciplines, paradigms, and—consequently— by the absence of science itself. It is a kind of category mistake to think of noncommunicative attitudes as just the unfortunate *effect* of sociohistorical contingencies. Far from being an obstacle on the path of cognitive activity, they provide a sort of protective and containing belt that makes cognition possible.[80]

Incommunicability and Incommensurability

To conclude, I want to suggest how incommensurability and the various forms of noncommunicative behavior described earlier may be tied together in the genealogy of scientific change.[81]

While incommensurability is a specific and usually uncommon linguistic phenomenon in science, noncommunicative behaviors, rhetorical tactics of nondialogue, and asymmetric access to bilingualism are much more common phenomena which seem more directly related to dynamics of identity development and preservation than to lexical structures. Moreover, as we have seen, one's adoption of noncommunicative behaviors or of bilingualism is related to one's legitimacy and power in a given environment. As a mathematics professor at Padua, Galileo was not publicly vocal about the incompatibility between his views and those of the Aristotelians (except under the pseudonym of "Cecco de Ronchitti") but he became so as soon as he migrated to court. However, the power and legitimation he obtained through that migration (which allowed him to adopt noncom-

---

80. Several philosophers have noted that, strictly speaking, theories are born refuted. I want to add that we cannot separate the process through which a theory is "resuscitated" or kept alive from the process through which people gather and develop a socioprofessional identity around it.

81. I am aware that this case study does not present a speciation from within one discipline or socioprofessional group, but rather it presents a challenge to the hierarchy between two related disciplines caused by the speciation of the subordinate discipline into a "higher" species. However, this does not mean that the lexical structures of the philosophers and the mathematicians were incommensurable to begin with and that, therefore, the example cannot provide evidence about the emergence of incommensurability in general. In fact, according to the received disciplinary hierarchy, the mathematicians' lexical structures were not supposed to extend into the philosophers' domain. More precisely, the mathematicians could deal with phenomena that were in the philosophers' domain, but they were not supposed to make philosophical claims about them. This does not mean that the mathematicians' claims were incommensurable with those of the philosophers: rather, they were subordinated. Thus there was no incommensurability to begin with, because the very possibility of comparison was ruled out in principle. The comparison between the two lexical structures (and incommensurability) emerged only when mathematicians began to consider themselves "philosophers" and to make claims about the physical dimensions of the world. In short, this may be a somewhat atypical example but not an anomalous one.

municative behaviors toward the Aristotelians) resulted also from his success in reinforcing his new socioprofessional identity through the discoveries of 1610. Power and legitimation are not external to the process of articulating one's paradigm and socioprofessional identity and, therefore, cannot be seen as the independent cause of noncommunicative behaviors. Similarly, I find it difficult to detect a consistent cause-effect relationship between incommensurability, noncommunicative behaviors, and communication breakdowns. Although incommensurability may precipitate communication breakdowns, it is also the case that noncommunicative behaviors may lead to incommensurability.

Rather than seek direct causal connections between these various phenomena and behaviors, an attempt that (given the mutual linkages among them) is likely to end up in a series of unanswerable questions, I prefer to sketch a possible picture of the process of scientific change which may indicate some of the relationships between these phenomena.

In the case of speciation within a socioprofessional group, a subgroup's unwillingness to talk to the rest of the group may initially be a rhetorical tactic of noncommunication aimed at maintaining the cohesion of the emerging group. At such an early stage of the speciation process, it would be quite likely that the linguistic grid of the subgroup would be still largely commensurable with that of the rest of the group.[82] But the cohesion of the subgroup would in turn help its members to articulate their new socioprofessional identity, which would allow for, and *commit* them to, the development of their new worldview. After a while, the new linguistic grid developed by the subgroup may actually become incommensurable with the old one, marking the emergence of a new "scientific species."[83] Self-fashioning and world-fashioning are not separable.

Therefore, although noncommunicative behaviors are not a necessary cause of incommensurability (there is nothing inherently incommensurability-producing in them), they can lead to its emergence by maintaining the cohesion of the group that is articulating the lexical structure. Although the relationship between incommunicability and incommensurability is not of cause and effect (in either direction) they are closely

82. One model that could relate (although incompletely) linguistic incommensurability and noncommunicative behaviors is offered by de Saussure's distinction between *langue* and *parole* (or Chomsky's competence/performance model). One could try to relate the lexical structure of the group responsible for incommensurability to *langue*, while considering the individual's possibly rhetorical statements of incommunicability as belonging to *parole*. Sometimes noncommunicative behavior would reflect an actual state of incommensurability (i.e., it would come from *langue*), while in other circumstances it would be uttered as private statements, reflecting personal perceptions or strategies rather than the group's lexical structure.

83. I think that this function of incommensurability-oriented forms of noncommunicative behavior is not unlike the role Feyerabend attributes to "propaganda" in the early phases of the development of a new world view (see Feyerabend, *Against Method*, esp. pp. 145–61).

connected by the articulation of new socioprofessional identity which, in turn, is at the basis of scientific change.

I have proposed a shift from a linguistic (and synchronic) view of incommensurability to one which sees it as the possible result of the diachronic articulation of a new socioprofessional identity. This contextualization helps analyze the place of bilingualism and noncommunicative behaviors within the process of scientific change and "scientific self-fashioning" and their relationship to the possible emergence of incommensurability. Finally, I hope to have shown that, once we place incommensurability in this framework, it ceases to appear as a mere problem but begins to look like a result (and a component) of the process of scientific change. Although one can still perceive it as representing a "cost," I would say that, like all processes that produce something, scientific change has certainly its own costs (incommensurability probably being one of the lesser ones).

Roma Theatrum Mundi

Patronage Pilgrims

"THE LODESTONE OF THE COURT" was the title of an oration given around 1625 in one of Rome's most fashionable academies—that of Cardinal Maurizio of Savoia.[1] What at first may seem another metaphorical image spun by a baroque literato turns out to be a quite accurate description of the pattern of migration of artists, literati, prelates, and ambitious courtiers toward Rome at the beginning of the seventeenth century. Galileo was one of these patronage pilgrims.

The perception of Galileo's interest in developing connections with Roman society and culture has often been retrospectively informed by the trial of 1633. As a Catholic Copernican, and with the legitimation of Copernicanism in a Catholic country being dependent on the reinterpretation of the Scriptures, Galileo had to develop connections with Rome because it was there that the people who could authorize such a reinterpretation (popes, cardinals, and theologians) were to be found. While this is quite true, such a perspective leaves out another crucial dimension of Galileo's interest in Rome. To him (as to many other clients) Rome was not only a place where the official keepers and interpreters of the Scriptures had their headquarters; it was also the seat of Italy's most powerful princely court.

Italy's political and economic importance had declined steadily since the end of the sixteenth century. The Duchy of Milano had long since lost its independence to become a territory controlled by Spain. Tuscany was still rich, but it was also turning into an increasingly provincial and agricultural state with little international relevance. Urbino—once the seat of

---

1. Agnolo Cardi, "La calamita della corte," in Agostino Mascardi, ed., *Saggi accademici* (Venice: Baba, 1653), pp. 242–64.

the prestigious court described in Castiglione's *Book of the Courtier*—experienced a sharp economic decline at the end of the sixteenth century and disappeared from the political map in 1631 when it was finally incorporated into the Papal State. After having been one of the most sophisticated Italian courts in the fifteenth and sixteenth centuries, Mantua (the court to which Galileo tried to migrate in 1604) was to lose its importance and eventually its political independence around 1629. Ferrara did not last even that long. By 1598 it was annexed to the Papal State, and what used to be the center of an elegant court which only a few decades earlier had hosted, among others, Ludovico Ariosto, quickly decayed to the status of a provincial city.

Although still powerful, Venice too was experiencing a sharp decline. Ironically, it was under the command of Zaccaria Sagredo (the brother of Galileo's friend) that in 1630 the Venetian army was badly defeated during the Italian episode of the Thirty Years' War, an event that marked the sharp decline of Venice's role in European politics. It was from the battlefield at Valeggio that Zaccaria wrote to Galileo authorizing him to use Giovanfrancesco's name in the *Dialogue*.[2] A few days after these letters were written, the Venetian army collapsed and Zaccaria was so fast in retreating from the battlefield that he arrived at the Peschiera camp four hours ahead of his soldiers.[3]

In the same period, Rome offered a much different picture. When Montaigne visited it in 1580, he found it "a city all court and nobility: everybody takes his share in the ecclesiastical idleness. There are no trading streets, or less than in a small town: it is nothing but palaces and gardens."[4]

But this process of concentration of power and wealth in Rome did not stop at the end of the sixteenth century. Also as a result of the decline of other Italian states, by the beginning of the seventeenth century Rome had become the most important political, cultural, and patronage center of baroque Italy and—until mid-century—of Europe.[5] In 1627, a much surprised Venetian ambassador said about Rome:

> in that city there are no gold mines, nor commercial exchanges with other nations, nor industry, nor artisanal manufactures through which other cities and kingdoms produce their wealth. Neverthe-

2. *GO*, vol. 14, pp. 95, 97.

3. As a result of this debacle, Zaccaria risked being sentenced to death (Gaetano Cozzi, *Il Doge Nicolò Contarini* [Rome-Venice: Instituto per la Collaborazione Culturale, 1958], p. 297). See also Maria Francesca Tiepolo, "Una lettera inedita di Galileo," *La cultura* 17 (1979): 66; and Romolo Quazza, "Il periodo italiano della Guerra dei Trent'Anni," *Rivista storica italiana* 50 (1933): 64–89.

4. E. J. Trechman, trans., *The Diary of Montaigne's Journey to Italy* (London: Hogarth Press, 1929), p. 149.

5. Paolo Prodi, *The Papal Prince, One Body and Two Souls: The Papal Monarchy in Early Modern Europe* (Cambridge: Cambridge University Press, 1987), pp. 46–49.

less, one sees gold flowing and being spent in Rome much more than in any other place because everybody who moves there brings with him his own wealth.[6]

Although Montaigne and the ambassador Contarini were quite selective and described only the life-style of those connected to the court, their remarks were basically on target. Galileo himself, in the preface to his *Dialogue*, referred to Rome as the "theater of the world."[7] Rome was a "land of opportunity."[8] Different from most Italian cities, Rome did not have a register of nobility.[9] While this encouraged frequent (and sometimes deadly) precedence disputes among aristocrats, it allowed lucky newcomers (including popes themselves) to gain titles and go up in the social scale very quickly.[10]

Rome's exceptional patronage opportunities were clearly perceived in Galileo's time. When, around 1614, the duke of Urbino, Francesco Maria II, offered Giovanni Ciampoli, Galileo's friend and supporter, a post at the Urbino court, Ciampoli's mentor, Giovanni Battista Strozzi, stepped in and "wanting to make sure that he would not end up with a declining prince, he offered to pay him a salary of three hundred scudi a year so that he could go to Rome and build his own fortune there."[11]

As shown by the remarkable percentage of non-Romans at the Ro-

---

6. Pietro Contarini in Nicolò Barozzi and Guglielmo Berchet, eds., *Relazioni degli stati europei lette al Senato dagli ambasciatori veneti del sec. XVII*, series 3, *Relazioni di Roma*, 10 vols. (Bologna, 1856–79), 2: 200.

7. Galileo Galilei, *Dialogue Concerning the Two Chief World Systems*, trans. Stillman Drake (Berkeley: University of California Press, 1967), p. 5.

8. Also, the salaries granted by the pope to his courtiers were remarkably higher than those at other Italian princely courts. On the stipends of the Medici court, see chapter 2. In the 1610s, a *cameriere secreto* at the papal court (the post Ciampoli would get in 1623) made 1,000 scudi per year (Girolamo Lunadoro, *Relatione della corte di Roma* [Rome: Frambotto, 1635], p. 4), while the corresponding role at the Medici court in the same years made about 150 scudi (*Archivio di Stato di Firenze*, "Miscellanea medicea 474"). In the case of other court roles, the difference was less sharp, but it is remarkable that Galileo's salary (exceptional by Florentine standards) was comparable to that of one of the many papal *camerieri*.

9. Laurie Nussdorfer, "City Politics in Baroque Rome, 1623–1644" (Ph.D. diss., Princeton University, 1985), p. 157; idem, *Civic Politics in the Rome of Urban VIII* (Princeton: Princeton University Press, 1992).

10. On the lack of a register of nobility and on the fluidity of social status in Rome, see Carlo Mistruzzi, "La nobiltà nello stato pontificio," *Rassegna degli Archivi di Stato* 23 (1963): 206–44; and Nussdorfer, "City Politics," 150–72. For an example of these precedence conflicts, see J. A. F. Orbaan, *Documenti sul barocco in Roma* (Rome: Società Romana di Storia Patria, 1920), vol. 2, pp. 275–76.

11. From the short biography of Ciampoli reproduced in Giovanni Targioni Tozzetti, *Notizie degli aggrandimenti delle scienze fisiche accaduti in Toscana nel corso di anni LX del secolo XVII* (Florence, 1780; reprint Bologna: Forni, 1967), vol. 2, part 1, (pp. 102–16), p. 105. In the same biography at p. 107 we find that "he had made up his mind to chase after his fortune in Rome, where even the most beggarly man has a chance to become a prince."

man court (and among Galileo's friends and foes there) Strozzi was not alone in his appreciation of the possibilities offered by Roman patronage.[12] The career of Cassiano dal Pozzo (a colleague of Galileo's at the Accademia dei Lincei) reflects similar patronage strategies. Rebuffing the offers coming from his father and family friends to accept good posts at the Medici court or marry into the northern Italian aristocracy, Cassiano remained in Rome leading the life of a court virtuoso until, more than ten years later, he was able to obtain a remarkably high post in the court of Urban VIII.[13]

However, Rome did not attract clients only because it was remarkably dense with patrons and well-paid jobs. True, the Curia, the papal court, the various cardinals' courts (some of them larger than those of many Italian secular princes),[14] the households of the Roman baronage, and the many *monsignori* and *cavalieri* patronized a remarkable number of artists, writers, physicians, and even some mathematicians; but what was most appealing about Rome to ambitious clients like Ciampoli or Galileo was the type of patronage that could be found there. While Cosimo II de' Medici was a powerful absolute prince and patron, the pope was the true epitome of that type of ruler. Perhaps Rome was the site where political absolutism and court culture best displayed the structure that would later become typical of the great European courts and modern states.[15] And it was precisely during the pontificate of Galileo's patron, Urban VIII, that political absolutism reached its peak in Rome.[16] Papal patronage was the

12. Panfilo Persico wrote in his *Del segretario libri quattro* (Venice: Damian Zenato, 1629) that the Roman court was a place where "everybody was a foreigner" (p. 82).

13. Giacomo Lumbroso, "Notizie sulla vita di Cassiano dal Pozzo," *Miscellanea di storia italiana* 15 (1874): 136–43. For a contextualization of Cassiano in Roman culture, see Francesco Solinas, ed., *Cassiano dal Pozzo* (Naples: De Luca, 1989).

14. For instance, in 1589 the familia of Cardinal Alessandro Farnese comprised 284 people, while that of his nephew of second degree, the duke of Parma and Piacenza, listed 226 courtiers (Gigliola Fragnito, "Parenti e familiari nelle corti cardinalizie del Rinascimento," in Cesare Mozzarelli, ed., *"Familia" del principe e famiglia aristocratica* (Rome: Bulzoni, 1988), 2: 565, 570. Although Farnese's court was particularly large, it was not unusual to find courts of cardinals with more than two hundred people (ibid., 569).

15. The result of this process of centralization was that, as noted by the Venetian ambassador Giovanni Mocenigo in 1611: "Cardinals do not have any part in state affairs, and if it sometimes happens that the pope shares some matter with the consistory, he does so to communicate his will to the cardinals. Nobody contradicts him any more as they used to do . . . so one can say that nowadays the government of Rome is one of supreme and absolute imperial rule" (Barozzi and Berchet, *Relazioni*, 2: 96). Giovanni Mocenigo reported a similar scenario: "And if sometimes—though rarely—he asks for advice, there is no one who dares utter anything but applause and praise" (ibid., 102). See also quotes from Francesco Contarini (ibid., 89) and Renier Zeno (ibid., 149), and Paruta as quoted in Prodi, *Papal Prince,* p. 37.

16. Prodi, *Papal Prince,* pp. 37–58.

most powerful tool for self-fashioning a client could aim for in early seventeenth-century Italy.

Between Libertines and Jesuits

Any ambitious client would have had plenty of reasons to go to Rome or to become connected to its patronage networks, but Galileo, as we have seen, had even more. When, in 1588, he was trying to get his career as a mathematician off the ground, he went to Rome to connect with Clavius. Rome loomed larger in his plans after his discoveries of 1609–10. With some help from Galileo, the Medici eventually realized that Rome was the place where his new discoveries needed to be finally certified. As Vinta put it in January 1611, the Jesuits' and the pope's endorsements would have secured the universal acceptance of his discoveries (and of the Medici celestial glory).[17] Rome's relevance in Galileo's career was increasing. If in 1588 Rome meant Clavius as an individual patron, in 1611 it meant the Jesuit mathematicians and the pope.

After 1611, thanks to Medici connections, Galileo's patronage networks began to branch increasingly toward Rome. Cardinals, prelates, Roman aristocrats, the Lincei, and the Jesuits were their nodes. At the same time, Florence was declining into an increasingly provincial cultural and political center, especially after Cosimo II's death in 1621. Galileo's shift of focus from Florence to Rome became clearer after 1623, when his friend Maffeo Barberini became Pope Urban VIII and several of his Lincei colleagues obtained high posts at the papal court. Galileo's Roman patronage connections became progressively stronger as the legitimation of his new socioprofessional identity and of his increasingly Copernican cosmological perspectives depended more and more on the papal prince.

In a sense, one may even say that Florence became for Galileo a somewhat sleepy but comfortable suburban base from which he could do business with Rome. Moreover, since he was the philosopher of the grand duke, Galileo could visit Rome as an official envoy of the Medici. In a status-bound society this was a tremendous advantage. Galileo would not be perceived and treated as a private gentleman but as a special "scientific ambassador" of the Medici. Usually, he would arrive in Rome with an impressive portfolio of letters of introduction for the Roman intelligentsia and be hosted at the Medici's Trinità dei Monti palace. As with the distribution of telescopes and copies of the *Sidereus nuncius* on which he built his international visibility, the Medici political networks allowed him access to Roman centers of power he could have never approached as a university professor of mathematics or as a private gentleman.

17. *GO*, vol. II, no. 464, pp. 28–29.

Although the courtly patronage system eventually propelled Galileo's career toward a dramatic ending, it also provided him with powerful resources along the way. Even when things took a bad turn for him, as when Copernicus was put on the Index in 1616, Galileo could have suffered much more serious consequences had he not been so well connected to so many cardinals, prelates, and Roman aristocrats.[18] That Galileo's name was not included in the condemnation of 1616 was a remarkable privilege he likely received because of his position as the grand duke's philosopher.

The nostalgia of an elderly, blind, and house-arrested Galileo for the "happiest eighteen years" of his life, spent in Venice and Padua with his gang of rich, libertine, anti-Jesuit, freethinking young patrician friends, is more than understandable. However, his statement should not be taken at face value, as the admission of a strategic mistake he might have committed when he left Padua for Florence in 1610.[19] Contrary to Giovanfrancesco Sagredo's proud statements, Venice was not exactly the Garden of Eden and the powerful protector of intellectuals he presented it to be in his letters to Galileo. We should not forget that Venice did not prevent Giordano Bruno's fatal extradition to Rome in 1592.[20] Also, it was Galileo's friend and combative theologian Paolo Sarpi who, when Copernicus's *De Revolutionibus* was put on the Index in 1616, advised the Venetian Senate that it was not worth challenging the edict because the Copernican doctrine "does not affect in any way the power of princes or favor them, and the temporal authority receives no benefit from it, nor does it touch the art of printing in the state, since it is sure that none of these books has ever been printed in Venice."[21]

Similarly, when in 1624 Fulgenzio Micanzio, Sarpi's friend and biographer, confidently told Galileo that he could have the *Two New Sciences* printed in Venice, he quickly had to come to terms with the unexpected limits of the Venetians' willingness to challenge the authority of the pope in the name of freedom of thought.[22] Despite quite understandable nostalgic myths spun by Galileo, his friends, and sympathetic historians in the light of the trial of 1633, it is not clear whether Venice would have been a much safer place for Galileo than Florence or Rome. Certainly, it

18. See also Richard S. Westfall, "Galileo and the Jesuits," in *Essays on the Trial of Galileo* (Vatican City: Vatican Observatory, 1989), pp. 31–57.

19. *GO*, vol. 18, no. 4025, p. 209.

20. Gaetano Cozzi, "Galileo Galilei, Paolo Sarpi e la società veneziana," *Paolo Sarpi fra Venezia e l'Europa* (Turin: Einaudi, 1969), p. 142.

21. Paolo Sarpi, "Sopra un decreto della congregazione in Roma in stampa presentato per l'Illustrissino Signor Conte del Zaffo a 5 maggio 1616," in *Opere,* ed. Gaetano Cozzi and Luisa Cozzi (Milan-Naples: Ricciardi, 1969), p. 604.

22. *GO*, vol. 16, no. 2903, p. 61; no. 3057, p. 193; no. 3075, p. 209; no. 3088, p. 230; no. 3098, p. 239.

did not offer him the powerful options for patronage and self-fashioning he could find in Florence and Rome.

As we have seen, Galileo tried to leave Padua and Venice as early as 1604. Moreover, during his stay in Padua he was also quite careful to remain on good terms with the Jesuits. The two moves could be related. As indicated by Gaetano Cozzi, Galileo's attitude toward the Jesuits was much less radical than those of Sagredo and Galileo's other young patrician friends.[23] Not only did Galileo obtain the position at Padua through patrons close to the Jesuits, but he then managed to balance himself between the more conservative, counter-reformist political faction of the *vecchi* and that of the *giovani,* the latter a group that stressed the need for Venice to regain political independence from the Church, Spain, and the Austrian Hapsburgs.[24] When Venice found itself in the *Interdetto* crisis, Galileo avoided taking sides and barely mentioned those events in his correspondence. Instead, in that same period, his good friend Sagredo (a former student of the Jesuits) was busy staging a farce at the expense of the former Venetian Jesuits, now exiled to Ferrara. Pretending he was an elderly Venetian noblewoman, Cecilia Contarini, Sagredo began to correspond with Father Barisone, rector of the College of Ferrara, asking advice on delicate spiritual matters. He then gathered the correspondence between Barisone and "Cecilia Contarini" and widely circulated it in Venice to deride the Jesuits' religious practices and political views. Apparently, the Contarini-Barisone correspondence was a local hit.[25]

Although Galileo greatly appreciated the wit and friendship of his patrician libertine friend, he never nominated him for election to the Accademia dei Lincei although he had much better scientific credentials than

23. Cozzi, "Galileo Galilei, Paolo Sarpi e la società veneziana," 135–234. While I find Cozzi's basic thesis about Galileo's oscillations between these two groups of friends and patrons convincing, I think that his sociopolitical taxonomy between "giovani" and "vecchi" sometimes becomes confusing. Also, occasionally he seems to blur the distinction between "giovani" and "vecchi" as political and generational categories.

24. Beginning with Guidobaldo (a good friend of Clavius) and continuing with Pinelli and many others, Galileo surrounded himself with Jesuit-friendly patrons like Benedetto Zorzi, Giacomo Contarini, Paolo Gualdo, Lorenzo Pignoria, and later Pietro Duodo. Further, as shown by Wallace, Galileo was on good terms with the Jesuits of the Paduan college (William A. Wallace, *Galileo and His Sources* [Princeton: Princeton University Press, 1984], pp. 269–72). For a detailed history of the formation of the "giovani" faction and on the political culture of the Venetian patriciate at the turn of the century, see Gaetano Cozzi, *Il Doge Nicolò Contarini* (Rome-Venice: Instituto per la Collaborazione Culturale, 1958), pp. 1–147.

25. On Sagredo's epistolary masquerade, see Antonio Favaro, "Giovanfrancesco Sagredo," in Paolo Galluzzi, ed., *Amici e corrispondenti di Galileo* (Florence: Salimbeni, 1983), 1: 208–10. Sagredo's action must have been well known if it was mentioned in a poem dedicated and read to him at a farewell party before his departure for Syria as the Venetian consul (Antonio Favaro, "Serie decimasesta di scampoli galileiani," *Atti e memorie della R. Accademia di Scienze, Lettere ed Arti in Padova,* new series, 22 [1905–6]: 10–13).

most (if not all) of the candidates Galileo ever nominated.[26] The political
choices behind Galileo's selection criteria become clear once we consider
his warm support of Welser's election to the Lincei. An important political
figure and banker from Augsburg, Welser was a financier and advisor to
Rudolph II, a strong supporter of the Jesuits, the probable author of a
strongly anti-Venetian pamphlet (the *Squitinio*) published in 1612, and
a friend of Clavius and of Galileo's pro-Jesuit friends Gualdo and
Pignoria.[27] As one might guess, Sagredo and Welser did not quite like
each other.[28] Galileo's attitude toward his longtime friend and collabora-
tor Paolo Sarpi reflects the same caution he used with Sagredo.[29]

Galileo's distancing himself from Sagredo, Sarpi, and the Venetian
giovani indicates that he did not want to compromise himself with associa-
tions that might have hurt him in the counter-reformist princely courts he
was aiming at: Mantua, Florence, and, indirectly, Rome. If it was true, as
Ciampoli argued, that the courtier should always be on good terms with
the Jesuits,[30] such advice was a prescription for a Catholic mathematician

26. Galileo's failure to nominate Sagredo is striking once we see that, in June 1612, Cesi
asked him explicitly whether he had somebody in Padua he considered worth electing to the
Lincei (*GO,* vol. 11, p. 312).

27. This attribution is made by Pierre Gassendi in his biography of Peiresc, *Viri illustris
Nicolai Claudii Fabricii de Peiresc, senatoris Aquisextiensis vita* (The Hague: Vlacq, 1655).
Welser's anti-Venetian feelings during the *Interdetto* and his disappointment in what he
viewed as Paul V's weakness against the Venetians are well documented in the letters to Faber
reproduced in Giuseppe Gabrieli, "Vita romana del 600 . . .", in *Atti del Primo Congresso
Nazionale di Studi Romani* (Rome: Istituto di Studi Romani), 1: 823–24.

28. *GO,* vol. 11, p. 314, p. 505; vol. 12, p. 45. Also, it is quite probable that Galileo (and
Filippo Salviati) were behind Welser's election, on 4 September 1613, to the Florentine Ac-
cademia della Crusca (Severina Parodi, *Catalogo degli Accademici dalla Fondazione* [Florence:
Sansoni, 1983], p. 56).

29. Around the time of the *Interdetto,* Galileo also began to move away politically from
Paolo Sarpi—the Venetian intellectual who symbolized the political and religious views of
the *giovani.* Although Galileo and Sarpi interacted again in 1609 regarding the development
of the telescope, nowhere in either the *Sidereus nuncius* or in the *Assayer* does Galileo mention Sarpi by name (Albert Van Helden, "Galileo and the Telescope," in Paolo Galluzzi, ed.,
Novità celesti e crisi del sapere [Florence: Giunti Barbera, 1984], pp. 149–53). Although this
may have largely to do with Galileo's trying to take as much credit for the invention as he
possibly could, it is also true that concerns for scientific credit and religious caution went
hand in hand in this case. Galileo was quite right in perceiving a potential political danger in
being represented as a close friend of Sarpi, since later developments indicate that his association with the Venetian theologian was repeatedly used as an innuendo about his alleged
religious unorthodoxy. For instance, Tommaso Caccini told the Holy Office that Galileo "is
held in suspicion on matters of faith because they say he is very intimate with Fra Paolo of the
Servi, so famous in Venice for his impiety, and they say that they still correspond" (*GO,* vol. 19,
pp. 309–10). Although Sarpi died in 1623, the letter Galileo wrote to him from Florence in
February of 1611 is the last item we have of their correspondence (*GO,* vol. 11, pp. 46–50).

30. "For hate or passion or spirit of contradiction, or because of natural inclination, one
should never try to antagonize a commonly esteemed religious order, like that of the Jesuits
today" (Giovanni Ciampoli, "Discorso di Monsignor Ciampoli sopra la corte di Roma," in

like Galileo, whose professional status was strongly dependent on their endorsement.[31]

The first letter we know Galileo wrote after his arrival in Florence in September 1610 was addressed to Clavius. It contained an apology for the long silence that, as a result of the *Interdetto,* he had to maintain while in Padua.[32] Galileo concluded by informing Clavius of the visit to Rome he was then planning and which eventually materialized in the spring of 1611. That visit was a crucial event in his career. Clavius and his students gave a public and enthusiastic endorsement of Galileo's discoveries (an event conspicuous enough to be reported in the *Avvisi di Roma*) which greatly boosted Galileo's status:[33]

> On Friday evening of the past week in the Collegio Romano in the presence of cardinals and of the Marquis of Monticelli [Cesi], its promoter, a Latin oration was recited, with other compositions in praise of Signor Galileo Galilei, mathematician to the grand duke, magnifying and exalting to the heavens his new observation of new planets that were unknown to the ancient philosophers.[34]

In his 1611 visit to Rome, Galileo began his connection with another institution that was to play a major role in his later career and patronage strategies: the Accademia dei Lincei, one of the first institutions dedicated to the new natural philosophy. The same *avvisi* in which we find the description of Galileo's triumph at the Collegio Romano reported that

> On Thursday evening [April 14], the Marquis of Monticelli [Federico Cesi], nephew of Cardinal Cesi, was the Maecenas of a

Marziano Guglielminetti and Mariarosa Masoero, "Lettere e prose inedite (o parzialmente edite) di Giovanni Ciampoli," *Studi secenteschi* 19 (1978): 237. An anonymous "guide" to the Florentine court, ("Avvertimenti per uno che entra in corte") makes the same point (*ASF,* "Miscellanea Medicea 502, fol. 317v).

31. Although the identification of the Jesuits with Counter-Reformation culture and politics is an oversimplification, it was a perception shared by Galileo's Venetian friends. To Sagredo (but also to several Venetian ambassadors), the political map of Italy could have been drawn as simply divided into two areas, Venice (or "free Italy") on one side and "Jesuitland" (or territories controlled or influenced by Spain) on the other. Sagredo's perception of the Jesuit hostility towards Venice was not idiosyncratic, but was explicitly shared by the Venetian ambassadors well after the *Interdetto* (Pietro Contarini in Barozzi and Berchet, *Relazioni,* 2: 183, 189). Galileo's cautious behavior in Venice shows that he did not want to lose access to "Jesuitland."

32. "In light of your prudence, it is not necessary for me to be specific about the reason why [I have not written] until today, that is, while I was in Padua" (*GO,* vol. 10, no. 391, pp. 431–32).

33. When he arrived in Rome that spring, Galileo found his elderly patron surrounded by a group of young mathematicians. This was Clavius's "Mathematical Academy," a group of mathematicians that would play a major role in Galileo's later career.

34. Orbaan, *Documenti sul barocco in Roma,* 2: 284.

banquet in the vineyard of Monsignor Malvasia outside the San Pancrazio gate in which he, the said cardinal, and Signor Paolo Monaldesco, his relative, participated. In this high and open site they met together with the aforementioned Galileo, Terrentio from Flanders, Signor Persio who is with Cardinal Cesi, Galla [Lagalla] who lectures in our university, Cardinal Gonzaga's Greek mathematician [Demisiani], Signor Piffari lecturer at Siena, and eight others. Some of them had come from elsewhere to participate in this observation. Although they remained until one o'clock in the morning, they still could reach no consensus.[35]

About the same time Galileo was getting acquainted with several of his future fellow Lincei and his discoveries were being hailed at the Collegio Romano, another person who would later become one of the leaders of the Lincei and Galileo's most influential broker was also trying to gain visibility among Roman patrons.

Giovanni Ciampoli was the precocious teenager described in another *avviso* of May 1611: "In the academy held Tuesday in the palace of Cardinal Deti, in the presence of seven cardinals, a young pupil of Signor Giovan Battista Strozzi gave a delightful discourse on silence."[36]

A few weeks later, Galileo and Ciampoli were riding back to Florence in a Medici carriage, much satisfied with the recognition they had received in Rome.[37] One had established his international reputation, strengthened his ties with the Jesuits, and gained membership in the Accademia dei Lincei. The other had impressed the Roman cardinals with his prodigious literary gifts which, a few years later, would take him to the highest level of the Roman court. Although Galileo and Ciampoli were by no means disciplinary colleagues, their later careers not only showed remarkable structural similarities and timing patterns but were largely based on the patronage contacts with the Roman court they had developed in those weeks.

Archipelagoes of Power

Rome was a crucial forum for Galileo but not one he was very familiar with. Cesi, Ciampoli, Cesarini, and his other brokers and supporters provided him with some of the necessary bearings. However, there were differences between the Florentine and Roman courts that could not easily

35. Ibid., 283.

36. Ibid., 284. Galileo may have been at Deti's on that occasion because, in a letter to Virginio Orsini, he described a previous meeting of the same academy (*GO*, vol. II, pp. 82–83). For a previous but similar performance of Ciampoli in Rome in 1607, see Guido Bentivoglio, *Memorie e Lettere*, ed. Costantino Panigada (Bari: Laterza, 1934), p. 75.

37. *GO*, vol. II, no. 538, p. 121.

be remedied by intelligence work. If in Florence Galileo had been able to use Medici dynastic mythologies to legitimize his discoveries and socioprofessional identity, he could not rely on anything comparable in Rome. Also, in Rome it was much more difficult to develop that type of exclusive patronage relationship that he had been able to establish with Cosimo in Florence. While all courts of absolute princes resembled each other to some extent, the Roman court displayed important peculiarities. These unique features were to play a major role in the development and conclusion of Galileo's later career.

The correspondence of Roman literati, the *avvisi,* and contemporary diaries are sprinkled with reports of meetings of academies and with other cultural and political events.[38] The topic and the speaker are not always reported, but the *avvisi* invariably notice the host of those academic gatherings, in whose place (or palace) they occurred, and which cardinals, aristocrats, ambassadors, and high prelates were present. Such notices were snapshots of the Roman elite.

An *avviso* of January 9, 1621, reports that "On Monday in the house of the Count of Novellara [Alfonso Gonzaga] a literary academy was announced, as had also occurred a week before. A wonderful lecture was given in the presence of Cardinals del Monte, Bandini, Bevilacqua, and d'Este. there were also many prelates and other nobles present."[39] The same indifference to the topic and speaker is displayed by an avviso of May 1609 reporting the activity of Cardinal Deti's academy, one that was visited by Galileo during his visit to Rome in the spring of 1611 and that saw the debut of precocious Ciampoli on the Roman scene.[40] "On Tuesday in the house of Cardinal Deti, the usual academy was held, attended by Cardinals Camerino, Bandini, Bellarmino, Ginnasio, Sannesio, and Delfini."[41]

---

38. On Roman academies, see the classic (but not always reliable) Michele Maylender, *Storia delle accademie d'Italia,* 5 vols. (Bologna: Cappelli, 1926–30); G. M. Garuffi, *L'Italia accademica, o sia le accademie aperte a pompa e decoro delle lettere più amene nelle città italiane* (Rimini: Dandi, 1688); Francis W. Gravit, "The *Accademia degli Umoristi* and Its French Relationships," *Papers of the Michigan Academy of Science, Arts, and Letters* 20 (1935): 505–21; Piera Russo, "L'Accademia degli Umoristi, fondazione, strutture e leggi: Il primo decennio di attività", *Esperienze letterarie* 4 (1979): 47–61; Luisa Avellini, "Tra Umoristi e Gelati," *Studi secenteschi* 23 (1982): 109–37; Renato Lefevre, "Gli Sfaccendati," *Studi romani* 9 (1960): 154–65.

39. Venceslao Santi, "La storia nella *Secchia rapita,*" *Memorie della Reale Accademia di Scienze, Lettere e Arti in Modena,* series 3, 9 (1910): 263. A similar *avviso* is reprinted in Orbaan, *Documenti sul barocco in Roma,* 2: 271. It was inaugurated by an oration by Agostino Mascardi included in *Prose vulgari* (Venice: Baba, 1653).

40. *GO,* vol. 11, no. 510, pp. 82–83; Orbaan, *Documenti sul barocco in Roma,* 2: 284.

41. Santi, "La storia nella *Secchia rapita,*" p. 262. Also, in an *avviso* of 29 April 1608 we find that "the other day for the second time the newly formed academy met in the house of Cardinal Deti, where many cardinals and a great number of prelates and gentlemen of the court intervened" (Orbaan, *Documenti sul barocco in Roma,* 2: 227).

INTERMEZZO

Evidently, the importance of these gatherings was judged by the quality of hosts and guests more than by that of the speakers. Interestingly, the *avvisi* did not seem to differentiate descriptions of literary events from those of banquets, comedies, weddings, jousts, receptions, and theatrical performances. For instance, on February 16, 1611, "On Monday evening in the palace of Cardinal Montalto the play called the Fable of Psyche was recited again. Present were Cardinals Cosenza, Monti [del Monte], Borghese, Montalto, and Peretti. Also present were Signor Francesco Borghese, Ambassador of Savoy, Constable Colonna, the Duke of Bracciano [Orsini], Altemps, and other nobles."[42]

The *avvisi*'s indifference to the genre of the event probably reflects the audience's own indifference. As with court spectacles, these events were occasions on which the Rome that counted displayed its hierarchies to itself. Where one was seated was more meaningful than the topic or genre of the event. Even the meetings of an unusual academy like the Lincei did not seem to differ from the ordinary banquets-cum-disputes.[43] What a historian of the Lincei has termed the "convivial academic meeting" of August 1613 was reported by an *avviso* as:

> On Sunday, the Duke of San Gemini had a banquet in his vineyard on the Pincian hill with Signori Cesarini, his nephews, and other relatives. . . . On Wednesday morning Prince Cesi offered a sumptuous banquet to the same Signori Cesarini and to other gentlemen and prelates in his palace near Saint Peter's, along with a handful of the foremost literary figures of the city. They engaged in disputations among themselves well into the night.[44]

It was probably during one of these events that Cesi managed to convince his cousin Virginio Cesarini to move from the Aristotelian camp into the Galilean one.[45]

Academies and public events were not organized by cardinals and Roman aristocrats only. Ambassadors and religious orders (especially the Jesuits) were also competing in sponsoring conspicuous events in order to develop or maintain close ties with the pope, the cardinals, and other political figures.[46] For instance, in 1625 the Venetian ambassador described

42. Ibid., p. 283.

43. Ibid., 278.

44. Ibid., 211.

45. Similarly, the banquet Cesi offered to Galileo, his uncle Cardinal Bartolomeo, and a number of Roman literati in the vineyard of Monsignor Malvasia in April 1611 fit a well-established social genre of vineyard-based events frequently encountered in the *avvisi*.

46. The pope's nephews were also engaging intensely in these types of patronage. Given their particularly powerful position, their academies were unusually well attended by cardinals. Sometimes the pope himself would be present, usually in a special niche attached to, but not completely integrated in, the academic space (Santi, "La storia nella *Secchia rapita*,"

his Spanish colleague's attempt to sway Urban VIII from his policy of equidistance between Spain and France by courting Urban's nephews, "often accompanying them hunting and in other similar recreations with great familiarity, to smooth the way for more important things. Having introduced comedies into his palace at Trinità dei Monti, he pulled many other cardinals and almost all the nobility into his circle."[47]

Comparable strategies were adopted by the French ambassador, who routinely organized comedies, operas, banquets, and tournaments to captivate cardinals, prelates, and Roman barons.[48] The Savoia too were determined to use *ricreazioni* to foster political interests. Since his arrival in Rome in 1621, Cardinal Maurizio di Savoia engaged in conspicuous spending and cultural patronage (dutifully reported in the *avvisi*) to show that his family was good at other things besides waging wars. The important Accademia dei Desiosi was part of this program.[49] Maurizio was so *magnifico* that, despite the abundant funds he received from the Turin court, by 1627 he went bankrupt and left Rome to avoid "the pressure of the creditors who crowded around him."[50]

Beside framing liturgical events (like the *Quarantore*) as theatrical spectacles through elaborate stage sets in the Chiesa del Gesù the Jesuits presented the graduation of their best (and usually noble) students as public events. For instance, "On the first of September [1614], D. Francesco Giuvara [Guevara?] defended his philosophical theses in the

---

vol. 9 (1910): 264–65). The academy met regularly to discuss religious subjects and became known as the Virtuosi (Ludwig von Pastor, *History of the Popes* [St. Louis, Mo.: Herder, 1938], vol. 27, pp. 69–70). Because of the close papal connection, the topics of discussion at these academies tended to be limited to commentaries on the Scriptures. Urban's nephew, the Lincean Cardinal Francesco Barberini, sponsored one of these academies. For example, on 17 July 1624: "Sunday after lunch the public academy of the Virtuosi was held in the rooms of Cardinal Barberini at Monte Cavallo. Monsignor Castracani gave a learned and elegant discourse on fortitude by taking inspiration from the Holy Scripture in the Book of Maccabbees. It was then responded to by a gentleman of the House of Rospigliosi [the future Clement IX]" (Santi, "La storia nella *Secchia rapita*," 9 (1910): 264).

47. Barozzi and Berchet, *Relazioni*, 231–32. This strategy must not have been a new one; in fact, twelve years earlier Virginio Cesarini had gone with Cardinal d'Este to see a comedy at the Spanish ambassador's house (Venceslao Santi, "La storia nella *Secchia rapita*," *Memorie della Reale Accademia di Scienze, Lettere e Arti in Modena*, series 3, 6 [1906]: 315).

48. Ibid., pp. 313–15.

49. Cardinal Maurizio's efforts must have been very successful, if two years after his arrival in Rome he was "admired in Rome not less for his piety and exemplary behavior, than for his nobility of blood. He united two things that one rarely sees together: grand courtly magnificence and splendor, and a life so innocent and of such marvelous customs that it impresses even those who profess to have a very chaste life-style. Therefore he is universally loved at court, and the pope has high esteem for all his deserving qualities" (Barozzi and Berchet, *Relazioni*, 165).

50. Francesco Luigi Mannucci, "La vita e le opere de Agostino Mascardi," *Atti delle Società Ligure di Storia Patria* 42 (1908): 155.

Collegio Romano in front of 24 cardinals and other important Signori."[51]
The popularity of these defenses and of the public debates (like the one
that started the dispute on comets) at the Collegio Romano may have encouraged the Jesuits to engage in more sophisticated cultural productions. During the carnival of 1623 and on the occasion of the canonization
of St. Francis Xavier, the tragedy *Il Primato,* performed in the theater of
the Collegio Romano, was

> recited several times, with the intervention of various cardinals,
> princes, and other important signori. It was a beautiful and pleasing
> event to all due to its excellent composition, staging, and the new
> and rich costumes of the speakers. Pleasing above all were the grace-
> ful machines employed, the clouds, the staging, the military games,
> the dances, and the musical performance, which caused many to
> turn to watch them.[52]

These types of events were quite common at the Collegio Romano.
The year before, a spectacular *Apoteosi* of Sts. Ignatius and Francis Xavier
was performed at the Collegio Romano. It was the highlight of the celebrations for the canonization of the two Jesuits. The writer of the play and
architect of its complex theatrical machines was Galileo's opponent during the dispute on comets and the architect of the new Chiesa del Gesù—
Orazio Grassi.[53]

Given their closeness to the pope, it is not surprising that his nephews
focused on more serious topics in their academies. One such academy met
in June 1622:

> Last Sunday after dinner Cardinal Ludovisi [Gregory XV's nephew]
> reconvened the academy in his rooms at the Quirinal, which is now
> held every 15 days during this summer season. Monsignor de' Rosis
> recited a learned and elegant letter in the vulgar tongue on flattery,
> which was discussed in a learned manner by signori Stefano Man-
> nara, secretary to Cardinal del Monte, and Girolamo Preti, gen-

51. F. Cerasoli, "Diario di cose romane degli anni 1614, 1615, 1616," *Studi e documenti di storia e diritto* 15 (1894): 280.

52. From an *avviso* quoted in Filippo Clementi, *Il Carnevale romano nelle cronache contem-poranee* (Rome: Tiberina, 189; reprint, Città di Castello: Unione Arti Grafiche, 1939), p. 364. The success was such that the pope decided to go to see it, therefore creating a true etiquette hurdle: "And the master of ceremonies had already gone to find a good spot from which he could see without being seen, but he was eventually dissuaded [from attending]" (ibid.). Another *avviso* on this play is reproduced in Santi, "La storia nella *Secchia rapita,*" 6 (1906): 315.

53. *Argomento dell'apoteosi o consagrazione de' Santi Ignatio Loiola e Francesco Saverio rap-presentata nel Collegio Romano nelle feste della loro canonizzazione* (Rome: Zanetti, 1622); and Carlo Bricarelli, "Il P. Orazio Grassi architetto della Chiesa di S. Ignazio in Roma," *Civiltà cattolica,* 2 (1922): 21, 24.

tleman of Cardinal Ludovisi. This occurred in the presence of His Most Illustrious Lordship [Cardinal Ludovisi] and of the Cardinals Bandino, Ubaldino, Santa Susanna, Sacrato, Gozzadino, and Aldo-brandino, as well as many prelates and other nobles. Because Our Lordship [the pope] had intervened, he remained retired in the small chapel in the same room of the aforementioned Most Illus-trious Ludovisi.[54]

The *avvisi* indicate that, within the limits imposed on them by their status, wealth, and religious affiliation, all those who counted in Rome did their best to gain or keep a place on the map of Rome's cultural field. Although cardinals, aristocrats, members of religious orders, and ambas-sadors had different resources and skills available, they all tried to get at-tention, distinction, and power by sponsoring some form of conspicuous and fashionable event. The picture that emerges (especially if one reads the *avvisi* extensively) is one of cultural proliferation, of the need for cul-tural distinction, of rapid change in trends, fashions, and of players.[55]

The remarkable proliferation of academies in Rome in this period was directly connected to this relatively fragmented and highly competitive power struggle which, in turn, reflected the peculiarity of the papal rule and court.[56] In fact, different from other Italian and European political

54. Santi, "La storia nella *Secchia rapita*," 9 (1910): 263–64. At the same academy, on 27 August 1622, from the small chapel attached to the apartment of Cardinal Ludovisi, Our Lord intervened privately with the usual academy in which, in the presence of seven cardi-nals, many prelates, and other nobles, Father Giovanni Battista Riccardi, called Father Mon-ster [later involved in the scandal of the *Dialogue*], gave a learned and elegant lecture in the vulgar tongue, and Signor Girolamo Aleandri (secretary of Cardinal Bandini) commented skillfully . . . on those words of Job in chapter 21: 'Dulcis fuit glareis Cocyti,' demonstrating that lay poets have utilized the Holy Scriptures" (ibid., 264). Another *avviso* of 13 August reports that "the usual ecclesiastical academy was held in the rooms of Cardinal Ludovisi with the intervention of many cardinals and prelates and nobles, in which Signor Rinuccini [nominated to the Lincei by Galileo] lectured . . . on those words from Isaiah the prophet: 'Tollite vobiscum verba et convertimini ad Dominum.' After this followed a second discourse by Signor Girolamo Maricucci, secretary to Monsignor Archbishop Volpio, . . . and Signor Francesco della Valle [brother of the Lincean Pietro], and everybody was praised highly" (ibid.).

55. Santi, "La storia nelle *Secchia rapita*," 6 (1906): 310–33, 9 (1910): 247–397; Cerasoli, "Diario di cose romane," 263–301; Orbaan, *Documenti sul barocco in Roma;* and Clementi, *Il Carnevale romano nelle cronache contemporanee,* vol. 1. Also useful are Maurizio Fagiolo dell'Arco and Silvia Carandini, *L'effimero barocco: Strutture della festa nella Roma del '600* (Rome: Bulzoni, 1978) 2 vols.; Marcello Fagiolo and Maria Luisa Madonna, eds., *Barocco romano e barocco italiano* (Rome: Gangemi, 1985); and Francis Haskell, *Patrons and Painters* (New Haven, Conn.: Yale University Press, 1980), pp. 3–166. Although it does not focus on the court and its culture, Nussdorfer, "City Politics in Baroque Rome," is very useful for a reconstruction of the broader Roman context. See also her *Civic Politics in the Rome of Urban VIII* (Princeton: Princeton University Press, 1992).

56. During the seventeenth century, no less than 132 academies were founded in Rome. This is an absolute record of natality for Italy and possibly for Europe (Amedeo Quondam,

centers, the Roman court was of a religious and nondynastic nature. Consequently, unlike in Florence where the culture of the academies tended to reflect a Medici cultural program, Rome's academic scene was both more intense and less directly supervised by the absolute prince.[57] These two features are closely connected.

A nondynastic prince, the pope did not have a specific "master narrative" to inform the activities of his court and academies. No pope had a family mythology like the one Galileo played into so well with his dedication of his discoveries to the Medici. Regardless of one's power, everyone was a tenant (or a short-term owner) in Rome. Although the Church obviously had a strong cultural tradition based upon the Scriptures and their interpretation, such a tradition did not determine the specific expressions of Roman court culture, which were often strongly influenced by a pope's specific tastes and idiosyncrasies. In particular, religious tradition was apparently not central to the culture of the cardinals' courts which, on average, do not seem to have been imbued with theological sensitivity. The discourse of the Roman courts was an intricate and shifting mix of religious and secular ingredients as well as mythologies about ancient Rome. In this, they reflected the ambiguities of papal rule, that is, of a prince who was both religious and secular.[58]

Moreover, as a religious prince, the pope could not get involved in too explicitly secular activities. Only very specific (usually liturgy-related) types of spectacles could be performed at his court. At least in principle, cardinals could not engage in secular forms of entertainment, could not go hunting, and were supposed to wear appropriate dress when participating in secular academies.[59] As a result, the other political constituencies present in Rome could occupy the cultural space left undeveloped

---

"L'Accademia," in Alberto Asor Rosa, ed., *Letteratura italiana*, vol. I, *Il letterato e le istituzioni* [Turin: Einaudi, 1982], p. 864).

57. However, at least at the level of policy, comedies and other forms of entertainment were to be carefully overviewed by the inquisitors. An edict of 1658 stated: "It is prohibited to anybody, even ecclesiastics, of whatever state, rank, or position, under pecuniary or corporal punishment at the will of His Most Illustrious Lordship, to perform in or have performed in whatever public or private place, even in one's own house, with doors closed or open, even only among friends and relatives, comedies, gypsy plays, or representations of any sort, also spiritual ones, without express license of His Most Illustrious Lordship" (Orbaan, *Documenti sul barocco in Roma,* 2: 282, note).

58. At best, the "master narrative" of the Roman upper class was very fragmented. Its continuity (besides that provided by scriptural interpretation) was provided by Rome itself, its history, and its monuments and ruins. This may explain some of the Roman courtiers' fascination with ruins and archaeology.

59. "Cardinals should never go to comedies or similar things. If they do, they should inform [the host], and they should not wear the *Berretta,* but rather the cap and the caped cassock. When he went to some comedy or similar party, the Cardinal of Florence would stay in a secluded spot . . . in order not to be seen" (Lunadoro, *Relatione della corte di Roma,* p. 55).

by the papal court and try to use it to their advantage. This may explain the remarkable proliferation of academies and other spectacles in Rome.

The Roman academic scene was not so much a prince-controlled space as a kind of political showcase featuring those who aimed at reaching the Holy See or at maintaining or developing strong ties to the present or future pope. In a sense, the space occupied by the many Roman academies was a jammed piazza located at the crossroads of the papal court, the cardinals' courts, the ambassadors, the Roman aristocrats, and the leading religious orders—a space that displayed continuously shifting patterns of people, alliances, and cultural tastes.

This shifting environment was Galileo's stage in Rome. Banquets in Roman palaces or suburban vineyards (as the one Cesi organized in honor of Galileo and his telescope in 1611), comedies featuring the Medicean Stars (as at the Peretti-Cesi wedding in 1614), the academies (public or private) were some of the spaces where the new natural philosophy was proposed and discussed. As shown by his correspondence and by descriptions of his performances in defense of Copernicus in Roman academies in 1616, these were the sites where Galileo promoted himself and the new natural philosophy, gained supporters, met challenges, and alienated opponents. And, we should not forget, he had access to these privileged spaces through his Medici patronage connections.

Rome's cultural and patronage scene could be compared to a volcanic archipelago subjected to rapid cycles of change. Like islands that emerged and sank, the cardinals' courts, the Roman barons, and (to a lesser extent) the religious orders tried to established themselves as centers of cultural activities and patronage. Most of these clusters appeared and disappeared with the people who established them. The papal court was the largest, most powerful (and therefore most dangerous) volcano of the archipelago. Clients moved through the peripheral islands trying to jump onto the pope's island at the right moment, that is, at the beginning of a new pontificate.

Ambiguity in one's allegiances was a structural need at the Roman court.[60] While a courtier's strong allegiance to his or her prince would have been rewarded in any dynastic court, it would have been a very simpleminded strategy in Rome; there, bonds had to be kept to the minimum so that one could be ready to jump onto another's faster-moving bandwagon on short notice. This does not mean that persons could shift from one patron to another with impunity. Clients needed a strong patron

60. Chronicles of the conclaves show that cardinals whose political allegiances were explicit did not have much chance of being elected. Instead, those who managed to steer between Spain and France without making too many enemies, and who did not have strong blood connections with Italian princely families, had a chance of becoming pope. Potential candidates needed to maintain a low profile. What held true for the pope also held for any courtier below him.

and they needed to be somewhat loyal to keep the patron's support. However, they were careful in not alienating other possible, future patrons. Moreover, because of the quickly shifting court power structure, it was unwise to antagonize someone who today might be your inferior but tomorrow could turn up above you. While *magnificentia* was the keyword in Rome, arrogance was not. It would have been stupid to insult people in Rome. In a handbook for secretaries of princes, Panfilo Persico advised them to be generous with the titles they used in letters because "in the court of Rome, and in the ecclesiastical republic there is no person of such low condition, that could not rise at some time to a grand state, as experience shows every day. Thus prudence teaches that one must pay attention to everyone and honor them more than they deserve."[61]

I do not want to say that Rome was a place in which factions and tensions did not exist and everything was solved by compromise and glossing over differences. On the contrary, the pattern of differences and tensions was impressive, but it was also continuously shifting. It is not just that people had specific identities, stances, views, that were then rhetorically masked. Rather, Rome was a place where, because of the dynamics inherent in the court structure, one's identity was continuously negotiated.

There may be a relationship between the volatility of Roman power scenarios, the lack of a prince-managed cultural master narrative, and the taste for eclecticism, for the ephemeral, and for literary wit that one found especially in Rome, the capital of the baroque. If cultural eclecticism characterized all baroque courts because of its symbiosis with the discourse of absolutism (and the culture of noncommitment that went with doctrines of *raison d'état*), in Rome we see that this cultural predicament was made extreme by the peculiar lack of master narratives and of a relatively stable cultural framework. If Roman courtiers appreciated unique "cultural gems" rather than complex and possibly coherent programs or philosophical systems, that was probably a reflection of their own identity.[62] In a sense, successful courtiers were like "gems" themselves, atomized individuals trying to go up in the social scale by displaying their singularity and uniqueness in continuously and unpredictably changing scenarios. If the "spirit of system" was alien to Roman culture, it was also because of the peculiar contingency that permeated the courtiers' lives, careers, and identities. It may not be accidental that what we call the baroque found its richest niche in Rome.

---

61. Persico, *Del segretario*, 171.

62. In a baroque court (and especially in a particularly unstable one like that of Rome), one's identity could not be safely anchored to a permanent system of reference. It was associated with one's *virtù*, something that (as court literature had been stressing since Castiglione) could not be defined. *Virtù* resided only in the presentation of one's uniqueness.

Gems, Flowers, and Other Fragments

More than anywhere else in Italy, Roman academies were antechambers to the court—places of entertainment and courtly training and recruiting.[63] Not only many of those who performed in academies had or were seeking a post at court or in cardinals' households, but—given the high turnover of popes and cardinals—the Roman job market was an active one, and cardinals or other major players were eager to monitor possible brilliant recruits. In Rome, academies were not institutions established by princes to domesticize their subjects by keeping them busy, but spaces where new talents were trained and discovered.

Such was the image presented in the inaugural oration to Cardinal Savoia's Accademia dei Desiosi in 1625.[64] It was delivered by Agostino Mascardi, Urban VIII's *cameriere d'onore* and a close friend of Cesarini and Ciampoli. In "That the Practice of Letters at Court is not Only Appropriate but Necessary," Mascardi argues that the courtier needs to learn how to control the exterior signs of his soul and that such a knowledge can be learned through academic exercises. Showing a clear understanding of baroque *raison d'état* Mascardi presents the academy as nothing but a school of courtly self-fashioning: "the prince and the courtier cannot and should not acquire the doctrine they need with length of study. Thus it is necessary for them to learn in some compendious fashion."[65]

Those who have to run a state cannot waste their time in lofty but useless thoughts.[66] Rather,

Like those Egyptian dogs that drink as they walk along the banks of the Nile without stopping, the courtier [*huomo civile*] should pick

63. This symbiosis was exemplified by orations and treatises on the role of the literato at court and on the prince's need to appreciate the pleasures of literature. See, for instance, Agostino Mascardi, "Che la corte è vera scuola non solamente della prudenza, ma delle virtù morali," *Prose vulgari*, pp. 46–63, which was well noticed and reported in the *avvisi*: "In the famous Academy of the Humorists that is held in the house of Signor Paolo Mancini, Signor Agostino Mascardi presented a learned and elegant lecture in the Tuscan language on the Court, criticizing the common opinion of those who were blaming it by demonstrating with lively reasons and arguments that it is a school where one teaches men to be prudent and honest" (reproduced in Clementi, *Il Carnevale romano nelle cronache contemporanee,* 433). Other examples are Cardi, "La calamita della corte"; and Matteo Pellegrini, *Che al savio è convenevole il corteggiare libri IIII* (Bologna: Tebaldini, 1624).

64. On the Desiosi, see Maylender, *Storia delle accademie d'Italia*, 2: 173–77; and Garuffi, *L'Italia accademica*. Some of the orations presented to the academy were gathered by Agostino Mascardi in *Saggi accademici*. See also Ildebrando della Giovanna, "Agostino Mascardi e il Cardinal Maurizio di Savoia," in *Raccolta di studi critici dedicati a A. Ancona* (Florence: Barbèra, 1901), pp. 117–26; and Mannucci, "La vita e le opere di Agostino Mascardi," pp. 139–76.

65. Mascardi, *Prose vulgari*, 9–10.

66. Ibid., 10–11.

from the gardens of the Muses those few flowers that offer themselves to his hand as he goes by. He should find a richer route that, by taking him away from the highways worn by the footprints of those who call themselves philosophers, will lead him through private paths to the great spirits' own way to the achievement of doctrine.[67]

The *compendiosità* of academic culture stressed by Mascardi was not just a convenient didactic package aimed at shortening the time required to acquire what was expected of a courtier. Mascardi was not proposing an evening MBA program. *Compendiosità* does not mean shallowness but refers to knowledge constituted as a rich, copious collection of "gems." Academic culture is a fragrant and selected bouquet of flowers from the garden of the Muses; one that the courtier can gather as he nonchalantly walks by. Also, it is the antithesis of pedantry. Academic culture is not learned through a long curriculum taught by boring technicians who call themselves philosophers; it is privately (almost secretly) absorbed in "private paths" directly from the "great spirits" (among whom Mascardi puts himself). All those centuries-old professorial glosses are discarded as "tacky." Without any mediation, the "great spirits" (who *de facto* know what counts and what does not) offer the pearls of knowledge to the Roman courtiers. What is being transmitted is *virtue,* not *technique.* What Mascardi can offer the courtiers is to "put them in touch" with their own "natural" knowledge. He facilitates the meeting of unique cultural gems and unique courtiers.

Eventually switching to a more mundane plane, Mascardi presents himself also as a keeper of a well-stocked warehouse:

> An academy is, gentlemen, like a well-stocked armory, and in it one finds every armament necessary to defend oneself from the blows of adverse fortune and to fight against the rebellion of one's feelings. It is an emporium rich in the finest goods of the Orient, in which some [wares] serve to delight, other to preserve health, and others to heal injuries of the soul. A politician [*huomo politico*] does not need only one type of teachings and teachers . . . because many and different are the negotiations that pass through his hands, as are the situations which require the prince's good judgment.[68]

Or, playing another favorite trope of baroque Rome, where many cardinals had a "distillery" (*stillaria*) where they produced (and then exchanged) essences and medical *secreti,*[69] Mascardi presents academic

67. Ibid., 12.

68. Ibid., 15.

69. The inventory of Cardinal del Monte's *stillaria* is reproduced in Christoph Luitpold Frommel, "Caravaggios Frühwerk und der Kardinal Francesco Maria del Monte," *Storia dell'arte* 9–10 (1971): 45, 47. Cardinal del Monte and Grand Duke Ferdinand I used to ex-

culture as a carefully distilled essence, something that "the less learned courtier should be delighted to sit down and receive . . . from others' mouths, the doctrine that others have gathered with incredible labor from the infinite volumes of the philosophers."[70]

Mascardi's image of the academy as a vast spice store (*drogheria doviziosa*) full of "oriental exotica" recalls another space where the cardinals enjoyed their culture: the museum, a space not yet technicalized by taxonomies but unsystematically (that is, nonchalantly) crowded with precious specimens.[71]

In another image, the academy was a gym where courtly selves were kept in shape. Knowledge and "honorable entertainment" could not be separated, just as one could not (or should not) find sharp boundaries between academic gatherings, comedies, banquets, operas. They were all court-related practices in which courtiers produced and reproduced their culture. As stated by Mascardi, academies were not meant to produce learned persons but *huomini civili,* and civility was the undefinable quality that epitomized the courtiers' identity and differentiated them from the noncourtly masses.

As we will see, the peculiar instability of the Roman courtly and academic scene, its shifting patterns of people and alliances, and its specific cultural attitudes described in Mascardi's oration played a major role in the last phase of Galileo's career.

---

change medical *secreti* and their recipes, sometime using ciphers (*ASF,* "Medlceo principato 3761," s.f.). It seems that del Monte's *secreti* could be lethal. One painter, Tommaso della Porta, died as a result of using a medicament given him by the cardinal (Luigi Spezzaferro, "La cultura del cardinal del Monte e il primo tempo del Caravaggio," *Storia dell'arte* 9–10 [1971]: 76). This must have been a fairly common event if Traiano Boccalini in his *Ragguagli di Parnaso,* jokingly wrote that "Apollo [the Pope] prohibits the princes from keeping distilleries or stills in their houses" because too many mysterious deaths had been occurring in Rome (Traiano Boccalini, *Ragguagli di Parnaso,* ed. Luigi Firpo [Bari: Laterza, 1948], 3: 255). Cesarini and Ciampoli were also very interested in iatrochemistry.

70. Mascardi, *Prose vulgari,* 17.

71. The literary and philosophical "gems" Mascardi wanted to offer the cardinals when they strolled along *sentieri riserbati* well befit a culture that gazed in wonder at ancient statues, mosaics, medals, pieces of petrified wood, and inscriptions elegantly displayed in museums, *casini,* or *antiquari* located somewhere in the gardens of the cardinals' villas. The global meaning of these fragments was usually lost, their place in a continuous narrative largely unknown, yet it was precisely their being "out of context" that transformed them into "gems"—into emblems of an unknown but necessarily grandiose past. In fact, it is interesting that whenever Cesi made some archaeological finding, he immediately assumed that it had to come from some emperor's villa or from some major temple conspicuously recorded in history. Like Mascardi's "flowers of the Garden of the Muses," these fragments came down to the Roman contemporary *mecenati* directly from the legacy of the past. They had simply surfaced, unmediated, uninterpreted, and untainted. Like flowers and plants, they were accessible to leisurely nobles, and yet the knowledge they represented was not shallow. Also, in a sense, the ruins were a metaphor of Rome's fragmented "master narrative" as well as of the "fragmented" identity of its courtiers.

FIVE

Courtly Comets

Contextualizing a Puzzle

THE *ASSAYER*, A WORK GALILEO PUBLISHED in 1623 in the midst of a long and bitter dispute on comets with the Jesuit mathematician Orazio Grassi, occupies an uncomfortable place in Galilean historiography. Some Galileo scholars have avoided this text altogether while others have glossed over much of its scientific content to praise it as a masterpiece of literary skill and dialectical argumentation.[1] In general, the *Assayer* has been presented as a virtuoso piece epitomizing the difference between Galileo's modern scientific method and Grassi's stubborn but hopelessly crumbling traditional discourse.[2] It has also been read very selectively. Some historians have focused on Galileo's fascinating speculations about the corpuscular nature of matter, heat, and light, or on the famous topos of the book of nature being written in geometrical characters and being open to anyone who can read that language. Others, instead, have paused on the few paragraphs where he discussed what would become known as the distinction between primary and secondary qualities.

---

1. The important exception is William Shea, "The Challenge of the Comets," *Galileo's Intellectual Revolution* (New York: Science History Publications, 1977), pp. 75–108, which remains the most balanced account of the dispute. Although Pietro Redondi's *Galileo Heretic* (Princeton: Princeton University Press, 1987) has focused only on certain aspects of the dispute on comets, it has provided a remarkably rich and insightful contextualization of this episode in the Roman cultural and political scene. This analysis is indebted to a number of Redondi's insights.

2. Galileo apologists—especially Italians—have glossed over most of the text, limiting themselves to praising it as "the most vigorous and cutting piece of writing that was ever produced by Galileo's pen" (Banfi), or "a fascinating work of cultural propaganda, a breaking off from old methods, of open denunciation of the compromising spirit that was hidden under the false modernity of the Jesuits' dialectics" (Geymonat).

The uneasiness of some Galileo scholars with the *Assayer* and their comments on only selected passages of it may indicate that this text does not fit the image of Galileo as the modern thinker deeply committed to the Copernican cause and to sound mathematical method that has informed many interpretations of his career.[3] Not only was the defense of Copernicanism not the main priority of the *Assayer,* but Galileo does not always come across as the "modern" thinker defeating the "traditional" Jesuit with empirically sound and logically coherent arguments. Although Galileo was remarkably good at exposing some of the logical flaws of Grassi's arguments, he seemed to commit some of the same faults himself. The empirical content of the book is no less puzzling. It displays a Galileo who revises Aristotle's theory of comets (by integrating it with that of the Pythagoreans) to deploy it against Grassi's Tychonic and empirically more plausible view that comets were pseudo-planets. In short, by modern standards, the *Assayer* contains a good dose of ad-hoc hypotheses, internal contradictions, and unjustified attacks on Grassi's positions.

One way to avoid confronting the puzzling features of this text would be to join Feyerabend in claiming that Galileo was right in being so unruly because it is only through opportunism and rule-breaking that fruitful scientific change can take place. Although Feyerabend's analysis of Galileo has been relevant to my reading of the *Assayer,* I propose here a more contextual reading of this text, one that places it within the culture of the Roman court and academies. By doing so, I hope to show that although the often puzzling discourse of Galileo's *Assayer* was highly opportunistic, Galileo's tactics did not rest on the systematic breaking of received rules of scientific discourse. Quite to the contrary, Galileo's apparently puzzling arguments were nicely framed by the options and constraints of contemporary court discourse. Galileo was neither the modern scientific methodologist depicted by later historiography nor Feyerabend's lawless opportunist. More simply, he was a competent courtier. Although his opportunism was remarkable, it was consistent with specific cultural codes.

Also, I want to argue that the clash between the Jesuits and Galileo over the comets was first initiated and then directed by patronage dynamics.[4] In particular, the virulence of the debate cannot be understood sim-

---

3. In fact, historians like Drake, whose main focus is on Galileo's mechanics rather than astronomy (and therefore does not assume a Copernican teleology in Galileo's career), are quick to notice that the *Assayer*'s argument does not have much to do with Copernicanism. For instance, in his *Galileo at Work* he says: "Hence, however strange as it may now seem, a contemporary basis existed for the Jesuits' contention that cometary orbits existed in the region of the planets. Just how this could damage Copernicanism is far from clear to me" (Stillman Drake, *Galileo at Work* [Chicago: University of Chicago Press, 1978], pp. 265–66).

4. On this issue, see also Richard S. Westfall, "Galileo and the Jesuits," in *Essays on the Trial of Galileo* (Vatican City: Vatican Observatory, 1989), pp. 31–57.

ply in terms of a collision between the new and the old worldview. The different styles of the natural philosophy of Galileo and the Jesuits were not only rooted in different intellectual traditions and cosmologies but they also reflected two different cultures interacting in Rome: that of the court and that of religious orders.

Mathematicians and Literati

Following the appearance of three comets in the second half of 1618, European mathematicians and astrologers received many inquiries about the nature, position, and astrological significance of those phenomena. Beginning in November, Galileo began to receive questions from patrons and friends.[5] Archduke Leopold of Austria (who had visited the bedridden Galileo just a few months before) was among the first to ask for his opinion.[6] Leopold was not the only European prince interested in comets. From the French court the Medici representative wrote to Galileo that

> Finding myself these past days in the company of some mathematicians who were discussing the comet that was seen and is still being seen, it was said by common consensus that none beside yourself could observe it. You are perfectly equipped to deal with these matters both because of the quality of your telescope and because the Grand Duke has excellent instruments to make this observation. Signor Aleaume [Jacques], the Royal Mathematician, said the same to the King, who had commanded him to observe it, and excused himself by saying that he did not have the right instruments and that only the Grand Duke could make it possible for Your Lordship to do so. I did not want to miss giving you this notice especially to rejoice in the esteem that is bestowed upon your person, and also to encourage you to satisfy the public expectation and curiosity.[7]

Letters expressing expectations of Galileo's forthcoming opinions about the comets arrived from Rome as well, especially from his fellow Lincei Stelluti and Cesarini. However, Galileo was bedridden and could neither observe the comets nor reply to his patrons and friends at this time.

The Jesuit mathematicians of the Collegio Romano also received plenty of queries but, unlike Galileo, they were able to conduct observations. Even more important, they were able to gather further data from other parts of Europe through the society's networks. Relying on a pool

---

5. *GO*, vol. 12, no. 1354, p. 420; no. 1355, pp. 420–21; no. 1356, pp. 421–22.
6. Ibid., no. 1369, p. 435.
7. Ibid., no. 1362, p. 428.

of observations that was possibly the best available at that time, the mathematicians of the Collegio Romano delivered a successful public oration which was well attended by the Roman intelligentsia and nobility.[8] The text was published in 1619 as *De tribus cometis anni MDCXVIII disputatio astronomica*.[9] Although his name did not appear on the frontispiece, Orazio Grassi, a former student of Clavius and the current professor of mathematics at the Collegio, was the author of the piece. The *Disputatio* claimed that the parallax computed from the observations put the comets well above the moon. A further argument for placing them above the moon was that the comets did not seem very enlarged even when seen through powerful telescopes, an argument similar to the one Galileo had made about the fixed stars in the *Sidereus nuncius*.[10] Also, the Jesuits claimed that comets were planets—although peculiar ones—and that, as argued by Tycho, they followed circular orbits. An unpretentious piece, the *Disputatio* was neither dogmatic in its claims nor disrespectful of Galileo (who was never explicitly mentioned in it). Complying with the genre of disputation, Grassi reviewed the available theories about comets and put forward the one that seemed to fit best the available observational evidence (which Grassi did not present as definitive).[11]

The *Disputatio* was not a dry piece of technical astronomy but a playful essay that blended together academic virtuosity, emblematics, qualitative astronomy, and a few geometrical propositions. Its initial *prolusio* (and the two introductory sonnets) were particularly witty, their point being to present comets not as terrifying omens but as a topic for learned and virtuous conversation.[12] At one point Grassi presents his task as that

8. Apparently, the disputation at the Collegio Romano was only one among the various academic orations on the comets: "The year 1618: Three comets appeared at the end of November, one of which was seen in India, in Persia, and in Japan . . . Many compositions of various sorts were made about these apparitions. In the salon of the Collegio Romano there was a dispute on these comets by a father of the Company of Jesus, a work of Father Oratio Grassi, professor of mathematics, which can be seen in press among the work printed by the aforementioned Father Grassi" (Girolamo Nappi, "Annali del seminario romano," parte 2, *APUG*, MS 2801; I would like to thank Father Lamalle for having shared this reference).

9. Reproduced in *GO*, vol. 6, pp. 19–34. An English translation is in Stillman Drake and C. D. O'Malley, trans., *The Controversy on the Comets of 1618* (Philadelphia: University of Pennsylvania Press, 1960), pp. 3–19.

10. Drake and O'Malley, *Controversy on the Comets*, p. 17.

11. For instance, Grassi used many conditionals in putting forward his claims: ". . . in order that we *may* now determine *almost* the true place of the comet, let us say that it *can probably* be placed between the sun and the moon" (ibid., p. 17 [emphasis mine]). Also, he was quite frank about the fact that the observations he was working from were good, but far from perfect. To make better ones, he claimed, he would have needed Tycho's instruments (ibid., p. 14).

12. Ibid., pp. 5–7.

of somebody writing the biography of an illustrious traveler engaged in some celestial grand tour:

> Since I believe that in this duty I ought not deviate from the masters of eloquence, in accordance with their practice, taking the first argument of my discourse from the comet's birth, I have sought its native land and parentage, and I have opened a pathway for myself through the illustrious circle of its subsequently very famous life to the far from obscure character of its death.[13]

In many ways, the *Disputatio* resembled the courtly style of Galileo's own scientific prose. Grassi was not just a competent mathematician, architect, painter, and inventor, but also an elegant writer and an accomplished playwright.[14] His cultural background and skills resembled Galileo's own more than those of some of Galileo's previous opponents such as delle Colombe, Capra, Mayr, or Magini. Also, the *Disputatio* marked the transition from the technical and scholarly style characteristic of the previous books by such Jesuit mathematicians as Clavius and Biancani to a more elegant discourse aimed at capturing an audience of courtly virtuosi.[15]

By adopting a courtly style Grassi was simply tuning himself to the cultural trends followed by his institution. As indicated by the work of the polymath Athanasius Kircher (and the success it met with in Rome), the Jesuits kept pursuing this court-oriented cultural politics with great determination. In fact, the Collegio Romano was trying to compete with the many Roman academies to establish itself not only as an elite university but also as a center of culture and edifying entertainment. Theatrical plays (some of them by Grassi himself), public disputations, poetry readings, and other forms of spectacle were routinely performed at the Collegio and always attracted a large number of cardinals and Roman aristocrats.[16] Reflecting the eclecticism of topics and genres we have seen

13. Ibid., p. 8.

14. Biographical information about Grassi can be found in Antonio Favaro, "Galileo Galilei e il P. Orazio Grassi," *Memorie del Reale Istituto Veneto di Scienze, Lettere e Arti* 23 (1887): 203–36; G. V. Verzellino, "Padre Orazio Grassi gesuita matematico eccellentissimo," *Memorie degli uomini illustri di Savona* (Savona, 1891), 2:347–51; Carlo Bricarelli, S. J., "Il P. Orazio Grassi architetto della Chiesa di S. Ignazio in Roma," *Civiltà cattolica* 2 (1922): 13–25; and Claudio Costantini, *Baliani e i Gesuiti* (Florence: Giunti, 1969). Copies of letters from the *Generale* of the Society of Jesus to Grassi are in *ARSI*, MED 23; MED 26; MED 27; ROM 17. They are useful to reconstruct Grassi's activities as an architect for the society.

15. Already in the *De iride disputatio optica* (Rome: Mascardi, 1617), the Jesuit mathematicians had tried to reach beyond the walls of the Collegio Romano into the fashionable academic environment by having the disputation printed and dedicated to the prince of the Accademia degli Umoristi—the most important and long-lasting literary academy in Rome.

16. These events are well described in the manuscript "Origine del Collegio Romano e suoi progressi" in *APUG*, MS. 143. On the Jesuits' theatrical activities, see Maurizio Fagiolo

in the Roman academic scene, the public disputations at the Collegio dealt with a wide range of arguments such as theology, sunspots, moral philosophy, optics, hydrostatics, and Copernican astronomy.[17] The public fête for Galileo and his discoveries in May 1611 fit well into this genre of events.

As members of a religious order, the Jesuits did not have the same cultural leeway enjoyed by lay literati and intellectuals. However, they were remarkably successful at giving modern and sophisticated appearance to traditional material and at competing successfully for cultural preeminence, especially in a place like Rome where the religious nature of the court limited the range of acceptable cultural productions. The Jesuits aimed at being the defenders of religious orthodoxy while avoiding the pedantry usually associated with such a stance. Such a sprinkling of *sprezzatura* over an otherwise technical body of knowledge was particularly necessary if they wanted to aim at the upper classes. In turn, their successful recruiting among the aristocracy legitimized their adoption of courtly cultural codes. Unlike other religious orders, the Jesuits were able to cross the boundary that had traditionally divided the culture of the cardinals (who usually knew some law but little or no theology) from that of the theologians and philosophers of the religious orders.

The Jesuits were unusually successful in shedding the cultural stigma courtiers attached to "friars," as courtiers tended to call all members of religious orders. Usually, members of religious orders were not seen as having that civility, freedom, culture (and life-style) that characterized courtiers. Some of the negative stereotypes we have seen used against philosophers were also employed against "friars." In his advice to Roman courtiers, Ciampoli claimed that:

> The friendship of friars is deleterious [to the courtier]. However, you need to know some of them who are well connected to various households and courts of princes so that they can praise you—for it is very important to be praised publicly. However, because those people are never esteemed very much, you should never get too close

---

Dell' Arco and Silvia Carandini, *L' effimero barocco: Strutture della festa nella Roma del '600* (Rome: Bulzoni, 1978), 2 vols.; and Per Bjurstrom, "Baroque Theater and the Jesuits," in Rudolf Wittkower and Irma B. Jaffe, eds., *Baroque Art: The Jesuit Contribution* (New York: Fordham University Press, 1972), pp. 99–110. On Grassi's theatrical activities, see Verzellino, "Padre Orazio Grassi gesuita matematico eccellentissimo," 348; and Bricarelli, "Il P. Orazio Grassi architetto," pp. 21–22.

17. See, for instance, Giuseppe Gabrieli, "Il Carteggio Linceo," *Memorie della R. Accademia Nazionale dei Lincei*, Classe di Scienze morali, storiche e filologiche, series 6; 7 (1938–42), part 2, section 1, pp. 267–68, 321; Cerasoli, "Diario di cose romane degli anni 1614, 1615, 1616," *Studi e documenti di storia e diritto* 15 (1894): 280.

to them but simply remain on good terms so as to reap those benefits mentioned above.[18]

As good as they were at being courtly, the Jesuits' move from the culture of the religious orders to that of the (often less religious) court could not be complete, and, as we will see, Galileo was able to capitalize masterfully on the "pedantic" leftovers of Grassi's intellectual style.

With the *Disputatio*, Grassi did in astronomy what his confreres had done in theater, poetry, and rhetoric. He demonstrated that the Jesuits were not just slavish thinkers but were willing to criticize Aristotle quite openly and engage in innovative natural philosophy.[19] At the same time, he steered away from Copernicus's unorthodox cosmology (that had been declared false by the Church just a few years before) and began to show his sympathies for Tycho's system—one that came to epitomize the Jesuits' blend of tradition and innovation. Grassi's *prolusio* typifies the Jesuits' ambiguous attitude toward novelties. As he put it, the comet was a terrifying omen of change only before the Jesuit mathematicians had explained it. Although the Jesuits were willing to entertain novelties, they did so only to domesticate them. Novelties could not be omens of radical worldly (or philosophical) change.

The Dispute Begins

As discussed previously, scientific disputes tended to resemble duels. Often, they were set off or propelled not by the disputants themselves but by their patrons' requests for their clients' opinion on a given phenomenon or on its interpretation as put forward by somebody else's client. Requests like these could not be safely ignored even if the clients might not have sound ideas about the phenomena. The dispute on comets fit this pattern well. In 1623, Galileo described his entry into the dispute in these terms:

> Perhaps you will tell me that I should have remained silent. To this I reply first that Sig. Mario and I were already too deeply obligated to let our ideas be known before Father Grassi's essay was published, so

18. Giovanni Ciampoli, "Discorso di Monsignor Ciampoli sopra la corte di Roma," in Marziano Guglielminetti and Mariarosa Masoero, "Lettere e prose inedite (o parzialmente edite) di Giovanni Ciampoli," *Studi secenteschi* 19 (1978): 236.

19. That the *Disputatio* was seen by the Jesuit mathematicians as an innovative piece that went against the beliefs of the philosophers of their own order may be gathered from Griemberger's surprise at Galileo's attack on it (Ugo Baldini, "Additamenta galilaeana I: Galileo, la nuova astronomia e la critica all'aristotelismo nel dialogo epistolare tra Giuseppe Biancani e i Revisori romani della Compagnia di Gesù," *Annali dell'Istituto e Museo di Storia della Scienza di Firenze* 9 [1984]: 22).

that to have remained silent would have been to invite upon ourselves contempt and almost universal derision.[20]

Also, neither Galileo nor the Jesuits seemed to have been particularly interested in comets until important patrons asked their opinion about them.[21] Grassi and Galileo *had* to speak and write about them—a patronage-driven move that was to propel them into an increasingly bitter dispute that ruined their previously good relationship. As Galileo put it,

> As soon as I learned this [that many astronomers were supporting Tycho's view of comets], I let it very plainly be understood that I deemed the reasoning quite empty. Many people made a joke of this, the more so when there appeared on their side authoritative support and confirmation from the mathematicians of the Collegio Romano, nor shall I deny that this caused me a bit of trouble. Finding myself placed under the necessity of defending my statement against so many opponents (who, strengthened by such assistance, rose up against me still more imperiously), I did not see any way to contradict them without including Father Grassi too. So it was through a necessary though fortuitous circumstance and not by choice that I pointed my opposition in the direction in which I least wished to.[22]

Galileo's description points to some important features of the patronage system in which he operated. Moreover, the debate did not remain limited to comets. By the time Galileo published the *Assayer* and Grassi responded with the *Ratio,* it had proliferated in many directions sometimes only indirectly connected to comets. The complex (if not messy) pattern of arguments and tactics of the five texts that punctuated the dispute from 1619 to 1626 does not seem to represent a clash between Copernican and anti-Copernican positions, but rather the play of patronage dynamics in a field turned much more insidious by the 1616 placement of Copernicus on the Index.

20. *GO,* vol. 6, pp. 227–28, as translated in Drake and O'Malley, *Controversy on the Comets,* pp. 178–79.

21. The *Disputatio* was not presented as an answer to questions posed by Roman or foreign notables about the nature and place of the comets (thought it was delivered as a public lecture). However, in the 1619 *Libra,* Grassi reminisced about the fall of 1618 and the context in which the Jesuits entered the debate on comets: "Hence it was decided that the academies of philosophers and astronomers ought immediately to be consulted. But why was it so readily believed that this Collegio of ours, renowned for the many interests of its academicians, should be considered as, among other things, the eyes of all, and that it ought especially to be consulted and its answers awaited?" Therefore, the Jesuits themselves were drawn into the debate by questions received from notables (Drake and O'Malley, *Controversy on the Comets,* p. 69).

22. *GO,* vol. 6, p. 226, as translated in Drake and O'Malley, *Controversy on the Comets,* pp. 178–79.

Until 1618, Galileo had dominated observational astronomy. The letter he received from the French court at the end of 1618 shows that he was seen as the undisputed leader of the field. As the royal mathematician told the French king, it was only Galileo who could observe the comets and answer his royal queries. By presenting Galileo as the unquestionable star, Aleaume was not just trying to get himself off the hook. In fact, not only all the remarkable astronomical discoveries that had taken place since 1609 were made possible by the telescope—an instrument that had become identified with Galileo—but it was Galileo himself who made those discoveries (or at least managed to get credit for them). It was on those discoveries (rather than on his support of Copernicanism) that Galileo had gained international reputation and built a remarkable career.

The comets of 1618 upset this situation. On that occasion, the king of observations remained silent (and in bed). As the beginning of the *Disputatio* indicates, the Jesuits were very happy to be finally able to observe something that Galileo had not yet studied: "Only comets have remained aloof from these lynx eyes. . . ."23

Also, the comets were the type of astronomical objects the Jesuits could study without having necessarily to enter into cosmological debates. After the placing of Copernicus's *De revolutionibus* on the Index in 1616, cosmological matters became a very sensitive issue for all Catholic astronomers but especially for progressive ones who (like the Jesuits) had realized that Ptolemaic astronomy had run its course. However, because comets had not been treated by either Ptolemy or Copernicus in their astronomical works, they were not immediately seen as objects whose interpretation was to lead into the domain of cosmology.24 Moreover, the only post-Copernican planetary model in which comets played some role (Tycho's) was geocentric and therefore theologically acceptable.25

Given this context, the Jesuits could approach the interpretation of the comets as cosmologically unproblematic; they had nothing to fear from the comets. If anyone brought up possible cosmological implica-

23. Drake and O'Malley, *Controversy on the Comets*, p. 6.

24. More specifically, Ptolemy did not mention comets in the *Almagest*, but did so briefly in his astrological text, the *Tetrabiblos*, in which he presented them as meteorological phenomena (Ptolemy, *Tetrabiblos*, trans. F. E. Robbins [Cambridge, Mass.: Harvard University Press, 1940], pp. 191–95, 217). Copernicus mentioned comets in passing only once in the *De revolutionibus*, referring to them as sublunary bodies. However, it is quite unclear whether he was expressing his opinion or he was reporting the common view on the matter (Nicolaus Copernicus, *On the Revolutions*, book 1, chapter 8).

25. Tycho had interpreted the trajectory of the comet of 1577 as ruling out the existence of the crystalline spheres (a belief that was posing some serious problems to the tenability of his system). In the Tychonic system, the spheres of Mars and of the sun intersect—a feature that would be unacceptable unless the spheres were immaterial. Because the comet was seen as having gone through various planetary spheres, Tycho argued that the spheres must have been immaterial.

tions, they could frame their claims (as they officially did in 1620) within the Tychonic system.[26] The study of comets also allowed them to enhance their visibility as a group. In fact, the *Disputatio* was not just the work of a single mathematician. The quality and quantity of the data on which Grassi relied were a tangible sign of the effectiveness of the society's networks. The study of comets had shown the resources of the Jesuits as a corporate entity.

Finally, we should consider a generational factor in the debate: the comets of 1618 marked the coming of age of Clavius's "children" as a group. If in 1611 they had collectively certified Galileo's discoveries (under Clavius's supervision), now Clavius was no longer alive and they were trying to get some credit for themselves. The debate on comets was not only between two individuals or institutions but also between two generations.[27]

Galileo may have felt that the scale was beginning to tilt against him. As an individual, Galileo was definitely much more famous than any of the Jesuits, but, as a network, they had many more resources than he (and not just more powerful connections with the Church). Moreover, the Jesuits had begun to display their wares in the courtly market that, until then, Galileo may have seen as his own. There is a sentence in Grassi's 1619 *Libra* that must have stung Galileo's ego. Recounting the initial phase of the Jesuits' involvement with the comets, he described virtuosi, aristocrats, and cardinals going to the Jesuit mathematicians to hear their opinion. In doing so, Grassi presented the Collegio Romano as the source of authority on these matters: "But why was it so readily believed that this Collegio of ours, renowned for the many interests of its academicians, should be considered as, among other things, the eyes of all, and that it ought especially to be consulted and its answers awaited?"[28] From what we have seen of the workings of the patronage system, being asked questions might turn into a nightmare, but not being asked questions was even worse; it meant that one's status was on the wane. Grassi's presentation of the Col-

26. The Jesuits publicly declared their acceptance of Tycho only in 1620 with Giuseppe Biancani, *Sphaera mundi, seu cosmographia* (Bologna: Bonomi, 1620). Because the Jesuits had a tight system of internal censorship, the publication of a book in which Tycho's system was clearly adopted amounted to a collective statement. Instead, the fact that Tycho is sympathetically treated in Grassi's texts on comets but is never clearly endorsed suggests that Grassi and his colleagues may have been using the debate on comets to test the waters with the society's censors.

27. By 1618 the scenario had changed for Galileo as well. Previously, his connection with the Jesuits meant Clavius. Galileo treated the elderly astronomer as a patron, that is, respectfully. But Clavius had died early in 1612, and the debt Galileo had toward him did not carry over to Grassi, Griemberger, and Guldin. Had Clavius been alive, Galileo might not have attacked Grassi as violently as he did.

28. Drake and O'Malley, *Controversy on the Comets*, p. 69.

legio Romano as the natural place for people of quality to go and ask questions about astronomy and mathematics amounted to a claim that—contrary to what people may still have believed in Paris—Galileo was no longer number one.

While the comets fit well the Jesuits' agenda (from patronage to cosmology) and helped them maximize their resources, they seemed to pose a serious threat to Galileo's own status as the astronomical courtly star. The puzzling harshness of Galileo's reply to a quite innocuous *Disputatio* can be understood in this context. Galileo's tough talk against Grassi may reflect his inability to accept what was happening. This was the first time he had not been first (or best) with observations. Furthermore, he could not do anything to win his primacy back in regard to the observation of the comets. Not only had the Jesuits already gone to print but, by the time he was out of bed, the comets had gone. It is also in this context of status anxiety that we can read Galileo's self-praise at the beginning of the *Discourse,* where he presented himself (or let Guiducci do so) as "that noble and sublime intellect who adorns the present age no less than this native land of his by having discovered so many marvels in the sky." Interestingly, he ended that passage by attacking the Jesuit Scheiner for having tried to steal from him the discovery of the sunspots.[29] Also telling are Galileo's sharp and surprising attacks on Grassi's use of emblems, puns, ironic turns of phrase, and the various rhetorical devices dispersed in the *Disputatio* and the *Libra*.[30] They too may be understood as expressions of bitterness by someone who saw his own ground being invaded with the use of his own tactics.[31]

If he wanted to reestablish his status vis-à-vis the Jesuits, Galileo had to come up with a new strategy. Moreover, he could not escape the expectations created for him by the patronage system. In particular, he had to demonstrate to the archduke of Austria, Leopold (who was the grand duke's brother-in-law), that he could always rely on Galileo for questions of natural philosophy. Galileo did not have many options. By his own admission, if he did not respond, he would seriously damage his credibility and patronage connections. If he endorsed the Jesuits' claims he would contribute toward reinforcing their position and, probably, encourage them to further trespass on his courtly domain. A third option was to present an alternative interpretation of the comets, possibly one that would undermine the lead the Jesuits had on him. Given his lack of observations,

29. Ibid., p. 24.

30. *GO*, vol. 6, pp. 88, 233–34. For Grassi's response, see ibid., pp. 116–17.

31. I suggest that the Lincei may have been sensitive to this dimension of Grassi's work as well. Consisting largely of courtiers, the Lincei (Cesarini and Ciampoli in particular) may have shared Galileo's perception that Grassi was trespassing on their turf.

Galileo decided not to try to refute Grassi on empirical grounds but to present new *hypotheses* on comets.[32]

Rather than respond in person to Grassi, Galileo communicated his views to Mario Guiducci, a close friend of his who was then consul of the Accademia Fiorentina. Together, they wrote a discourse which Guiducci presented to that academy in 1619. It was then printed as the *Discourse on Comets* under Guiducci's name (but with explicit credit to Galileo as the source of it).[33] The *Discourse* was dedicated to Leopold of Austria, who was credited for directing Galileo's intellect toward the comets.[34] For simplicity's sake, I will refer to Galileo as the main author of the *Discourse*.

Galileo's response to Grassi was, to put it mildly, provocative. He claimed that, before using the parallax method, Grassi should have made sure that comets were real physical objects rather than optical artifacts like rainbows.[35] Unless Grassi could show that comets were real objects, all his evidence would melt into air.[36] As an alternative view, Galileo suggested that comets might not be physical objects at all but only the result of vapors rising perpendicularly from the surface of the earth.[37] Accordingly, what was seen as the comet's tail might have been nothing more than the refraction of the sunlight through these vapors.[38] In short, comets might

32. Galileo's predicament in 1618 was not unlike that he had experienced in 1612 when he received Welser's letter asking him to respond to Apelle's work on the sunspots. At the time Galileo received that letter, he had not conducted systematic observations of the sunspots, but he wanted nevertheless to reclaim the priority for their discovery while impressing his new patron. As with the *Discourse*, the first letter on the sunspots had a definite hypothetical character.

33. Mario Guiducci, *Discorso delle comete* (Florence: Cecconcelli, 1619); reproduced in *GO*, vol. 6, pp. 39–93. An English translation is in Drake and O'Malley, *Controversy on the Comets*, pp. 21–65.

34. "And finally, my action has been determined above all by your desire (as shown by your gracious letters to Galileo) to learn his opinion upon this matter" (Drake and O'Malley, *Controversy on the Comets*, p. 22).

35. *GO*, vol. 6, pp. 65–71.

36. Also, Galileo sharply criticized his claim that an auxiliary argument for placing the comets above the moon could be provided by the slight enlargement they received through the telescope. Although what I think Grassi was conveying with an improper wording was that one did not *see* comets much enlarged through the telescope, Galileo presented him as an ignoramus in optics for having claimed that the telescope did not enlarge all objects with the same ratio of magnification, but enlarged close objects more then those far away (ibid., pp. 72–82). However, Grassi was by no means an ignoramus in optics, as shown by his 1617 *De iride disputatio optica*.

37. It seems that Galileo's work on the nova of 1604 may have provided him with a model by which to interpret the comets (Willy Hartner, "Galileo's Contribution to Astronomy," in Ernan McMullin, ed., *Galileo, Man of Science* [New York: Basic Books, 1967], p. 185).

38. This view of the comet's tail goes back at least to Gemma Frisius. See Peter Barker, "The Optical Theory of Comets from Apian to Kepler," *Physis*, forthcoming.

not be planets and might not move in circular orbits. More probably, they moved away from the earth along a straight line perpendicular to the earth's surface. To Galileo this was the only viable hypothesis to account for the rapid variation of the comets' luminosity.[39] That at some point comets seemed to follow a curvilinear trajectory might have to be explained away as an optical illusion.

Galileo's response to Grassi addressed several problems at once. First, by attacking Grassi's views on comets with aggressive and paradoxical arguments he could show that he, rather than Grassi, was the real courtly (that is controversial, original, and creative) natural philosopher. Ciampoli thought that "The discourse looks absolutely marvelous and, to me, miraculous. New stuff [*roba nova*]; propositions that are paradoxical to the philosophical plebs and are argued with such a clarity that it is impossible not to marvel."[40] This aggressive intellectual style was likely to please somebody like Leopold, a great patron whom Galileo needed to answer.[41]

Second, by delegitimizing Grassi's use of the parallax argument, Galileo managed to get rid of Grassi's (and the Jesuits') greatest asset: the observations they had gathered all over Europe. Through this move, Galileo had turned a situation in which he had to counter the resources of several Jesuit observational astronomers into one in which he faced Grassi alone, and not just as a mathematician but also as a *natural philosopher*. Galileo thus conveniently shifted the discussion to less empirical and more philosophical grounds on which he thought he could more easily outdo Grassi. Then, by stressing that his arguments were presented only as "gracious conjectures," Galileo managed to stay clear of theologically sensitive areas, something he was expected to do as a result of the admonition he had received in 1616 not to hold or defend the Copernican doctrine.[42] Finally, by relying on the courtly taste for "gems" that we have seen articulated in Mascardi's oration, Galileo could present his hypothetical approach not as the result of his contingent difficulties but as

39. *GO*, vol. 6, pp. 51, 90–91.

40. *GO*, vol. 12, no. 1399, p. 466.

41. We have some later evidence that it had been quite important for Galileo to answer Leopold's query. In the spring of 1621, shortly after the death of Cosimo II, Galileo wrote Leopold asking him to send a letter of recommendation on his behalf to Leopold's sister, the Archduchess Maddalena. A few weeks later, Leopold wrote Galileo that he had taken care of it (*GO*, vol. 13, no. 1494, p. 61; no. 1503, p. 70). Evidently, with the death of Cosimo, Galileo's position was not necessarily safe. His post had been created by the deceased grand duke and his successor may not have confirmed it. Knowing that a new regency would be installed (one in which Leopold's sister was to play a major role), Galileo wanted to make sure that his contract would be confirmed. It is interesting to speculate on what could have happened had Galileo not answered Leopold's question on the comets via Guiducci's *Discourse*.

42. For examples of Galileo's hypothetical approach, see *GO*, vol. 6, pp. 47, 51, 73, 99.

something dictated by his own virtue. He made this clear at the very beginning of the *Discourse:*

> Since freedom to extract any jewel of quality from so rich a treasury [the cosmos] is thus very limited, those who were able to bring us back a few should be held in high esteem as fortunate men and magnificent benefactors. Likewise, they should be pardoned if the brevity of the time they have been permitted to remain in such a place has prevented them from singling out the better things from the poorer, and if sometimes, in place of the cause of some effect which we have requested from them, they have brought us something else. But just as they amply deserve to be excused, so we must not be blamed if after careful examination of such causes we do not equally approve them all.[43]

Galileo's views on comets were not to be seen as a systematic treatise but as a few brilliant insights the readers had been lucky to receive. The scarcely glamorous image of a Galileo sick in bed and unable to conduct observations was magically transformed into the topos of Galileo's superior mind traveling (through for a very short time) into the unknown spaces of the cosmos to come back with a few gifts for his virtuosi friends, gifts they were expected to be delighted with or, at least, to appreciate for their novelty. While this methodological choice may have puzzled other mathematicians, it seemed to work quite well with the courtly audience Galileo was addressing.[44] In the *Assayer* Galileo extended the use of the "gem" trope to dismiss further difficulties of his views and to characterize his book as a piece of courtly natural philosophy.

Galileo versus Tycho

The legitimation of a hypothetical approach to comets was only one aspect of Galileo's tactics. He also needed to show that his hypotheses were the best available. To do so, he produced a systematic (if not always fair) critique of Grassi and of everybody else who had previously written on comets. Galileo's overall strategy is reminiscent of some of the tactics of the 1615 "Letter to the Grand Duchess," There, he tried to argue that the heliocentric hypothesis should not be struck down but given breathing room simply because the Ptolemaic alternative was plainly refuted. More precisely, by showing that his adversaries' claims were refuted, Galileo claimed he did not need to prove his own hypotheses but that they ought

43. Translation adapted from Drake and O'Malley, *Controversy on the Comets,* p. 23. The original is in *GO,* vol. 6, pp. 45–46.

44. For Kepler's view of Galileo's position on comets (as expressed in the *Assayer*), see the appendix he attached to his *Tychonis Brahei Dani Hyperaspistes* (Frankfurt, 1625), translated in Drake and O'Malley, *Controversy on the Comets,* pp. 339–55.

to be taken seriously simply because they were not yet refuted. They were better than anything else available and it was up to his adversaries to disprove them if they wished to do so. This tactic reemerged in the *Discourse on Comets* and was expanded in the *Assayer*. There, Galileo began to claim that he was not obliged to prove his hypotheses about comets. Rather it was Grassi (whose views he claimed to have refuted) who had to do the job of refuting Galileo's hypothetical claims, if he cared to.

If Galileo's systematic and somewhat harsh critique of Grassi may be contextually understandable, the attacks on Tycho in the *Discourse* seem, at first, uncalled for, given the very minor role Tycho's doctrine has in Grassi's *Disputatio*. Grassi did not mention Tycho's cosmological views (he actually avoided explicit references to cosmological issues altogether) but cited him only once in connection with his great instruments which, had the Jesuits possessed them, would have provided them with much better observations of the comets.[45] The only implicit link between Tycho and Grassi is that the Jesuit presented comets not as sublunary phenomena but planet-like celestial bodies orbiting in circles. Unlike Tycho, who claimed that the sun was the center of the orbit of comets, Grassi (probably to leave cosmological issues out altogether) did not address this point explicitly.[46]

Some historians have seen Galileo's attacks on Tycho as a preemptive strike. According to this view, Galileo feared that a comet-based anti-Copernican argument Tycho had presented in a letter to Christopher Rothmann could be picked up by the Jesuits and deployed against him.[47] There, Tycho had presented the comets' behavior as confirming his belief in the stability of the earth. He argued that the earth did not move; if it did, one would have been able to notice some reflection of the earth's revolution around the sun projected on the comets' trajectory. In particular, Tycho had expected comets, like the other planets, to retrograde when in opposition to the sun, something that he had not been able to observe.[48]

For a number of reasons, I think that the power of this argument has been overstated. Nowhere in the heated exchanges with Galileo did Grassi

45. Drake and O'Malley, *Controversy on the Comets*, p. 14.

46. Ibid., p. 16.

47. Tycho Brahe, *Epistolarum astronomicarum libri* (Uraniborg, 1596), in *Tychonis Brahe Dani Opera Omnia*, ed. I. L. E. Dreyer (Amsterdam: Swets and Zeitlinger, 1972), 6: 179. See also Tycho's letters to Magini (*Tychonis Brahe*, 7: 289–99, esp. 295) and to Peucer (7: 127–41, esp. 130).

48. Christine Jones-Schofield, *Tychonic and Semi-Tychonic World Systems* (New York: Arno, 1981), p. 74. Other useful works on early modern cometary theory are Peter Barker and Bernard R. Goldstein, "The Role of Comets in the Copernican Revolution," *Studies in History and Philosophy of Science* 19 (1988): 299–319; C. Doris Hellman, *The Comet of 1577: Its Place in the History of Astronomy* (New York: Columbia University Press, 1944); and Roger Ariew, "Theory of Comets at Paris during the Seventeenth Century," *Journal of the History of Ideas* 53 (1992): 355–72.

ever bring up Tycho's interpretation of comets as an argument against Copernicus. During the dispute, we find only one mention of the possible use of comets against Copernicus and it is to be found not in printed texts but in Galileo's correspondence. In a letter about the publication of the *Disputatio,* Giovanni Battista Rinuccini reported that somebody *outside* of (and not within) the Jesuit order was spreading rumors that the comets could undermine Copernicus.[49] That was in March 1619, before Grassi's publication of two additional and increasingly hostile responses to Galileo on comets in which he never deployed such a comet-based anti-Copernican argument.

He did not do so even when, in the *Libra,* he entered cosmological grounds by arguing that Galileo's view of comets as moving along a rectilinear trajectory might require the assumption of the earth's motion, a hypothesis unacceptable to Catholics.[50] In the *Assayer,* Galileo responded

---

49. Westfall has translated this passage incorrectly and argued: "Thus when Giovanni Rinuccini reported the lecture to Galileo, he added the Jesuits . . . were claiming that the lecture destroyed the basis of Copernicanism" (Westfall, "Galileo and the Jesuits," p. 45). Instead, Rinuccini's letter reads: "The Jesuits discussed the comets in a public lecture now in press, and they firmly believe that it is in the heavens. Some *outside* the Jesuit Order are spreading the rumor that this is the greatest argument against Copernicus and that it knocks it down" (*GO,* vol. 12, no. 1378, p. 443, as translated in Shea, "Challenge of the Comets," p. 75; emphasis mine).

The only other reference of Tycho's comet-based anti-Copernican argument found in Galileo's correspondence is in Ludovico Ramponi's 1611 letter to Galileo (*GO,* vol. 11, no. 561, pp. 161–62). However, as far as I know, this is the only letter addressing that specific point in Galileo's entire correspondence—and Galileo corresponded with many competent sympathizers of the Tychonic system (like Baliani). Moreover, there is no mention of comets in relation to Copernicanism in the correspondence between Galileo and the Lincei in that period. This silence is significant because, as mentioned earlier, the Lincei frequently corresponded with Galileo about the "Operation Grassi," its form, goal, and possible dangers.

50. In the *Discourse,* Galileo had admitted that if comets move away from the earth along a rectilinear trajectory perpendicular to it, then they should be seen as eventually approaching but never passing the zenith. Instead, he conceded, evidence showed that they did pass the zenith and decline toward the north. As Galileo put it: "This forces us either to change what has been said or else to retain it, but to add some other *cause* for this apparent deviation. I cannot do the one, *nor should I like to* do the other" (Drake and O'Malley, *Controversy on the Comets,* p. 57 [emphasis mine]). Grassi translated Galileo's "cause" as "motion" and "nor should I like to" as "I dare not" (ibid., p. 97), implying that Galileo's unspoken additional cause was the earth's motion—a hypothesis that Catholics were not allowed to consider: "But at this point I hear something or other softly and timidly whispered in my ear about the motion of the earth. Away with the word dissonant to the truth and harsh to pious ears! It had better be whispered with lowered voice. But if the matter were so, Galileo's opinion would have proclaimed that which is considered as a false foundation by others than he. For if the earth is not moved, this straight motion does not agree with the observations of the comet; but it is certain that among Catholics the earth is not moved . . . Nor do I believe that this had ever come to the mind of Galileo whom I have always known as pious and religious" (ibid., p. 98). Elsewhere, Grassi deployed the same theological argument against Kepler's (and by association Galileo's) view that comets follow straight trajectories (ibid., p. 75).

to this oblique remark by challenging Grassi to himself prove or answer his own insinuations, something which the Jesuit did not do.[51] Given that Grassi seemed to use whatever resources he could muster against Galileo, he doubtless would have used Tycho's argument had he thought it was a powerful one.

I find it hard to believe that Galileo would have gone through the entire dispute just to strike back against an anti-Copernican argument that nobody deployed, an argument that, when posed to him elsewhere, he actually met. That was in the 1624 "Reply to Ingoli."[52] Monsignor Ingoli was one of the literati with whom Galileo debated on the Copernican hypothesis at a public gathering during his visit to Rome in 1616.[53] Later, Ingoli wrote down his argument in the form of a letter to Galileo. In it, he briefly referred to Tycho's comet-based anti-Copernican argument.[54] At first Galileo ignored this letter altogether, but then he revived it eight years later when, before going ahead with the writing of the *Dialogue,* he wanted to circulate some pro-Copernican arguments in Rome to see how people would react to them. In the "Reply to Ingoli" Galileo claimed that

> [Ingoli's] fourth argument is an arbitrary invention of Tycho based on something which, in my opinion, he never observed and could not have observed. I am referring to the motion of comets when they are in opposition to the sun. Now, if it is true, as I most certainly believe, that their tail always points away from the sun, then it is impossible for us to see any of them when they are in opposition to the

51. "This argument is superfluous and vain, since neither Signor Mario nor I ever wrote that the cause of such deviation depended upon any other motion, either of the earth or of the heavens or of any other body. Sarsi has introduced this as a caprice, so let him answer it himself" (ibid., p. 262). Elsewhere, Galileo responded to Grassi's remarks about the possible heliocentric and therefore impious underpinnings of Kepler's (and his own) views on comets by saying that "since this opinion was deduced by Kepler as a consequence of the earth's motion, a proposition that cannot be piously and religiously held, Sarsi has therefore deemed it worthless. Why this must be so much the more motive for destroying it and demonstrating its impossibility; surely it is not a bad idea to prove also by physical reasons, if possible, the falsity of those propositions which have been declared repugnant to the Holy Scriptures" (ibid., p. 192). One would expect that, had Grassi believed in the soundness of Tycho's anti-Copernican argument, he could have deployed it to respond to these two challenges and refuted *both* Galileo's (and Kepler's) rectilinear hypothesis *and* the heliocentric assumptions that allegedly constituted its background. Grassi's bland replies to these two challenges in the *Ratio* are in *GO,* vol. 6, pp. 401, 453.

52. The "Reply to Ingoli" had circulated in Rome since the fall of 1624, and Grassi was personally told of it and, probably, of its pro-Copernican agenda. In fact, on 2 November 1624, Guiducci wrote Galileo that he had told Grassi of the "Reply to Ingoli" and that he was going to show it to him (*GO,* vol. 13, no. 1678, p. 224). I suggest that although Guiducci then changed his mind and did not give it to Grassi, the Jesuit (already knowing the topic of the reply) may have tried to intercept a copy of it in Rome. Given the Jesuit networks in Rome, it should not have been too difficult for Grassi to obtain it and include a reply in his 1626 *Ratio,* which, instead, he did not.

sun since in this case their tail would be invisible. Furthermore, what does Tycho know with certainty about a comet's own motion, as to be able confidently to assert that, when mixed with the earth's motion, it should produce some phenomenon different from what is observed? Regarding himself as the arbiter and ruler of all astronomical affairs, he finds true and right only those things which correspond to his observations or imaginations; thus, having invented a very implausible theory of comets, he saw nothing in comets that could support the Copernican hypothesis, and he preferred to deny and reject the latter than to abandon his own conceited whim.[55]

Although the first part of Galileo's answer may be a bit specious, he was right to point out that Tycho's argument was highly contextual, assuming his own theory about comets as the paradigm according to which one should interpret the apparent lack of retrograde motion in the comets. It was a theory-laden interpretation that could be used by Tycho to confirm his own interpretation to himself but that would have not carried much weight with a nonbeliever.

As Tycho shows, people holding different cosmological paradigms managed to perceive comets as perfectly fitting their views. If the comet of 1577 had helped Tycho become Tychonic, it had also been crucial in convincing Maestlin of the truth of the Copernican system.[56] Similarly, in his 1619 *De Cometis,* Kepler presented the comets as providing arguments for the earth's motion around the sun.[57] As he put it, "there are as many arguments (in addition to those deduced from the motion of planets) for the

53. That Galileo responded to Ingoli's 1616 letter in 1624 does not reflect, I think, Galileo's difficulty in responding to the comet-based anti-Copernican argument (only one of many that Ingoli presented), but rather it indicates the predicament in which he found himself after 1616. It would not have been wise for Galileo to defend Copernicus a few months after it had been put on the *Index*. Instead, he did so in the much more friendly environment resulting from Urban's election in 1623.

54. Francesco Ingoli, "De situ et quiete terrae disputatio," in *GO,* vol. 5, pp. 403–12. At p. 410 we find: "Quarta est ex Tychone in libro Epistolarum Astronomicarum, pag.149, ubi asserit, cometas caelitus conspectos, et in Solis opposito versantes, motui Terrae annuo minime obnoxios est, cum tamen esse deberent, quia respectu ipsorum evanescere motum huiusmodi non est necesse, sicut in fixis syderibus, cum cometae praedicti illam maximam fixarum a Terram distantiam non habeant."

55. *GO,* vol. 6, p. 554, as translated in Maurice Finocchiaro, *The Galileo Affair* (Berkeley: University of California Press, 1989), p. 191.

56. Robert S. Westman, "The Comet and the Cosmos: Kepler, Mastlin and the Copernican Hypothesis," in Jerzy Dobrzycki, ed., *The Reception of Copernicus' Heliocentric Theory* (Dordrecht: Reidel, 1972), pp. 7–30; and idem, "Michael Mastlin's Adoption of the Copernican Theory," *Studia copernicana* 13 (1975): 53–63.

57. A useful overview of Kepler's changing views about comets is offered in Alan James Ruffner, "The Background and Early Developments of Newton's Theory of Comets" (Ph.D. diss., Indiana University, 1966), pp. 94–118.

annual motion of the earth about the sun as there are comets in the sky. Farewell Ptolemy, I return to Aristarchus through Copernicus."[58] In short, given the complexity of the matter, the extreme variability of the phenomena, and the paucity of reliable observational evidence, the interpretation of comets could go either way, depending on one's cosmological beliefs.[59]

To summarize, we have no direct evidence that the Jesuits took Tycho's argument as a particularly good one or that Galileo was particularly worried by it. Also, Grassi did not seem eager to bring up cosmological issues unless attacked. The Jesuits were probably still cautious at going public with their post-1616 Tychonic stance on planetary theory (something they did only in 1620). When Grassi employed cosmological arguments (or threats) it was either to reply to Galileo's severe criticism of his support of Tycho, or to undermine Galileo by insinuating that he was a closet Copernican. In a sense, the heat of the debate (generated by Galileo) brought out Grassi's Tychonic sympathies in the same way the debate on the discoveries of 1610 helped consolidate Galileo's Copernicanism. Although cosmological arguments did appear in the later part of the dispute on comets, they did so more as auxiliary weapons then as foci of the debate.

A better key to understanding the motives of Galileo's attack on Tycho may be found in the possible symbiosis he saw between Tycho's planetary theory and the Jesuits' predicament and resources. This may explain Galileo's puzzling misrepresentation of the extent of Grassi's reliance on Tycho. Although the *Disputatio* mentioned Tycho only once, Galileo criticized Grassi for "subscribing to every claim of Tycho," an endorsement that Galileo presented as perplexing.[60] For instance, he presented Tycho's work on the comet of 1577 as nothing more than "a very diligent history" and chastised his interpretation of those data as "imaginations."[61] These dismissive remarks about Tycho's interpretive skills would become more conspicuous in the *Assayer*, where Galileo went so far as to claim that Tycho should be taught the rudiments of mathematics.[62] These attacks were strategic. By presenting Tycho as a good observer but a

58. Johann Kepler, *De cometis libelli tres* (Augsburg, 1619), p. 98, as translated in Ruffner, "Background and Early Developments of Newton's Theory of Comets," 113–14.

59. The vast variability of comets was well known to everyone. For instance, in the *Discourse* Galileo had stressed the differences between the last comet of 1618 and that of 1577 (Drake and O'Malley, *Controversy on the Comets*, p. 49).

60. *GO*, vol. 6, pp. 64–65.

61. Ibid., pp. 86, 92, 93.

62. "God forbid that Father Grassi should have imitated Tycho in these without noticing that Tycho, in his manner of investigating the distance of a comet by observations made from two different places on the earth, shows himself needful of attention to the very first elements of mathematics" (Drake and O'Malley, *Controversy on the Comets*, pp. 180–81).

CHAPTER FIVE

bad interpreter, Galileo eventually argued that Tycho's system was no system at all. It was a piece of astronomical patchwork that could not be compared to what Ptolemy or Copernicus had done. As we know, Galileo's *Dialogue on the Two Chief World Systems* did not bother to analyze Tycho's model.[63]

Galileo's surprising attack on Tycho and on Grassi's allegedly total reliance on him did not stem primarily from Galileo's fear of a possible comet-based refutation of Copernicus by Tycho and the Jesuits. Instead, Galileo wanted to discredit Tycho's model as a system so that it would not become the canonical position of Catholic astronomers after the condemnation of 1616. This should not be read just as a pro-Copernicus move. Rather, as we have done earlier, we should consider the homology between Galileo's increasing commitment to Copernicanism and his patronage predicament. Galileo was not against Tycho to defend Copernicus and against the Jesuits to save his own status as the star of astronomy. *These two tactical levels should be collapsed into one.* What Galileo worried about was the symbiosis between the Jesuits and Tycho. He wanted to prevent the Jesuits from becoming the new authorities in astronomy and Tycho from turning into their patron saint. Coupled with Tycho's authority, the Jesuits' already remarkable resources of colleges, mathematicians, and observatories in many countries probably would have produced an undefeatable package. As Galileo told Ingoli, Tycho regarded "himself as the arbiter and ruler of all astronomical affairs," and the Jesuits seemed now interested in following in his footsteps.

Symmetrically, by defending Copernicus, Galileo was also defending his own "uniqueness" as an astronomer and as a high-visibility client. Quite probably, Galileo thought of himself (rather than Tycho) as the greatest astronomer since Copernicus. Since his discoveries (especially the phases of Venus) had deposed Ptolemy, Galileo had probably imagined himself taking his place on the pedestal. It was bad enough that the condemnation of 1616 was preventing him from pursuing the high-visibility and controversial work that had brought him so many rewards. To see the Jesuits trying to put Tycho (and themselves) on the pedestal probably was too much. Given the limitations imposed on him by the condemnation of 1616, Galileo's best hope was to have people suspend their judgment on cosmological matters altogether. He did not want any decision or any new canon introduced (and institutionally backed by the Jesuits), trusting that the situation might change (or that he might find the final proof of Copernicus). For Galileo, it was much better to be a "suspended" star than simply number two.

This explains why in the *Assayer* Galileo argued that because Ptolemy

---

63. On Galileo's exclusion of Tycho from the *Dialogue,* see Howard Margolis, "Tycho's System and Galileo's Dialogue," *Studies in History and Philosophy of Science* 22 (1991): 259–75.

was refuted by Galileo's own observation of the phases of Venus, Copernicus was declared false by the Church in 1616, and Tycho had put forward a *nonsystem,* no decision could or should be taken about planetary theories.[64] *Mutatis mutandis,* this was a replay of his tactics in the "Letter to the Grand Duchess." The important difference was that now Galileo was in a weaker position. Copernicanism, having been declared false by the Church in 1616, could no longer be presented as a tenable hypothesis. Yet, he could still hold on to a version of the same tactic. Rather than try to have his adversaries come around to Copernicus by being unable to disprove him, this time he could try to stall a decision on matters of planetary theory.

Also, Galileo could conveniently attack Tycho and Grassi together by casting them in the unflattering cultural stereotype of the pedant. Although both Tycho and the Jesuits might be given credit for their diligent observations, they could not be taken too seriously as interpreters. Galileo tried to suggest that neither the Jesuits nor Tycho shared his "philosophical virtuosity." As much as the Jesuits tried to disguise it, they were closet pedants.[65] Otherwise, why did they feel such a strong need to rely on authorities? Galileo's misrepresentation of Grassi as an authority-bound pedant was effective. A stung Grassi responded by trying to muster all possible arguments to defend himself and Tycho from what he perceived as Galileo's unfair attack on him. However, Grassi's defense of Tycho (and his attempt to get back at Galileo by pointing to his lingering Copernicanism, i.e., his lack of respect for the Church's *authority*) provoked Galileo's counterattack on Grassi as a pedant who could not argue without bringing in authorities. Grassi's saying "Whom then should he have followed?" gave him away.[66] In a sense, Galileo brought out the pedant in Grassi with strategic attacks in the *Discourse* and then poked fun at him in the *Assayer.*

64. After criticizing Grassi for his tendency to look for authorities on cometary and astronomical matters, he claims: "I do not see why he [Grassi] selects Tycho and sets him before Ptolemy and Nicholas Copernicus, for both of these have constructed and followed out to the end, with the greatest skill, complete systems of the universe. This I cannot see that Tycho has done, unless indeed Sarsi thinks it is enough for Tycho to have rejected the other two systems and promised us a new one, though failing afterward to carry out his promise . . . Thus the two systems being surely false and Tycho's being null, Sarsi should not reprehend me for desiring with Seneca the true constitution of the universe" (Drake and O'Malley, *Controversy on the Comets,* pp. 184–85).

65. In this regard, I want to point to Altieri-Biagi's claims that Galileo's use of common (rather than technical) language in the *Assayer* was a strategy to present Grassi as a stuffy user of academic, technical jargon (Maria Luisa Altieri-Biagi, *Galileo e la terminologia tecnico-scientifica* [Florence: Olschki, 1965], p. 34).

66. Drake and O'Malley, *Controversy on the Comets,* pp. 71, 183 (emphasis mine). More specifically, Galileo argues in the *Assayer* that Sarsi should not have been so sensitive about his claim that, in the *Disputatio,* Grassi was endorsing Tycho in everything. True, Galileo did say that in the *Discourse,* but what he meant was "in everything *about comets.*" In short, Galileo suggests that Sarsi is overreacting and in doing so he is not serving well his master (Grassi),

Basically Galileo tried to portray the Jesuits as pedants in need of an authority figure while, at the same time exposing their bad judgment in picking their new source of authority. Obviously, this was not a technical argument that would have been appreciated by many professional astronomers, but the cultural connotations Galileo was imposing on the Jesuits and Tycho were not lost to the audience of courtiers and virtuosi that Galileo and the Jesuits were trying to attract.

Galileo's attack on Tycho was neither unreasonable nor exclusively motivated by his commitment to Copernicanism. A more complex and satisfying picture emerges once we move from an intellectualist approach that focuses primarily on issues of proper method, empirical adequacy, cosmological arguments and stakes, and post-1616 censorship, to consider a broader context that includes issues of status, resources, patronage dynamics, and different cultural styles. Once contextualized, Galileo's apparently puzzling use of the hypothetical argument and the apparently arbitrary lumping together of Tycho and Grassi turn into very smart tactics. Not only were these effective ad-hoc devices to get out of a contingently difficult situation, they also helped him maintain his patronage stature while holding off the closure of the cosmological debate. In particular, by proposing interesting hypotheses ("gems") rather than authority-based arguments, Galileo could present himself as a sophisticated courtly author while labeling Tycho and the Jesuits as pedants. It was a remarkable way of turning a serious defensive predicament into an elegant and courtly response. Once more, Galileo's commitment to Copernicanism and to his own image and patronage concerns went hand in hand.

Comets on the Roman Scene

Grassi's reply to the *Discourse* arrived promptly. Before the end of 1619 he published the *Libra astronomica et philosophica* under the pseudonym of "Lotario Sarsi."[67] In it, the polite manners of the *Disputatio* gave way to a

---

who is definitely too intelligent to endorse Tycho in everything. In fact, Galileo goes on to show some of Tycho's problematic claims (claims that, he is sure, Grassi would never think of endorsing). But everybody knows that Sarsi is Grassi. Therefore, behind the rhetorical defense of Grassi from the stupidity and authority-prone thinking of his student, Galileo is actually showing that Grassi felt pressured when he should not have and that his overreaction confirmed his "enslavement" to Tycho. Moreover, by confirming his overall endorsement of Tycho (something that Galileo had not requested), Grassi only invites further trouble because Galileo can then challenge him on more grounds. Yet to the contrary, Galileo presents himself as an individual with no hidden agendas, who was not bringing in any authority (either Tycho or Copernicus) but was just hypothesizing about comets (*GO*, vol. 6, pp. 228–33).

67. That Grassi adopted a pseudonym may have reflected his superiors' concern for the society's image. In a letter from General Vitelleschi to Father Rettore dated 6 August 1619, we find: "If Father Grassi should respond to the writings of this person, Your Reverence can

more aggressive and polemic style comparable to Galileo's own.[68] The 1619 *Libra* did not put an end to the debate. Grassi's response further upset Galileo, who after a long delay published the *Assayer* in 1623, which in turn provoked Grassi's final 1626 *Ratio*.

If the early phase of the dispute did not have a specific geographical center, its conclusion became an almost exclusively Roman affair. Archduke Leopold drifted out of the scene and his role was taken up by the Accademia dei Lincei. As Galileo was a member of that academy and had been obliquely addressed as such in Grassi's *Libra,* the dispute on comets was quickly transformed by the Lincei into a duel between their champion and that of the Collegio Romano.[69] Ciampoli nicknamed the *Assayer* the "Sarseide."[70]

When the *Assayer* was in press, a remarkable patronage conjuncture occurred. Maffeo Barberini, a good friend of Galileo (to whom he had dedicated a poem, the *Adulatio perniciosa,* three years earlier), reached the Holy See as Urban VIII.[71] Cesarini became his lord chamberlain, Ciampoli his secretary, while the pope's nephew, Francesco Barberini, was elected to the Accademia dei Lincei shortly before becoming a cardinal in October. Another Lincean, Cassiano dal Pozzo, became the secretary of Cardinal Barberini. In a matter of weeks, the Lincei found themselves closer to the center of power than any other cultural faction in Rome.

Galileo had written the *Assayer* as a long letter to Cesarini. In the light of the new events, the Lincei decided to dedicate it collectively to the newly elected pope. Urban, who loved to think of himself as a great poet and an intellectual on his own, was glad to accept the dedication by a star of Galileo's status. Like the *Sidereus nuncius,* the *Assayer* was one of those gifts that helped to establish a prince's image when it was most needed: at the beginning of his rule. In October 1623, during a presentation ceremony in the Vatican Palace, Prince Cesi delivered the *Assayer* to Urban.

---

rest assured that, having to publish, he will issue a well founded and considered response displaying religious modesty" (*ARSI, ROM* 17, 2, fol.305v; I would like to thank Father Lamalle for sharing this reference).

68. At the beginning of the *Libra,* Grassi presented himself as very surprised by the tone of the *Discourse,* especially given the friendship, support, and cooperation Galileo had always received from the Collegio Romano (Drake and O'Malley, *Controversy on the Comets,* 70–71). He also voiced his puzzlement that such a "courteous gentleman" was so rude and unappreciative of the *Disputatio*'s light and witty style. In short, Grassi tried to present Galileo as a grumpy, rude man (ibid., 72).

69. On this aspect of the dispute, see Pietro Redondi, *Galileo Heretic* (Princeton: Princeton University Press, 1987), pp. 68–106.

70. *GO,* vol. 13, no. 1518, p. 84.

71. In the *Adulatio,* Maffeo Barberini celebrated Galileo for the discovery of the Medicean Stars, of the unusual appearances of Saturn, and of the sunspots. Maffeo signed the letter that accompanied the poem "as brother" (*come fratello*)—an unusually informal title for a cardinal (*GO,* vol. 13, no. 1479, p. 49).

Presentation copies were also distributed to the cardinal-nephew and to other important cardinals who "had asked for them with much interest."[72] Grassi was less pleased. It was said that when he saw a copy of the *Assayer* at a Roman bookstore, he changed color, then grabbed the copy and walked away.[73]

By accepting the dedication, Urban VIII became the implicit arbiter of an increasingly Roman dispute. The Lincei's timely turning of Urban into the arbiter of the dispute was not just a way of cashing in on their newly strengthened connections to him. The dispute on comets had partially turned into a competition between two parties closely connected to the Roman court and competing for visibility on the Roman cultural scene.

Very likely the Roman audience saw the debate on comets as another literary *querelle*. The courtly literary style adopted by both parties, the frequent digressions, and the limited technical content of the texts contributed to such a perception. From a contemporary point of view, the Jesuits and the Lincei would have appeared as two factions among the many others competing in the chaotic Roman cultural scene. What differentiated them from most of the other factions or academies was that they were interested in natural philosophy instead of straight rhetoric or poetry. However, the audience they were trying to appeal to was not the professional astronomers (who in fact did not seem to pay much attention to what was going on between Grassi and Galileo) but the Roman intelligentsia, that is, those cardinals, prelates, and literati we have seen hopping from one academy to another or from one banquet or comedy to the next. Some of these people did notice the *querelle*. Poems on comets were circulated and commented upon with reference to the ongoing dispute, while the tension between the *novatori* and the conservatives in philosophy (likely to have been increased by the *Assayer* and the *Libra*) became the topic of an oration at the Savoia academy.[74] Also, as Urban was a literary

72. *GO*, vol. 13, no. 1590, p. 141.

73. Ibid., no. 1595, p. 147.

74. Agostino Mascardi, "Sopra un componimento poetico intorno alla cometa. Al Signor Conte Camillo Molza," *Prose vulgari* (Venice: Baba, 1653), pp. 151–67. The reference to the dispute is on p. 152. In a letter of February 1625 Mario Guiducci informed Galileo: "This past Thursday, in the academy that is held every week in the house of Cardinal Savoia, Signor Giuliano Fabrizi . . . gave a very nice lecture, and he criticized all the Peripatetics, and particularly those who take great stock in the authority of the writers" (*GO*, vol. 13, p. 253). Redondi has interpreted Fabrici's oration as referring directly to the *Assayer* (Redondi, *Galileo Heretic*, 74). The *avvisi* reported the event as the usual oration *cum* banquet: "Thursday evening, on the occasion of the academy held in the house of the Cardinal Savoia, Doctor Giuliano Fabritij of Spoleto gave a nice lecture on ambition that was very well received by Cardinals Barberini and Magalotti, who had intervened together with the Most Excellent Signor Don Antonio Barberini and Carlo Magalotti. All stayed for dinner with the same Cardinal Savoia" (quoted in Venceslao Santi, "La storia nella *Secchia rapita*," *Memorie della*

personality eager to demonstrate intellectual openness as much as political independence (especially vis-à-vis the Spanish crown), the *Assayer* was perceived by some as the manifesto of the innovators against the more traditional culture epitomized by the Collegio Romano.[75]

As we have seen, between 1619 and 1623 the Lincei put pressure on Galileo to respond to Grassi, reminding him that, if he failed to do so, his honor and that of the Lincei would be tainted. As Cesarini put it,

> On this occasion, I take the liberty to encourage you to publish the response to Sarsi that, for many reasons, you owe to the world, but in particular to regain from the ignorant that false sense of victory they have obtained from those writings [the *Libra*]. The Lord Prince and all the Lincei are asking you most warmly . . . and although—being in no need of further glory—you may disdain these too easy disputes, nevertheless you are obliged to the public name of the Lincei that has been offended by Sarsi and other malignants.[76]

The Lincei's animosity toward Grassi and the Jesuits in general can be better understood by looking at the cultural identity of the Lincei and at the changing relationship between them and the mathematicians of the Collegio Romano.

Different from the many literary academies competing for visibility in Rome, the Lincei were much more elitist, met much less often, and did so rarely in public.[77] The reserved nature of the Lincei reflected Cesi's own identity. Although a member of the highest Roman aristocracy, Cesi was an unwilling courtier. Like all the so-called Roman barons (the handful of top aristocratic families such as the Orsini, Colonna, Savelli, Cesarini, Conti, etc.) the Cesi were being slowly ruined by the costly life-style of the court.[78] As was typical of the development of political absolutism, the Ro-

---

Reale Accademia di Scienze, Lettere e Arti in Modena, series 3, 9 [1910]: 265). Fabrici's oration "Dell'ambitione del letterato" is included in Agostino Mascardi, ed., *Saggi accademici* (Venice: Baba, 1653), pp. 97–121.

75. On the Roman cultural scene and the Lincei's place in it, see Redondi, *Galileo Heretic*, pp. 28–136.

76. *GO*, vol. 13, no. 1523, p. 89.

77. Actually, academic meetings were not the *locus* of the academy's activity, which took place mostly through correspondence and publishing.

78. Usually, the Cesi (with the Cesarini) were put in the second class of Roman barons (the first class being reserved to the most ancient families like Colonna, Orsini, Savelli, Conti). On the financial decline of the Roman baronage, see Carlo Mistruzzi, "La nobiltà nello stato pontificio," *Rassegna degli Archivi di Stato* 23 (1963): 206–44; and (with reference to the Cesi family) Jean Delumeau, *Vie économique et sociale de Rome dans la seconde moitié du XVIe siècle* (Paris: De Boccard, 1959), 1: 153–55, 434–38, 467, 471–72. As Enrico Stumpo put it: "Although some of the most important Roman families such as the Colonna, Orsini, Cesarini, and Caetani were involved, the stocks of those 'monti' did not meet much luck in the title market precisely because those families were already in serious financial crisis. The in-

man aristocracy received worthless but high-sounding titles to compen-
sate for its declining political power. For instance, by the time he was a
young man Cesi was prince of San Polo and Sant' Angelo, duke of Aquasparta, and marquis of Monticelli. However, most of these places were little more than villages that were dignified with impressive names.[79] The
Roman barons had little political power and survived socially by letting
themselves be married into by the new (less noble but much richer) papal
families, or by trying to keep a cardinal in their family so that the necessary
face-saving privileges could be obtained.[80] The financial difficulties that
occupied Federico Cesi throughout his life and prevented the Lincei from
becoming more firmly established were typical of his social group. But if
this crisis of finances and identity eventually crippled the development of
the Lincei, it may also have been a cause of its initial establishment.

As a teenager, Cesi saw the life of the new natural philosopher as an
alternative to the life of the courtier. In his maturity, especially through
the training he received from his uncle, Cardinal Bartolommeo, he became a skilled political operator sensitive to all the subtle variations in the
Roman power structure.[81] However, he still maintained a sharp disdain

comes from their estates, while rich, were not sufficient to allow for and to maintain the very
high standards of living necessary in the Rome of that age, standards that could have been
assured only by the benevolence of the various popes" (Enrico Stumpo, *Il capitale finanziario
a Roma fra Cinque e Seicento* [Milan: Giuffré, 1985], p. 268). This decline was clear to contemporaries. As Traiano Boccalini sarcastically put it, "the poppies, high as cypresses already"
had been reduced to "the humble lowness of dwarfish violets" (Traiano Boccalini, *Ragguagli
di Parnaso,* ed. Luigi Firpo [Bari: Laterza, 1948], 3: 83).

79. Celestino Piccolini, "Ricevimenti ai feudatari nel Seicento," *Atti e memorie della Società Tiburtina di Storia e Arte* 7 (1927): 217–37 (this refers to the reception of Federico's
brother, who took over his title at Federico's death); idem., "Federico II Principe de' Lincei,
Marchese de Monticelli," *Atti e memorie della Società Tiburtina di Storia e Arte* 9–10 (1929–
30): 197–207; Giuseppe Gabrieli, "Memorie Tiburtino-Cornicolane di Federico Cesi fondatore e principe dei Lincei," *Atti e memorie della Società Tiburtina di Storia e Arte* 9–10
(1929–30): 230–47; idem., "Il Palazzo Cesi a Tivoli," *Atti e memorie della Società Tiburtina di
Storia e Arte* 8 (1928): 262–68; and Edoardo Martinori, *I Cesi* (Rome: Tipografia Compagnia
Nazionale Pubblicità, 1931), pp. 87–98.

80. For instance, it is quite evident that Cesi's financial troubles became severe after the
death in 1621 of his uncle, Cardinal Bartolomeo, who had managed to obtain the authoriza-
tions to establish and then renew the *monti* that financed the debt of the Cesi family. After his
uncle's death, Federico scrambled to use the two cardinals of the Caetani family (Federico's
grandmother was Beatrice Caetani) (Giuseppe Gabrieli, "Cesi e Caetani," *Rendiconti della
Reale Accademia Nazionale dei Lincei,* Classe di Scienze morali, storiche e filologiche, series 6,
13 (1937): 255–69). However, the eventual failure of Federico's efforts was not untypical.
As Stumpo put it: "In the first half of the seventeenth century, real estate properties worth
hundreds of thousands of scudi passed from the old group of nobles like the Colonna, Orsini, Cesi, or Caetani families to the new group of the Aldobrandini, Borghese, Barberini, or
Chigi families, often . . . from the relatives of the dead pope to those of the newly elected"
(Stumpo, *Il capitale finanziario a Roma,* p. 268).

81. For instance, see Cesi's very insightful account of the conclave from which emerged
Gregory XV. It seems that Cesi was able to write this report because he had accompanied his

for court etiquette, self-display, noisy disputations, and all the other typical features of Roman courtly and academic life. Although, out of necessity, he routinely engaged in all of these activities (he and his family kept a high profile in the gossippy *avvisi*), Cesi always did his best to spend long periods away from Rome on his country estates, writing and botanizing. His plans for the Lincei as presented in the unpublished "Lynceographum" reflect this identity and the culture that went with it.[82]

Cesi's establishment of the Lincei was also an attempt to fashion a new identity and a new mark of distinction for himself in a social context that was threatening his traditional social role. Although he was well aware of the decline, not only of his family but of the high Roman aristocracy in general, he also felt he was above engaging in the courtly rat race. That was the domain of aggressive social climbers like his friend Ciampoli. People like Cesi were uncomfortable in a court populated and often ruled by nouveaux riches (like some of the popes themselves). Still, he was no hermit. He knew too well that the Roman court was the source of all power, status, and privilege; if he wanted to keep his finances afloat, he needed to remain constantly in touch with the court either in person or through his various brokers (several of them members of the Lincei).[83] The new natural philosophy provided him with an alternative to what he experienced as a sense of social decline, identity crisis, and his unwillingness to engage in the routine of an aggressive court life.[84]

Because of his high social status, Cesi was expected to become neither

uncle, Cardinal Bartolomeo, into the conclave (Giuseppe Gabrieli, "Relazione del Conclave di Gregorio XV," *Archivio della Reale Società Romana di Storia Patria*, 50 (1927): 5–32.

82. "Lynceographum quo norma studiosae vitae Lynceorum Philosophorum exponitur," *Archivio linceo*, MS. 4. The manuscript has never been published, but a summary is found in Baldassare Odescalchi, *Memorie istorico critiche dell'Accademia de' Lincei e del Principe Federico Cesi* (Rome: Salvioni, 1806), pp. 204–42. Crucial for the understanding of Cesi's intellectual and moral outlook is his essay (unpublished in his time) "Del natural desiderio di sapere et Institutione de' Lincei per adempimento di esso," reproduced in Gilberto Govi, "Intorno alla data di un discorso inedito pronunciato da Federico Cesi fondatore dell'Accademia de' Lincei," *Memorie della Reale Accademia Nazionale dei Lincei*, Classe di Scienze morali storiche e filologiche, series 3, 5 (1879–80): 244–61. The essay that has best captured the relationship between Cesi's identity and the Lincei's agenda is Giuseppe Olmi, "'In essercitio universale di contemplatione e prattica': Federico Cesi e i Lincei," in Laetitia Boehm and Ezio Raimondi, eds., *Università, accademie e società scientifiche in Italia e in Germania dal Cinquecento al Settecento* (Bologna: Il Mulino, 1981), pp. 169–236. See also Mario Biagioli, "Declining Patrons and Pindaric Upstarts: The Lincei," in *Atti del convegno "La concezione del mondo in Europa: Religione, scienza e modernità attraverso l'opera di Alexandre Koyré (1892–1964),"* forthcoming.

83. On the reliance of Cesi on Francesco Barberini, see Giuseppe Gabrieli, "Il carteggio Linceo," *Memorie della Reale Accademia Nazionale dei Lincei*, Classe di scienze morali storiche e filologiche, series 6, 7 (1938–41): 853–54, 860–61, 883–84, 918–19, 934, 935–36, 948, 1206.

84. On the issue of the ethos and identity of aristocratic patrons of early modern science, see also Mario Biagioli, "Filippo Salviati: A Baroque Virtuoso" *Nuncius* 7 (1992)(in press). It may be interesting to explore similarities and differences between Cesi's identity and atti-

CHAPTER FIVE

a flamboyant courtier (one who engaged in noisy academic debates) nor a
pedant (like a traditional philosopher, slave to a system).[85] His commitment to freethinking in matters of natural philosophy was a sign of his
social status. Consequently, his academy was neither a tiresome gathering
of literati, nor a modern scientific institution, nor some sort of monastic
order. Rather, it was something closer to a knightly order, say a philosophical version of the Knights of Malta. As a social gesture, the Lincei
were a bit like Tycho's Uraniborg. Just as Tycho's feudal ethos led him to
conceive (and construct) Uraniborg as the castle of an astronomical lord,
so the Lincei were the expression of the ethos of a declining feudal lord
who (like Tycho) shunned the court and wanted to establish his own
realm, in this case an order of philosophical knights.[86]

The aristocratic connotation of the Lincei emerges well in the description of the rituals that marked its institution on Christmas of 1603.
Since the beginning, the Lincei were not going to be a "republic of letters." As Eckius put it, "to you lord, to you who shine with such heroic
virtues and are adorned with such truly princely qualities, suits not the
title and rank of brother, but rather that of prince of us other brothers. We
are only brothers, but you are, oh lord, our prince."[87] Stelluti confirmed
the need for this hierarchical distinction among them: "no one thing
could be more justly agreed upon than that our role is to be true brothers
among us while your role is that of the prince. Love belongs to us, to you
the empire to which you were born, and to which you were destined from
heaven. To you we give the sceptre, and you will govern us other
brothers."[88]

What followed was a ritual resembling that of knighting:

[Cesi,] after having dressed in long robes decorated with purple, as-
cended to the cathedra, and made each of the Lincei come to him
and, having read to them some new constitutions, he asked each of

tudes about natural philosophy and those of other major aristocratic virtuosi like Tycho and
Boyle.

85. Cesi's reflections on his philosophical program, as well as on his own aristocratic status, are sketched in a manuscript reproduced in Giuseppe Gabrieli, "L'orizzonte intellettuale
e morale di Federico Cesi illustrato da un suo zibaldone inedito," *Rendiconti della Reale Accademia Nazionale dei Lincei,* Classe di Scienze morali, storiche e filologiche, series 6, 14
(1938): 663–725, esp. pp. 689, 691–94.

86. On the relationship between Tycho's aristocratic ethos and science, see Owen Hannaway, "Laboratory Design and the Aim of Science: Andreas Libavius versus Tycho Brahe," *Isis*
77 (1986): 585–610.

87. Odescalchi, *Memorie istorico critiche,* p. 28 (Odescalchi's source here is the manuscript
"Gesta lynceorum," now *Archivio linceo,* MS 3). On the intriguing figure of Eckius, see Elisja
M. R. van Kessel, "Joannes van Heeck (1579–?), Co-founder of the Accademia dei Lincei in
Rome," *Mededelingen van het Nederlands Instituut te Rome* 38 (1976): 109–34.

88. Odescalchi, *Memorie istorico critiche,* p. 28.

them if they were willing and able to obey them. First among all of them, Eckius said he would and could observe them, and putting his open right hand over his chest, he took his oath. The prince then, opening his robe, showed a gold chain that hung around his neck on his chest, with a lynx suspended from the middle of the chain. He gave Eckius a similar one, saying, receive this symbol of brotherhood common to you and me: this is not only a sign of virtue and of brotherliness, but also a prize for future and present labours. All the others approached him and he hung on the neck of each one a similar chain. By so doing, he established solemnly the academy of the Lincei and the order of the studious lincei on the most august day of the birth of Christ.[89]

As young Cesi told his (probably perplexed) family when they questioned him on the nature of the Lincei and on why he was spending most of his time with them rather than with the other scions of the Roman baronage: "I carry a lynx of gold suspended from a golden chain as emblem of my studies and of my literary projects. I gave a similar one to my friends and to those in my *familia* whom I know to be most faithful to me; many princes give those persons whom they want to honor similar friezes and ornaments."[90]

Although the academy was established in 1603, it became much more active after 1610. By that time, Cesi had developed a less exclusive and more realistic image of the Lincei, but he still abided by the basic principles laid down in 1603.[91] In 1611 Galileo joined its ranks and, after that, the Lincei's status and visibility grew with his. Given the nature of the Lincei and of the cultural space they were trying to carve out for themselves, their competitors were not the various literary academies so much as the Collegio Romano. In particular, they were as sensitive as Galileo to the Jesuits' successful trespass into courtly natural philosophy. As much as Cesi disliked the court, it was there that his finances could be kept afloat, and it was at court that he eventually wanted the uniqueness of his academy to be recognized. Moreover, several of the less ascetic but most active members of the Lincei like Ciampoli, Cesarini, and Cassiano dal Pozzo

89. Ibid., p. 29.

90. Ibid., p. 68.

91. See Joannes Faber, *Praescriptiones Lynceae Academiae curante Joanne Fabro Lynceo Bambergensi* (Terni: Guerrero, 1624). On Cesi's ongoing interest (crippled by his financial difficulties) in establishing Lincean institutions outside Rome, see Antonio Favaro, "Di una proposta per fondare in Pisa un Collegio di Lincei (1613)," *Archivio storico italiano*, series 5, 42 (1908): 137–42; Giuseppe Gabrieli, "Il Liceo di Napoli," *Rendiconti della Reale Accademia Nazionale dei Lincei*, Classe di scienze morali, storiche e filologiche, series 6, 14 (1938): 499–564; and idem, "Marco Welser Linceo augustano," *Rendiconti della Reale Accademia Nazionale dei Lincei*, Classe di scienze morali, storiche e filologiche, series 6, 14 (1938): 74–99.

CHAPTER FIVE

were deeply engaged in courtly life.⁹² To them (especially to Ciampoli) the Lincei represented not so much Cesi's ideal of secluded philosophical life but a way of distinguishing themselves in the oversaturated cultural environment of the Roman court.

Despite the inherently competitive nature of the Roman scene, contrasts between the Roman Jesuits and the Lincei were a fairly new development. In the past, the interests of the two groups had often coincided or, at least, overlapped. Furthermore, Cesi's high aristocratic status and his closeness to his uncle, Cardinal Bartolommeo, helped to maintain a friendly relationship between the Lincei and the Collegio Romano. In 1611 the Jesuit mathematicians had supported Galileo (and, indirectly, the Lincei) because his astronomical discoveries helped them strengthen their position vis-à-vis the natural philosophers within the society. The Jesuits also supported Galileo's anti-Aristotelian work on buoyancy and seemed quite displeased by the events of 1616.⁹³ However, in the long run, both

92. On Ciampoli and Cesarini, see Giuseppe Gabrieli, "Bibliografia Lincea: II, Virginio Cesarini e Giovanni Ciampoli," *Rendiconti della Reale Accademia Nazionale dei Lincei*, series 6, 8 (1932): 422–62; idem, "Due prelati lincei in Roma alla corte di Urbano VIII: Virginio Cesarini e Giovanni Ciampoli," *Atti dell'Accademia degli Arcadi* 3 (1929–30): 171–200; idem, "Una gara di precedenza accademica nel Seicento fra Umoristi e Lincei," *Rendiconti della Reale Accademia Nazionale dei Lincei*, series 6, 11 (1935): 235–57; Domenico Ciampoli, "Un amico del Galilei: Monsignor Giovanni Ciampoli," *Nuovi studi letterari e bibliografici* (Rocca S. Casciano: Cappelli, 1900), pp. 3–169; Marziano Guglielminetti and Mariarosa Masoero, "Lettere e prose inedite (o parzialmente edite) di Giovanni Ciampoli," *Studi secenteschi* 19 (1978): 131–257; Maurizio Torrini, "Giovanni Ciampoli filosofo," in Paolo Galluzzi, ed., *Novità celesti e crisi del sapere* (Florence: Giunti Barbèra, 1984), pp. 267–75; Antonio Favaro, "Giovanni Ciampoli," in Paolo Galluzzi, ed., *Amici e corrispondenti di Galileo* (Florence: Salimbeni, 1983), vol. 1, pp. 135–89; Iustus Riquius, *De vita Virginii Caesarini* (Padua: Thuilii, 1629); Agostino Mascardi, "Per l'esequie del Signor D. Virginio Cesarino," *Prose vulgari*, pp. 349–67; Mario Costanzo, *Critica e poetica del primo Seicento* (Rome: Bulzoni, 1970), 2 vols.; Augusto Favoriti, "Virginii Caesarini Vita," in Virginii Caesarini, *Carmina* (Rome: Bernabò, 1658); Ezio Raimondi, *Anatomie secentesche* (Pisa: Nistri-Lischi, 1966); and Ianus Nicius Erythraeus, *Pinacotheca* (Leipzig: Gleditschl, 1692), pp. 59–60, 63–72.

93. Giovanni Bardi, a noble Florentine who had joined the Society of Jesus, gave a public demonstration on buoyancy at the Collegio Romano that was largely based on Galileo's work (*GO*, vol. 12, no. 1021, pp. 76–77; no. 1024, p. 79). Cesi and the Linceans Valerio, Faber, and Stelluti were invited together with the usual crowd. The sixteen-page text, dedicated to Cesi, was then published as *Eorum quae vehuntur in aquis experimenta* (Rome, 1614). Because extremely little is known about Bardi, I want to indicate that many letters to him are found in *ARSI*, ROM 17, 1; ROM 18; ROM 20, 1; ROM 21; ROM 22; ROM 23. Information about his studies and roles within the society are found in *ARSI*, ROM 55 ("Catalogi triennales 1616–1622"). In ROM 56 ("Catalogi triennales 1625–1633"), fol.337r, he is evaluated as "ingenium et iudicium medium, complexione ignea et cholerica." He died on 14 June 1635 (*ARSI*, HIST. SOC. 43, fol.11r). The Jesuit interest in buoyancy was not limited to Bardi. In 1614 Giuseppe Biancani (who had submitted *Brevis tractatio de iis quae moventur in aqua* to the society's *revisori*) was denied permission to publish on grounds that the piece was not original but only reiterated the claims of Galileo's *Discourse* (Baldini, "Additamenta galilaeana I," 18, 31). Finally, another Jesuit mathematician, Marino Ghetaldi, had done work on buoyancy in-

the Lincei and the Jesuits mathematicians established themselves in their respective niches and the conditions that had enhanced their collaboration slowly dissolved. Eventually, as shown by the debate on comets, the two groups collided. From Galileo's point of view, the publication of the *Assayer* marked a patronage clash between two networks he had carefully developed over the years. As we have seen, this patronage clash was propelled by patronage itself.

Strategic Detours

Less than four hundred copies of the *Assayer* were printed in the fall of 1623 (as compared with about two thousand copies of the *Letters on the Sunspots* and a thousand of the *Dialogue*).[94] Evidently, the debate had grown both nasty and local. By 1623 the larger public seemed to have forgotten the comets of 1618.[95]

Although Galileo wrote the *Assayer* as a line by line refutation of Grassi's *Libra,* he often used that text as a point of departure for long detours. Grassi's imprecisely worded claim that the comets must have been above the moon because they did not appear visibly enlarged by the telescope gave Galileo the chance to ridicule Grassi's alleged ignorance of optics and to enter into a long digression which allowed him to go back to the invention of his famous telescope and display his knowledge of optics while presenting Grassi as an ignoramus.[96] Another of these digressions was triggered by Galileo's critique of Aristotle's views of comets. In the *Discourse* he argued that Aristotle was wrong in claiming that comets result from the ignition of dry and hot vapors located immediately under the

---

formed by Archimedes (Pier Daniele Napolitani, "La geometrizzazione della realtà fisica: Il peso specifico in Ghetaldi e in Galileo," *Bollettino di storia delle scienze matematiche* 8 [1988]: 139–237).

94. On 8 September (*GO*, vol. 13, no. 1575, p. 129) Stelluti wrote Galileo that they had completed the printing, which consisted of 12,000 sheets. The *Assayer* is 236 pages of text and 14 of portraits, frontispiece, dedicatory letters and poems, etc., bringing the total page number to 250. Because the volume is a quarto, we derive an edition of 384 copies. For the size of the edition of the "Letters on Sunspots," see *GO*, vol. 11, no. 845, p. 482. There Cesi wrote Galileo that he had 2,000 copies printed. However, in *Archivio linceo*, MS 2, fol.133r, we find an account sheet in which only 1,400 copies are mentioned. On the size of the edition of the *Dialogue*, see *GO*, vol. 14, no. 2188, p. 281.

95. As shown by the very heterogeneous content of the *Assayer,* comets were no longer the exclusive focus of Grassi and Galileo. The few reports we have about the reception of the *Assayer* suggest that what attracted the Roman audience was not the strength of Galileo's own views on comets, but the spectacular display of argumentative skills that he employed to ridicule Grassi and the several brilliant speculative detours dispersed in the text.

96. *GO*, vol. 6, pp. 245–61. Grassi was not alone in using the telescope-based argument. The Lincean Stelluti wrote Galileo in December 1618 making the very same point (*GO*, vol. 12, no. 1365, pp. 430–31).

sphere of the moon and carried around the earth by the motion of that sphere.[97] From the problem of the ignition of sublunar vapors, Galileo drifted into a brief speculation about the nature of heat. That excursion was then expanded into those well-known passages from the *Assayer* where he discussed our perception of hot bodies as a result of heat particles impinging on our senses, which eventually led him to introduce what would become known as the distinction between primary and secondary qualities.[98]

These detours were not accidental. Not only did they provide Galileo with spaces in which to display his philosophical originality, they also reflected the patronage and courtly framework in which the debate developed. As we have seen, much of the natural philosophy produced within a courtly or patronage environment stemmed from questions posed by patrons. The question-answer structure did not apply to the patron-client relationship only, but it informed the interaction between the disputants as well. Especially in those cases in which the two parties did not share a comparable paradigm, a debate tended to develop into a barrage of often divergent questions, answers, and counterquestions. In a sense, debates were conglomerates of mini-duels. Questions could drift in almost any direction, often one in which a disputant would have some knowledge or argumentative resource. Also, by assaulting the critic with a range of counterquestions, a disputant could try to dispel the sense that he had not satisfactorily answered a question put to him previously.

As indicated earlier, disputes taking place within the patronage system did not usually reach closure because of the economy of the patrons' status. Now, I would add that disputes taking place within a patronage environment show that natural philosophers did not have a distinct argumentative etiquette but abided by the traditional practices of literary or philosophical *querelles*. The only difference is that the new natural philosophers (but also courtly writers with a humanistic background) expressed their frustration at the litigiousness of the disputational practices usually associated with traditional philosophy.[99] However, such expressions of dissatisfaction seem to be little more than gestures of discursive politeness.[100]

97. *GO,* vol. 6, pp. 52–3.

98. Ibid., pp. 54–56, 347–52.

99. For instance, in his oration on "Dell'ambitione del letterato," Giuliano Fabrici presents the texts of the Scholastic philosophy as "litigious" (Mascardi, *Saggi accademici,* 105). Similarly, in the *Praescriptiones lynceae* (p. 7), Cesi had Faber stress that the philosophy sought by the Lincei was against "verbose contentions." Galileo made a similar point in the *Assayer (GO,* vol. 6, p. 236).

100. Sometimes these claims seemed sincere. For instance, the new philosophers would occasionally claim to resent a disputation-based reward system (this is the reason behind Cesi's plans for the Lincei); however, they realized that such a system was the only one they

In a patronage context, disputes not only tended to avoid closure because of the patrons' reluctance to pass a final judgment; they also proliferated in all directions, making it difficult to decide even the grounds on which to designate a winner or loser. Obviously, this dynamic fit the patrons' interest very well. Until the development of standardized protocols of argumentation or the introduction of experiments as a way of circumscribing what could be talked about (and the ways in which one could produce acceptable evidence) disputes remained framed by the patrons' status economy: they remained entertaining performances. The dispute on comets is a good example of this pre-institutionalized etiquette of argumentation and of its results.

During the seven years it lasted and the five texts it produced, the dispute traveled a lot, and Galileo proved more skilled than Grassi at directing its migration. Two slim booklets had generated three respectable tomes. By the end, Galileo had succeeded in moving out of a very dark corner in which he had been put by the *Disputatio* to gain a position from which he could call the shots. That move had been made possible by his reliance on courtly aesthetics. By the time the *Assayer* was published, the topics of contention had multiplied to the point of eclipsing Galileo's initial trouble. By the same token, Grassi had been forced out of his safe initial enclave into the jungle of Galileo's dialectical traps and philosophical virtuosity. As a result, it seems that whenever Grassi felt he did not have strong empirical counterarguments to Galileo's, he was not comfortable in following and arguing with him on speculative grounds. Instead, he found refuge in citations from ancient authors. As a result, while the *Disputatio* was an empirical and well-argued piece (one that could have been written by Galileo himself), the *Libra* was populated by arguments from authority which provided Galileo with rich material to deride Grassi's pedantry.

For example, one of Galileo's interesting speculative digressions in the *Discourse* concerned the corpuscular nature of heat. After trying to criticize Galileo's hypothesis on empirical grounds, Grassi had compiled a list of quotations (from Aristotle, Ovid, Virgil, Lucan, and contemporary artillerymen) about projectiles becoming hot and sometimes melting as a result of the air's friction. In doing so, he wanted to defend Aristotle's

had and they adapted "reluctantly" to it. In fact, as Cesi put it, while in disputations there is no guarantee that the best will win, it certainly helps if the best has good disputational skills. In short, in voicing their frustrations, the new philosophers were also voicing their desires about alternatives—alternatives that, in their work, they did not (or could not) adopt. The tropes they employed in criticizing the contentiousness of the philosophers were often a replay of the usual humanist condemnation of the professional philosophers' disputational practices—something that we find even in Petrarch (Neal W. Gilbert, "The Early Italian Humanists and Disputation," in A. Molho and J. Tedeschi, eds., *Renaissance Studies in Honor of Hans Baron* [Florence: Sansoni, 1971], pp. 203–26, esp. 219–20.

claim that although heat is *caused* by motion through friction, it is not motion itself. More specifically, heat should not be thought of as a specific type of moving matter as Galileo had claimed.[101] After a long list of quotations from ancient poets (probably to show off his humanistic skills) Grassi ended with an amusing statement from the *Histories* of Suidas (a tenth-century Greek lexicographer): "The Babylonians whirling about eggs placed in slings were not unacquainted with the rude hunter's diet, and by this method which the solitude of the army required, by that force they also cooked the raw eggs."[102]

Galileo did not waste the chance to aim a funny mock-syllogism at Grassi's flying eggs:

> If Sarsi wants me to believe from Suidas that the Babylonians cooked eggs by whirling them rapidly in slings, I shall do so; but I must say that the cause of this effect is very far from that which he attributes it to. To discover the truth I shall reason thus: "If we do not achieve an effect which others formerly achieved, it must be that in our operations we lack something which was the cause of this effect succeeding, and if we lack but one single thing, then this alone can be the cause. Now we do not lack eggs, or slings, or sturdy fellows to whirl them; and still they do not cook, but rather they cool down faster if hot. And since nothing is lacking to us except being Babylonians, then being Babylonians is the cause of the eggs hardening." And this is what I wished to determine.[103]

Although Galileo's mocking of Grassi is quite devastating, we should not be swept away by his wit and forget that the view of heat Galileo was defending was as speculative (and, by present standards, incorrect) as the Aristotelian one defended by Grassi. The main difference between Grassi and Galileo was in argumentative style rather than in the plausibility of their positions. Actually, as we will see later, Grassi scored several empirical points against Galileo. We should not buy uncritically Galileo's masterful representation of Grassi as a pure pedant.[104] That was the result of Galileo's tactic, one he later adopted against the Simplicio of the *Dialogue on the Two Chief World Systems*. Although Grassi had a good potential as a pedant, he certainly did not develop it, and certainly not in the *Disputatio*.

---

101. Drake and O'Malley, *Controversy on the Comets*, pp. 115–22.

102. Ibid., p. 119.

103. Ibid., p. 301.

104. In fact, Grassi's attitude toward ancient authorities was not one of straightforward acceptance as Galileo claimed. Although Grassi did present quotes from the ancients as sources of evidence, he did not present those reports as indicating normal states of affairs. Rather, he claimed that those things did happen at some point in time (although they might no longer happen in the present). In short, he presented them as *singular* refutations of Galileo's claims (which he represented as having been put forward as lawlike statements).

Quite the contrary, he went a long way to shed the dogmatism usually attached to the authority-based discourse of the Aristotelians and to adopt a more elegant courtly style of argumentation. Similarly, in the *Libra* he did not always rely on authorities but used plenty of empirical arguments. Then, in his defense of Aristotle's claim that the movement of the moon's sphere could carry vapors around with it, he even introduced experimental evidence based on models simulating the processes being debated.[105] Pedantry was not an essential quality of Grassi's. Rather, it usually emerged when he was short of options which, in turn, resulted also from Galileo's opening up the debate well beyond its initial boundaries. It was Galileo who (to get out of his initial corner) provoked Grassi's transition from potential to actual pedantry and then laughed at it with his courtly friends.

The Fable of Sound

In the fall of 1623, Cesarini wrote to Galileo that the pope loved the *Assayer* so much that he was having it read to him during meals.[106] A few days later, Galileo was told that Ciampoli kept reading parts of the book to the pope, who seemed particularly fond of the so-called "fable of sound."[107] Those were the pages in which Galileo told the story of a man who hears a certain sound and tries to discover its origin. Each time he thinks he has found its real cause, he hears that same sound again and realizes that there was yet another way by which nature produced it. By following this same sound through a number of different causes, this man eventually finds a cicada. He thinks that, finally, he has a chance to find the *real* cause of the sound and decides to perform a "crucial experiment":

> At length, lifting up the armor of its chest and seeing beneath this some thin, hard ligaments, he believed that the sound was coming from a shaking of these, and he resolved to break them in order to silence it. But everything failed until, driving the needle too deep, he transfixed the creature and took away its life with its voice, so that even then he could not make sure whether the song had originated in those ligaments.[108]

Urban liked this parable not only because it was the literary high point of the *Assayer* but because it was the epitome of court culture itself. It showed that pleasure was in the appreciation of the virtuosity of nature, that is, in the multiplicity of the causes by which nature (and, therefore,

---

105. Ibid., pp. 105–15. They are presented as a response to Galileo in *GO*, vol. 6, pp. 53–54.

106. *GO*, vol. 13, no. 1589, p. 141.

107. Ibid., no. 1593, p. 145.

108. Drake and O'Malley, *Controversy on the Comets*, p. 236.

God) could produce a given sound. By trying to find the real, unique cause of the sound, one not only fails, he also kills the cicada and the pleasure of the inquiry with it. By seeking necessary causes rather than enjoying the novelties encountered along the way, one is a deluded philosopher *and* one who does not know how to play the courtier. By killing the cicada, one displays lack of philosophical virtuosity, courtliness, and respect for God's infinite power.

We have seen a similar view of court culture in Mascardi's 1625 oration to the academy of Cardinal Savoia. Like Galileo's virtuoso who is expected to appreciate the novelties and marvels he finds in nature without seeking their necessary causes, Mascardi's courtier (as a good humanist who appreciates *copia*) is supposed to pick the flowers he finds in the garden of the muses without asking for or wasting time developing dogmatic philosophical systems.[109]

The fable of sound does not only reverberate closely with Mascardi's oration but also with some of the dynamics of court patronage discussed earlier. At court and in patrician salons patrons were interested in brilliant performances and original interpretations rather than in dry syllogisms. What mattered was the show more than the end product. However, that the patrons privileged the performative aspects of disputes over the evaluation of the truth-value of the claims being debated was not just a sign of their superficiality or inability to engage with the details of the arguments. Great patrons could not put their status and power on the line by siding with a client whose claims might be judged to be wrong or too controversial. Consequently, strong claims might also have been represented as unfashionable because they threatened the economy of the entire social system of the court. Those who made strong claims were not represented as court virtuosi but as technicians—uncivil people who did not appreciate the elegant play of alternative views. They were not presented as threatening but simply as boring. The way they thought and argued demonstrated that they had servile intellects. To be mentally enslaved by a philosophical system was similar to belonging to the lower classes, that is, to being a slave to one's material conditions. Being pedantic was like being mechanical. As we have seen, court literature tended to portray university philosophers in these terms.

To sum up, there is an interesting link between court taste and the dynamics of patronage. The aesthetic of gems presented in Mascardi's oration or Galileo's celebration of the irreducible bountifulness of nature is not just the epitome of courtly urbanity and the opposite of pedantry: it is also the aesthetics that fit best the economy of the patrons' power. In fact, this type of discourse reflected the interests of great patrons like Cosimo

109. On *copia* see Terence Cave, *The Cornucopian Text* (Oxford: Clarendon Press, 1979), pp. 3–34.

de' Medici, Leopold of Austria, and the pope. This refined eclecticism typical of the baroque court allowed for elegant debates while providing the patron with a philosophical justification (actually a prescription) for not taking a stand. It was a type of discourse that helped patrons maximize their visibility and play safe at the same time. Nature's ability to produce a given effect in a number of different ways, celebrated in Galileo's *Assayer,* allowed Urban to enjoy Galileo's spectacular discoveries and controversial hypotheses without having to take a stand about how they fitted the Scriptures. This sort of "court nominalism" (or baroque eclecticism) was the discourse that fit best the needs of political absolutism.[110] The courtly disdain of pedantry, the impropriety of seeking causes and speaking *ex professo,* the celebration of the bountifulness of nature and of God's unlimited power, the appreciation of witty hypotheses, inventions, and discoveries, the unfashionableness of dogmatic philosophical systems, and the patron's noncommittal attitude were all intertwined in that remarkable cultural edifice typical of baroque courts.

The fable of sound was not only an epitome of court culture but of the argumentative structure of the *Assayer* as well. Galileo introduced it almost halfway into the text at a point where he was trying to justify his use of an argumentative style that did not involve hard proofs but original hypotheses and local interpretations. It was explicitly presented to ridicule Grassi's alleged pedantry and his inability to enjoy the "play" of novelties. The ways in which Galileo articulated this basic tactic to meet a wide range of specific arguments and problems are quite remarkable.

One of the *Assayer*'s most serious challenges was to undermine the credibility of Grassi's mathematical propositions about the position and trajectory of the comet. Galileo's main strategy was to say that—as he had already argued in the *Discourse*—before one deploys a geometrical method like parallax measurement, one should make sure one is dealing with real physical objects. According to Galileo, Grassi had failed to do so. This was not just a mere technical mistake. The flaw in Grassi's argument derived from a too literal application of his mathematical tools. He was dogmatic even in handling parallaxes. He lacked philosophical finesse.[111] As Galileo put it, the geometrical method should be used carefully because "it is a challenge too dangerous for those who cannot handle it."[112] The problem was not with the mathematical approach but with Grassi's stubbornness.

Another problem repeatedly mentioned by Galileo was that Grassi

110. I am using "nominalism" because of the apparent analogy between this position and that of Ockham. In fact, in both cases, the critique of the possibility of finding necessary cause for phenomena derives directly from God's omnipotence.

111. In a sense, Grassi comes across as someone who tried to learn how to become a courtier by reading handbooks of etiquette but failed to develop a "feel" for it.

112. *GO,* vol. 6, p. 296.

routinely misread his claims. Where he had presented a hypothesis, Grassi would see a real claim and would try to fault him for not having produced the necessary demonstration.[113] Rooted in Grassi's dogmatism, such a wrong-headed critique had to be dismissed.[114] Grassi was a "philosophical brute," like the man in the fable of sound who did not want to admit that nature can produce the same effect through a variety of causes.[115] Grassi did not seem to understand that, in certain cases, not searching for ultimate causes was not a sign of lacking intellectual nerve but rather a mark of finesse. If you insist in finding *the* cause, not only do you not find it, you end up killing the cicada as well. Consequently, if Grassi did not know how to play and took everything too seriously, he should not have expected a virtuoso like Galileo to play by his rules.[116] If Grassi wanted proofs, he should have gone and got them. In a sense, Galileo had turned the debate into a parlor game. When Grassi failed to abide by the rules of the courtly game, he had to do penance by being stuck with the burden of proof.

When Grassi criticized Galileo for saying that comets are a refraction of vapors while failing to explain why the motion of the comets was not linked to that of the sun as, say, in the case of rainbows, Galileo answered: "It would be quite adequate here to respond by saying that there is no necessity for the comet to behave like a rainbow, a halo, or other illusions, because the comet is different from a rainbow, a halo, and from the others [the illusions]."[117] Without bothering to specify in what ways comets were different from other optical artifacts produced by refractions, Galileo implied that there are other ways by which nature may produce comets, ways that are different from those it employs to produce rainbows and other optical phenomena. A courtly audience should have been satisfied with this. While the tactic of putting the burden of proof on his adversaries was nothing new to Galileo (for that is what he had already tried—though not very successfully—in the "Letter to the Grand Duchess"), now he had the powerful cultural trope of the "gem" to back up that tactic.

There is another way in which court discourse provided Galileo with a powerful resource to use against Grassi. Both in the debate on comets and in much of his work in astronomy and mechanics Galileo never presented

113. Ibid., pp. 225, 235–37, 273, 276–77, 278–79, 281–82, 289, 297, 303, 306, 316, 343–44. Galileo also claims that Grassi is consciously misrepresenting his hypothetical claims as empirical in order to attack them (ibid., pp. 294, 303, 305, 310, 314, 316).

114. At *GO*, vol. 6, p. 236, Galileo relates dogmatism to aggressiveness and to one's desire to win an argument by rhetoric rather than sound arguments.

115. Ibid., p. 289.

116. Drake and O'Malley, *Controversy on the Comets*, pp. 227–28; and *GO*, vol. 6, pp. 240, 279, 333–35, 341–43.

117. *GO*, vol. 6, p. 297 (translation mine).

a comprehensive philosophical system. Although isolated metaphysical considerations were common in his texts, he never put them together in any systematic fashion. The lack of a philosophical system put Galileo in a difficult position when it came to arguing with natural philosophers, who could refer their claims back to a system and who thought of explanation in terms of fitting a given phenomenon into its place within that coherent framework. As we have seen in the debate on buoyancy, the philosophers criticized Galileo for lacking a system that would have anchored the physical principles he employed in his arguments. Moreover, the range of phenomena and problems Galileo could address with his mathematical approach were minimal compared to the wide-ranging qualitative explanations made available through the Aristotelian system. Although Galileo tried to make it a point of intellectual pride that he felt free to say he did not have answers to many questions, such an approach would not have gone far in convincing the Aristotelian philosophers. As argued by Feyerabend, the lack of an articulated alternative philosophical system was a major handicap for Galileo.[118] Without a philosophical support system, his worldview was, so to speak, born refuted, and it could be sustained only by surrounding it with a range of ad-hoc hypotheses and auxiliary theories. Such an ad-hoc shield would have protected the new worldview for the time necessary for its articulation. Court discourse provided Galileo precisely with this type of shield and support system.

The homology between Galileo's argumentative structure and court discourse helped him delegitimize Grassi's authority-based arguments while legitimizing his own new natural philosophy. Galileo presented his hypotheses not as claims that referred to an all-encompassing philosophical system (one he did not have) but as free-standing novelties. While novelties or novel hypotheses might have been threatening to supporters of dogmatic philosophical systems, they turned into pleasurable spectacles of the boundless variety of nature once they were viewed through the eyes of a sophisticated, undogmatic court virtuoso. By adopting this discourse and audience, Galileo tried to turn the problem of being able to demonstrate only a few things into a virtue by saying that only undiscerning, noncourtly individuals would favor systems that claimed to provide all answers although providing no true ones. By relying on the tropes of court elitism, distinction, and singularity he claimed that *real* virtuosi should appreciate a few true propositions much more than many half-baked ones.[119] As with Mascardi's characterization of court culture, a few well-chosen flowers were all that one needed and wanted. Galileo no longer

118. Paul K. Feyerabend, *Against Method* (London: Verso, 1975), pp. 69–120.

119. *GO*, vol. 6, p. 237. This shows that Galileo was not an innovator in the sense of someone who favored new world-views in general. To the contrary, he was not in favor of new philosophical systems like those Campanella or Bruno would have liked. As stated in the *Assayer*, he was neither an ancient nor a modern in terms of what philosophical systems he

needed to criticize the Aristotelians for the ad-hoc nature of their system or for its empirical problems. He had an a priori argument against them: if you were part of *le monde* you did not want to build a system to begin with.

Courtly virtuosi appreciated novelties because, in a sense, the novelties were emblems of the intellectual freedom they possessed by virtue their status (or rather of the ideal representation of it). Nobles were represented as objective because they were disinterested, and they were disinterested because, by being economically independent, they did not have vested interests in what was being debated. For the same reasons, nobles (or those who wanted to represent themselves as such) could not let their intellects be enslaved by a philosophical system. To a noble, being eclectic in philosophy was a natural choice, almost a matter of *noblesse oblige*. And an image of knowledge as an eclectic collection of singular gems was precisely what Galileo's fable of sound or Mascardi's oration had presented. It did not really matter if these gems were new empirical discoveries or novel hypotheses. What mattered was their singularity and originality.[120]

This is the context in which we should place Galileo's famous image of the book of nature:

> Philosophy is written in this grand book—I mean the universe—which stands continually open to our gaze, but it cannot be understood unless one first learns to comprehend the language and interpret the characters in which it is written. It is written in the language of mathematics, and its characters are triangles, circles, and other geometrical figures, without which it is humanly impossible to understand a single word of it; without these, one is wandering about in a dark labyrinth.[121]

Although this passage has been usually interpreted as a straightforward declaration of Galileo's allegiance to the mathematical method (or even to Platonism), it does so only in a quite roundabout way.[122] Galileo's audience for the *Assayer* was not twentieth-century historians and philosophers but early seventeenth-century courtiers. The image of the open book of nature appealed to them because of the sense of *unmediated*

---

preferred (ibid., p. 235). Rather, he presented himself as being against the use of arguments from authority based on *any* philosophical system.

120. Ibid., pp. 236–37.

121. Drake and O'Malley, *Controversy on the Comets,* pp. 183–84.

122. In fact, although the passage is a general attack on authority in natural philosophy, the immediate target here was Tycho and not Aristotle. In the short term, Galileo was trying to argue that one should not canonize Tycho (and therefore rule out Copernicus) because there should be no need of authority in astronomical matters now that Copernicus had been declared false and Ptolemy had been refuted.

knowledge it conveyed.[123] True, one had to learn how to read those characters, but to learn a language was *not* like enslaving oneself to a philosophical system. Once that linguistic ability was acquired, the book was open and the interpretation free.[124] Actually, I do not think it really mattered that most of the courtiers could probably not read those characters. What was more important was that Galileo's *idea* of knowledge fit their own. The potentiality was more important than the actuality.

Not only could they read it by themselves: they could also choose what page or paragraph they wanted. As with the *Assayer,* one did not need to absorb an entire philosophical system (or follow a long narrative) before understanding what the book was about. Like Ciampoli when he read it to Urban, they could open any page at random and find some "gem." One did not have to read the entire *Organon* of Aristotle in the appropriate order before being able to talk philosophy. Although Galileo was definitely trying to legitimize the mathematical method, the trope he was relying on was not mathematical realism or Platonism, but the courtly desire for nonpedantic knowledge.[125]

There is a last tactic of Galileo's that I want to present. One would expect that by introducing a quite open hypothetical approach, Galileo would give up on any parameter to demarcate between proper and ad-hoc hypotheses. That does not seem to have been the case. He seemed to consider all his hypotheses as acceptable while criticizing Grassi for introducing ad-hoc ones. Again, this puzzling asymmetry is, I think, rooted in the courtly disdain for systems. A consequence of Galileo's linking novelties and hypotheses to the courtly ethos was the presentation of system-holders as intellectually unethical. If nobles were objective because they did not have vested interests, then system-holders were not objective be-

123. In his "Dell'ambitione del letterato," Giuliano Fabrici (probably taking off from the *Assayer*) wrote: "Philosophy ought to study the great text written by God, a volume which is the world, and the characters are the experiences. It should not be subjected to the rule of a litigious text that, after two thousand years of interpretation, is not yet understood." In short, the book of nature signified not only a type of knowledge that bypassed philosophical systems, but also one that was no longer litigious. With the "vested interests" of philosophical systems went also their tendency to trigger disputes (Mascardi, *Saggi accademici,* 105).

124. The image of the book of nature open before everyone's eyes resembles other "unmediated" practices of aristocrats: archaeology or botanizing.

125. Much interesting and often convincing work has been done on the relationship between the practices of the early modern craftsmen and notions of progress and openness to conceptual change. Although the culture of the baroque court is certainly not that of an artisan's workshop, I want to indicate that, for very different reasons, these two cultures provided different resources for the legitimation of the new knowledge. In particular, change, novelty, and singularity were crucial elements of the culture of the baroque court. They were intrinsically tied to one's need for distinction and to one's representation as a nonservile individual.

cause they had axes to grind. Not only the very fact of having a system indicated a servile, non-noble mentality, but (as argued by Bacon as well) having a system induced vested interests (idols) in one's mind. To have a philosophical system was like having a bad conscience; it was as much a sign of uncourtliness as of intellectual "impotence." Ultimately, it signified a lack of intellectual power (and therefore of objectivity).

As a result, one's reaction to discoveries (or novel hypotheses) functioned as some sort of lie detector. If you were disturbed or threatened by novelties, it meant that you were hiding something—maybe a philosophical system.[126] As Galileo put it, Grassi's tendency to attack his hypotheses instead of simply registering them as such was a "sign of a soul altered by some passion."[127] In a sense, Galileo's bringing up Grassi's pedantry was like exposing his bad conscience. In the *Disputatio* Grassi had managed to pass as a free-playing courtier, but the *Libra* (with Galileo's help) had brought his vested interests to the surface.

It is by manipulating the cultural connotations of the appreciation of philosophical systems and novelties that, I think, Galileo tried to justify his asymmetric stance on hypotheses. When Grassi criticized him for the major gaps Grassi perceived between his hypotheses and the available observational evidence, Galileo did not even attempt to deny the gaps but responded that there were many ways—unknown to us—by which nature could fill them. Instead, Galileo accused Grassi of trying to save his claims by introducing patently ad-hoc hypotheses. I think the difference here is that, while Galileo's hypotheses were "pure," Grassi's were "interested."[128] Galileo was playing; Grassi was cheating.

Let me give an example of one of Galileo's ad hoc (but "innocent") rescues of his claims. At one point, Grassi tried to refute Galileo's claim that heat was produced by subtle material particles that separate from solid bodies under friction and enter the pores of our skin, thus producing that sensation that we call "heat." Grassi said that he had carefully checked the weight of a body before and after heating it through hammering, but he did not detect any variation whatsoever. Grassi's apparently reasonable remark was sharply dismissed by Galileo as methodologically untenable. To him it was quite *possible* that the particles released by hot bodies had a

126. Galileo's critique of pedants and system-holders as impotent intellectuals resonates well with the negative stereotype of the repressed friar. In a sense, given his socioprofessional identity, the friar is the perfect system-holder. I am not suggesting a psychohistory of Jesuit mathematicians, but rather want to call attention to the sharp distinction between the culture of the cardinals and the court and that of the religious orders in Rome—a distinction on which Galileo tried to capitalize.

127. *GO*, vol. 6, p. 236.

128. Basically, Galileo claims that novelties are good because they help criticism of received views, and—although they do not necessarily lead to truth—they at least help to produce the best possible argument (ibid., p. 282).

much lower specific weight than the medium surrounding them and that—following the Archimedean laws of buoyancy—there was no *neces-sary* reason for the body to become lighter after having been heated.[129] What to Grassi seemed (quite justifiably) to be an ad-hoc move was not so to Galileo. True, his hypotheses were additional but not ad hoc. Galileo was disinterested and, therefore, his hypotheses had no hidden agenda.[130] Nature was ready to intervene with its bountifulness and to fill the gaps in Galileo's demonstrative chain. Nature sided with the good guys.

I hope to have shown that, when placed in the context of court culture and patronage, the *Assayer* ceases to be a puzzling text and begins to appear as a remarkable courtly artifact. With it, Galileo managed to turn serious handicaps into resources. The result was a text that did not try to demonstrate anything specific about comets or Copernican astronomy but tried—by adopting the discourse of the court—to legitimize his way of doing natural philosophy and delegitimize that of his adversaries. In the short run, the strategy paid off very well. With the important exception of the Jesuits, Galileo's cultural manifesto was very well received and praised by the pope and by the Roman courtly and academic community.[131] The fact that Galileo did not respond to Grassi's 1626 *Ratio* indicates that his honor was perceived as having been firmly restored by the *Assayer* and that nothing else was needed. As Cesi put it,

> I believe that everybody knows very well that Your Lordship is outside of the joust, and that you are not obliged to enter any arena or enclosed field [for duels]. . . . Mons. Ciampoli is of the same idea as are other courtiers and literati who love and esteem the work of Your Highness as it deserves. . . .[132]

---

129. Ibid., p. 334. This argument is very similar to that about the comets as refraction. In both cases, Galileo denies that measurement can be legitimately used on these phenomena. It is interesting to note that the way in which Galileo dismissed weighing as a proper probe of his hypothesis resembles his defense of the hypothesis that comets were refractions and that, by being so, their position could not be determined through a type of measurement—the parallax—that could not discriminate between real and fictitious bodies.

130. In fact, legally speaking, since the 1616 condemnation of Copernicus, Galileo could not have had any hidden agenda. As we know, he could not uphold the heliocentric cosmology anymore.

131. However, in a letter to Galileo in April 1625, Guiducci mentioned that an accusation had been filed against the *Assayer* with the Holy Office (*GO,* vol. 13, no. 1720, pp. 265–66). From what Guiducci had gathered, it focused on the alleged Copernicanism of the text. The accusation did not go anywhere (probably because the reviewer selected by the Holy Office, Father Giovanni di Guevara, was quite sympathetic to Galileo); but an anonymous accusation was actually filed against the atomistic doctrines included in the *Assayer.* This is the document that was uncovered by Redondi and became central to his thesis as presented in *Galileo Heretic.*

132. *GO,* vol. 13, no. 1902, p. 448.

Galileo's standing in Rome can also be read in the unusually warm reception he received from the pope (who granted him six audiences in six weeks) when he visited a few months after the publication of the *Assayer*. It was in these meetings that Galileo hoped to have the pope give conditional approval to his writing of the *Dialogue on the Two Chief World Systems*. Although Urban did not commit himself on that issue, he gave positive indications to Cardinal Hohenzollern, who relayed them to Galileo.[133] Eventually, years later, Urban gave a conditional approval to the publication of the *Dialogue*. Actually, Urban's request that Galileo should conclude the *Dialogue* by emphasizing that God, because of his omnipotence, could have structured the cosmos in many different ways was a direct reference to Galileo's argument in the fable of sound. The homology between Galileo's trope and Urban's theological views is spelled out in a dialogue between the two as reproduced in *De Deo uno,* a treatise published in 1629 by Agostino Oreggi, Urban's theologian:

> Having agreed with all the arguments presented by that most learned man [Galileo], [Urban] asked if God would have had the power and wisdom to arrange differently the orbs and stars in such a way as to save the phenomena that appear in heaven or that refer to the motion, order, location, distance, and arrangement of the stars. If you deny this, said Sanctissimus, then you must prove that for things to happen otherwise than you have presented implies a contradiction. In fact, God in his infinite power can do anything which does not imply contradiction; and since God's knowledge is not inferior to his power, if we admit that he could have done so, then we have to affirm that he would have known how. And if God had the power and knowledge to arrange these things otherwise than has been presented . . . then we must not bind divine power and wisdom in this manner. Having heard these arguments, that most learned man was quieted, thus deserving praise for his virtue no less than for his intellect.[134]

Quite probably, Urban expected another pious and entertaining book he could have read to him over meals. However, precisely because of Urban's election, the success of the *Assayer,* and the lack of opposition to a Copernican manuscript he had circulated in Rome, Galileo saw himself in

---

133. Ibid., no. 1637, p. 182. More signs of Urban's vague openness to the Copernican question were reported in March 1630 by Father Campanella, who entered in the matter with the pope (*GO,* vol. 14, no. 1993, pp. 87–88).

134. Agostino Oreggi, *De Deo uno* (Rome, 1629), pp. 194–95, as translated in Maurice Finocchiaro, *Galileo and the Art of Reasoning* (Dordrecht: Reidel, 1980), p. 10. Given the dates of Galileo's visits to Rome, it is likely that this dialogue took place in the spring of 1624.

a much stronger position than the one he had found himself in 1618.[135] His tactics changed accordingly. In the *Assayer* he had wholeheartedly embraced "court nominalism" because it represented the best strategy, given the several handicaps and problems he was facing. Things were different with the *Dialogue*. He knew very well what claims he wanted to push, understood that Urban's pontificate was a unique conjuncture, and thought he had some very strong evidence on his side (the explanation of the tides)—a particularly nontechnical argument that could be easily grasped by a courtly audience. As a result, the *Dialogue* was a much more ambiguous text than the *Assayer*. Although it used a fictional genre and formally remained within the boundaries of hypothetical discourse, it did its best to convey a message that was not hypothetical.[136] If the *Assayer* had presented Galileo as a philosopher in the sense of being a speculative and courtly freethinker, with the *Dialogue* Galileo wanted to go back to being a "philosophical astronomer," one who studied and argued about the physical structure of the cosmos.

As shown by the debate on comets, hypotheses, even when explicitly introduced as such, could be interpreted as hard claims. However complex and well planned, a text could not prevent the reader from viewing it through more or less hypothetical or realist lenses. Especially in the case of an inherently ambiguous text like the *Dialogue*, its interpretation was bound to be contextual. As we will see, the peculiar dynamics of the Roman court set this context in a way that Galileo may not have expected.

135. This is the so-called "Reply to Ingoli," reproduced in *GO*, vol. 6, pp. 509–61. It had a limited circulation in Rome beginning from the fall of 1624.

136. On the hypothetical form of the *Dialogue*, see Finocchiaro, *Galileo and the Art of Reasoning*, pp. 3–26.

SIX

Framing Galileo's Trial

V ERY FEW EVENTS IN THE HISTORY OF SCIENCE have received
more attention than Galileo's trial of 1633.[1] However, while much
literature has focused on the trial's conceptual (theological, cosmological, methodological) dimensions and on the specific personal interactions between Galileo and his friends and foes, it is only in the
recent work of Westfall that the role patronage might have played in the
events of 1633 has been finally addressed.[2]

Given the major gaps in the available documentary evidence, what I
propose here cannot be a comprehensive narrative about "what really happened" but only a possible alternative interpretive framework based on
the analysis of patronage and courtly dynamics presented earlier.[3] In particular, I will argue that Galileo's career was propelled and then undone by
the same patronage dynamics. I will try to show that the dynamics that led
to Galileo's troubles were typical of a princely court: they resembled what
was known as "the fall of the favorite."

---

1. The most important documents related to the trial have been recently made available in
English by Maurice Finocchiaro, *The Galileo Affair* (Berkeley: University of California
Press, 1989). The secondary literature on the trial is so vast that I will cite only those texts that
have been directly relevant to this analysis. The most significant secondary sources are listed
in the bibliography of Finocchiaro's volume at pp. 365–73.

2. Richard Westfall, "Patronage and the Publication of the *Dialogue*," *Essays on the Trial of
Galileo* (Vatican City: Vatican Observatory, 1989), pp. 58–83. Related issues are discussed in
idem, *"Galileo Heretic:* Problems, As They Appear to Me, with Redondi's Thesis", ibid.,
pp. 84–103.

3. Although Sergio Pagano's publication of *I documenti del processo di Galileo Galilei* (Vatican City: Archivio Vaticano, 1984) is a much-welcome collection that presents new material,
it does not fill the documentary gap on the trial of 1633. The only important new document it
presents was discovered and published by Pietro Redondi (see his *Galileo Heretic* [Princeton:
Princeton University Press, 1987]). It is the anonymous denunciation of the *Assayer* so central to Redondi's thesis.

My goal is not to present patronage as an auxiliary cause of Galileo's trial while leaving the clash between two incompatible worldviews as its main cause. Rather than talking about primary, secondary, necessary, or sufficient causes of the trial, I will try to reconstruct the specific patronage dynamics a client encountered in a baroque court like that of Urban VIII and to see how these may have affected Galileo's later career and trial. Although we do not know much about what was said in the meetings of the Congregation of the Holy Office in 1632–33, we can reconstruct the processes by which careers were launched and destroyed every day at the Roman court. Although we may never know who "terminated" Galileo and for what reason, we can understand how patronage dynamics set the stage for such a termination.

Conjunctions and Retrogressions at the Roman Court

Arriving in Rome from Paris in 1632, precisely at the beginning of Galileo's troubles, Alvise Contarini was struck by the peculiar character of the papal court. An experienced visitor at many European courts himself, he nevertheless found this one

> very different from any other court. Constituted as it is by a capricious mixture of men, nationalities, and interests, the Roman court is very difficult to figure out, to work within, and even more to describe. In Rome, ambition and the wings of hope take men to worship fortune. In the process, courtiers tend to undergo a strange metamorphosis. They forget about their princes and even about their own fatherland and concern themselves only in refining their skills and cunning while indulging in the excesses of courtly life and in its vices.[4]

Although this description tells a lot about Contarini's own "republican" values, the Venetian ambassador spotted well the unusual instability, the exceptional competitiveness, and the peculiar "habitus" that distinguished the Roman courtiers. However, what he observed did not come from the courtiers' genes but was the product of the peculiar power structure and generational cycles of the papal court.

Unlike other princely courts, Rome was not the seat of a dynasty. Also, one did not need to be of aristocratic origin to become pope. As shown by the election of Maffeo Barberini as Urban VIII, it could happen that a socially undistinguished subject of the Medici could, for the lifespan that was left to him after the election, rule (with great pleasure) over

---

4. Alvise Contarini in Nicolò Barozzi and Guglielmo Berchet, eds., *Relazioni degli stati europei* . . . (Bologna, 1856–79), 2:353.

his former ruler.[5] Similarly, in the late 1620s, Federico Cesi, the marquis of Monticelli, duke of Aquasparta, and prince of San Polo and Sant' Angelo, depended on cardinal-nephew Francesco Barberini (somebody who, until his uncle's election to the pontificate a few years before, had been socially much inferior to Cesi) for the privileges necessary to keep his finances afloat.[6]

Ciampoli's biographer stated that Rome was the place where "even a bum could become a prince," while Persico remarked that "as shown by our daily experience, at the Roman court there is nobody of so low a condition that he may not climb to a great position at some point."[7] After Urban's election in 1623, a distinguished cardinal commented ironically that

it was miraculous that somebody of less than private [nontitled] background could be lifted up to the sublime throne and made lord of the Soul, of the State, and of everything else. It was a miracle of the Lord God that somebody who could barely have supported himself would become respected by all the Christian princes.[8]

As stated by this cardinal, the political role the pope came to play was so unheard of that it must have been the result of a miracle. Other people used to say that the bees of the Barberini's coat of arms used to be horse flies. In any case, these "miracles" propagated downwards, allowing skillful and lucky courtiers to have "miraculous" careers as well. As expressed by Alvise Contarini's moralistic condemnation of the specific habitus of the Roman courtiers, this peculiar demographic pattern turned the papal court into a place of extreme competition and of extremely fragile patronage allegiances. It was a place where "nobody is well enough supported and connected to be sure of not falling under any circumstances,"[9] but

5. At the time he was elected, Barberini's patrimony was valued at 15,000 scudi. The sum is not negligible, but it indicates that when he became cardinal Maffeo was far from being as wealthy as most of his colleagues (Romolo Quazza, *L'elezione di Urbano VIII nelle relazioni dei diplomatici mantovani* [Rome: Reale Società Romana di Storia Patria, 1922], p. 43). However, within five years of his election he allowed his brother to pile up 1.5 million scudi, and he bought for 750,000 scudi an estate with title for his nephew Taddeo from the Colonna (Barozzi and Berchet, *Relazioni,* 2:262).

6. Giuseppe Gabrieli, ed., "Il carteggio della vecchia accademia di Federico Cesi," *Memorie della Reale Accademia Nazionale dei Lincei,* Classe di scienze morali storiche e filologiche, series 6, 7 (1938–41): 853–54, 860–61, 883–84, 918–19, 934, 935–36, 948, 1206.

7. Giovanni Targioni-Tozzetti, *Notizie degli aggrandimenti delle scienze fisiche accaduti in Toscana* . . . (Florence: Bouchard, 1780; reprint, Bologna: Forni, 1967), vol. 2, part 1, 106–7. A very similar statement is found in Barozzi and Berchet, *Relazioni,* 354; and Panfilo Persico, *Del segretario libri quattro* (Venice: Damian Zenato, 1629), 2:171.

8. Barozzi and Berchet, *Relazioni,* 2:246.

9. Alvise Contarini in Barozzi and Berchet, *Relazioni,* 2:353.

one where identities and roles (even ones as unusual Galileo's) could be fashioned (or ruined) much easier than in any other court.

Because of the lack of papal dynasties, the election of a new pope involved a drastic redistribution of power at the Roman court. Together with a new pope, a large group of nephews and close clients (a new "administration") would come to power.[10] Then, because the popes tended to reach the pontificate in their old age and because they did not have biological offspring to benefit from the products of their patronage, they tended to use their power and resources very quickly to achieve a vicarious progeny through patronage and in constructing an image of themselves that could last without the support of a dynastic mythology.[11] As shown by Frommel's study of popes' "egocentric" urban planning in Renaissance Rome, the peculiar demographic cycles of the Roman court had a profound impact even on the way the city looked.[12] These cycles, I think, also explain something about the obsession with *fama* and *fortuna* (and their dangers) apparent in Roman court literature and poetry.[13]

Quite literally, the popes' careers (or those of less aristocratic clients who tried to climb with them) were a race against time. Unlike secular princes, who might reach power in their early years, popes usually reached

10. Moreover, the physical stress of the conclave (which was particularly severe during the summer when malaria made its annual comeback) proved fatal to many of the very elderly cardinals. Six of them died as a result of Urban's conclave. As noticed by the Gonzaga ambassador in Rome, these deaths opened up career opportunities for younger prelates, therefore contributing to the reshuffling of cadres mentioned before (Quazza, *L'elezione di Urbano VIII*, p. 42).

11. As Contarini put it: "The more the Popes realize their mortality, the more they are generous with everybody, they distribute favors abundantly, and they do not temporalize as they would if the papacy were hereditary. Finally, the court in general finds employment and fortune only in the frequent changes [of popes]" (Barozzi and Berchet, *Relazioni*, 2:366–67). Contarini also comments: "From the private (and sometimes very low) state, the Popes ascend to one that is most eminent in dignity, authority, and wealth. Their affection toward their family interests increases with the same proportion, and it grows so much that they cannot bear the idea that their house and their descendants would return to their private state after their death. But they concentrate on making them remain princes and great lords, and therefore their study and attention focuses on this issue more than on anything else" (ibid., 215–16).

12. Christoph Luitpold Frommel, "Papal Policy: The Planning of Rome During the Renaissance," *Journal of Interdisciplinary History* 17 (1986): 39–65.

13. Agostino Mascardi, *Le Pompe del Campidoglio* (Rome: Zanetti, 1624), pp. 213–14, 219–20; and Giovanni Ciampoli, *Lettere di Monsignor Giovanni Ciampoli* (Venice, 1676), pp. 33, 45, 52, 58, 60, 63, 68, 78, 92, 102, 106, 108. Fama is also a preeminent figure in Maffeo Barberini's poems (see his *Poemata* [Paris, 1625]). See also Mascardi's "Discorso o invettiva fatta in una accademia intorno alla iniquità della fortuna," *Prose vulgari* (Venice: Baba, 1653) pp. 501–17. On the trope of "fortuna" at the Roman court toward the end of the century, see Renata Ago, "La cultura della carriera," *Carriere e elientele nella Roma barocca* (Bari: Laterza, 1990), pp. 104–13.

their "exaltation" late in life. And it was not uncommon for a pope to die soon after having reached the pinnacle of power. As Persico put it in 1629:

> Sudden and miraculous exaltations are more common at the Roman court than elsewhere because of the turnover of princes, the shortness of life, and other conditions. There being plenty of occasions, frequent mutations, the wheel of fortune spinning continuously, everybody would try to take advantage of those careers that are open to all either through religion or virtue. However, some are excluded because of clientele and envy, and others are taken aback by seeing virtue being less rewarded than malice and power. Although this is true of other courts, it is not as conspicuous as here because elsewhere the mutations are not as frequent, the goals and interests not as divided, and the form of government as changing and arbitrary.[14]

I think that the patronage frenzy of somebody like Urban VIII, of his secretary, Ciampoli, or of Galileo himself, may be better understood in this context.

Galileo demonstrated he had a good understanding of the periods of the revolutions of Roman patronage. In October 1623 (right after Maffeo Barberini's election to the pontificate) he wrote Prince Cesi that the election of Urban VIII was a *mirabil congiuntura*. This marvelous conjuncture, Galileo said, made him think about the possibility of important changes "within the republic of letters." Well aware of his age (he was then about sixty years old) Galileo claimed that if these changes "do not materialize in this marvelous conjuncture, then they will never come about because—at least as far as I am concerned—there is no point in hoping that a similar situation will come around again."[15] The changes that Galileo was considering were those connected with the legitimation of Copernican astronomy.

Because of these peculiar power cycles, Rome was a place where exceptional social acceleration, aggressive clients, conspicuous expenditure, and great patronage were the norm. In a sociohistorical period in which birth was the greatest determinant of one's social standing and career, Rome was a "carnevalesque" place where the legitimation of most unlikely social identities was possible. Private gentlemen like Maffeo Barberini could become popes, and mathematicians like Galileo could try to displace philosophers and theologians.

However, the heavy movement of clients to and away from Rome at the beginning of each pontificate indicated that the frequent cycles that made Roman patronage so powerful were also responsible for the extreme

14. Persico, *Del segretario,* p. 6.
15. *GO,* vol. 13, no. 1581, p. 135.

instability of Roman court life. For instance, writing in 1624 to Strozzi, Ciampoli—who was then Urban VIII's *segretario de' brevi* and *cameriere secreto*—described his position at the Roman court as powerful yet precarious: "Now, it could happen at any moment that I could find myself out of this Grand Hotel—as many here call the Pontifical Palace—and would have to get out of Rome for some time and live in a hut. . . . These are events that happen to all sorts of people at every new pontificate."[16]

If the presence of popes, cardinals, and church officials attracted clients to Rome from all over, the demographic pattern of the papacy reshuffled the hierarchies of clients and modified their composition, especially at the highest levels.

The careers of Galileo's contemporaries like the mathematician Benedetto Castelli, the Lincean virtuosi Cassiano dal Pozzo, Giovanni Ciampoli, and Claudio Achillini, the poets Agostino Mascardi, Gabriello Chiabrera, Alessandro Tassoni, Giovanni Battista Marino, Antonio Querenghi, and many visual artists display these migration patterns, which had characterized the Roman scene in the sixteenth century as well.

For instance, in the 1530s the mathematician Federico Commandino came to Rome from Urbino in the entourage of Cardinal Niccolò Ridolfi. Through Ridolfi, he entered the household of Pope Clement to become his *cameriere secreto*. But the pope died soon after and, having lost his patronage connection, Commandino moved to Padua.[17] Years later, he entered the household of Cardinal Ranuccio Farnese and traveled with him to Rome where he met, and made his qualities known to, Cardinal Marcello Cervini. As a result, when Cervini became pope, Commandino was called back to Rome. But when Federico eventually arrived in Rome, the pope was seriously ill. He died shortly after, and his newly arrived client had to go back to Urbino again, this time with a psychological depression that stayed with him for the rest of his life.[18]

Although Maffeo Barberini's career was a smooth one, his ambitious hopes risked collapse when his patron, Clement VIII, died in 1605. At that time, Maffeo was the papal ambassador (or nuncio) to the French king and it was expected that the new pope, Leo XI, would relieve him of that prestigious post (which usually led to the cardinalate) to give it to a closer client of his. Maffeo's secretary tells us that when news of Clement's death reached Paris "Monsignor Barberini wept and sighed most bitterly for a long time because of the tragedy of the pope's death, but even more for

---

16. Marziano Guglielminetti and Mariarosa Masoero, "Lettere e prose inedite (o parzialmente edite) di Giovanni Ciampoli," *Studi secenteschi* 19 (1978): 190.

17. Bernardino Baldi, "Vita di Federico Commandino," in Filippo Ugolini and Filippo Polidori, eds., *Versi e prose scelte di Bernardino Baldi* (Florence: Le Monnier, 1859), pp. 516–17.

18. Ibid., pp. 519–20, 527–28.

seeing his career wrecked."[19] Fortunately for Maffeo, Leo XI died in a few weeks and his successor, Paul V, was a good patron of Maffeo's.[20] As his secretary politely put it: "How much Nuncio Barberini regretted the Pope's death we will leave it to the reader to judge."[21] The sudden death of Leo XI saved Maffeo's career but it also "put an end to the fortune of thousands right when it [Leo's life] was on the verge of delivering them some benefit"; the Lincean Achillini described the "great rain of tears of his many clients, lifted up only to be dramatically dropped immediately after," a fate characteristic of the Roman court in those days.[22]

Even artists who had a relatively safe patronage connection with some cardinal—like the famous poet Giambattista Marino with Cardinal Maurizio di Savoia—experienced the vacant see as a time of frenzy and uncertainty. They spent those days anxious to see whether their patron would gain or lose power in the new conjuncture.[23] Marino's letters written just before the election of Urban VIII in August 1623 became increasingly short and populated by statements such as "I have time only for a short note because in this period of the vacant see I find myself most busy,"[24] or "I am so busy in these days that I do not think I will be able to send you that Discourse."[25] In those days, similar concerns are conveyed to Galileo by Ciampoli, who wrote to him that "we find ourselves entangled in the most busy occupations. . . ."[26] But Marino—together with the Lincei, Galileo, and many artists and poets—received a nice surprise in August 1623. As a much relieved Marino told Bernardo Castello: "Enough! Thank God, after so many turbulences during the vacant see, we have a poet pope—a virtuoso and a close friend of ours."[27] In those

19. "Il conclave di Urbano VIII," Isidoro Carini, ed., in *Spicilegio vaticano di documenti in editi e rari* (Rome: Loescher, n.d.), 1:345.

20. Ludwig von Pastor, *History of the Popes* (St. Louis, Mo.: Herder, 1938), 28:29.

21. Carini, *Spicilegio vaticano*, 1:346.

22. Giambattista Marino, *Epistolario seguito da lettere di altri scrittori del Seicento*, ed. Angelo Borzelli and Fausto Nicolini (Bari: Laterza, 1912), 2:115.

23. On the uncertainties of conclaves and vacant sees, see Quazza, *L'elezione di Urbano VIII*, and a contemporary report (probably written by Barberini's secretary) reproduced as "Il Conclave di Urbano VIII," in Carini, *Spicilegio vaticano*, 1:333–75. See also the description of the conclave of Gregory XV in the report of the Venetian ambassador in Barozzi and Berchet, *Relazioni*, 2:115–16. Petrucelli della Gattina's *Histoire diplomatique des conclaves* (Paris: Librarie Internationale, 1865), 3:1–94 gives a very detailed coverage of the conclaves of Gregory and Urban. On the rituals of the conclave, see J. Davies, trans., *The Ceremonies of the Vacant See or a True Relation of What Passes at Rome Upon the Pope's Death* (London: H. L. and R. B., 1671).

24. Marino, *Epistolario*, 2:25. Similar remarks are contained in another letter at p. 26.

25. Ibid., pp. 26–27.

26. *GO*, vol. 13, no. 1562, p. 119.

27. Marino, *Epistolario*, 2:27.

days, the correspondence of a literato like Mascardi or of a Lincean virtuoso like Stelluti delivered very similar messages to their friends, Galileo among them.[28]

Although the patronage crises that followed the election of a new pope often ruined previously established careers, they were eagerly awaited by younger, upward-moving clients. To move from one patron to another was no easy task in early seventeenth-century Italy. Because exclusive patronage relationships were represented as personal bonds, to drop a patron could turn out to be an insulting gesture and one which could prove very dangerous for the client.[29] Given this situation, the frequent patronage quakes that shook the Roman court were looked forward to because they provided relatively safe occasions for changing patronage allegiances.[30] With his usual lucidity, Persico commented that, although shifting patronage relationships could be found anywhere, they were especially common "at the court of Rome where, because of the so frequent revolutions, and because people have no other goal than their own profit, everybody turns himself toward the rising sun, leaving the setting one."[31]

Consequently, courtiers looked forward to new pontificates (unless, of course, their patron was in power). In 1609, the Venetian ambassador, Francesco Contarini, claimed that the Roman court was not too happy with Paul V "because they worry he may live long," an outcome contrary to the desires of those who "would like to see new pontificates frequently."[32] Similarly, the diarist Gigli saw the government of the city of Rome as "desiring changes and bothered by the long pontificate of Paul V."[33] The emergence of such a popular sentiment had been correctly predicted by the ambassador of the Republic of Lucca who, visiting Rome in June 1605 right after the election of Paul V, remarked:

> The Court is much saddened, and the Roman people have not demonstrated much happiness for the election of this pope. And that

28. See Mascardi's letters on the *congiuntura* reproduced in the appendix to Francesco Luigi Mannucci, "La vita e le opere di Agostino Mascardi," *Atti della Società Ligure di Storia Patria* 42 (1908): 494–95. Stelluti's letter on the happy ending of the vacant see is in *GO*, vol. 13, p. 121.

29. See Baldassarre Castiglione, *Book of the Courtier,* trans. Charles Singleton (Garden City, N.Y.: Anchor Books, 1959), book 2, no. 22.

30. For instance, Castelli—a client of the Barberini in Rome—declined the grand duke's invitation (delivered to him by Galileo) to go back to teach mathematics at Pisa, saying that "I have no way to free myself from this place . . . without the danger of ruining my affairs in a way that I could never reverse" (*GO,* vol. 17, pp. 361–62). Castelli's patronage dilemma has been discussed by Westfall in "Galileo and the Jesuits," *Essays on the Trial of Galileo,* p. 36).

32. Barozzi and Berchet, *Relazioni,* 2:87.

33. Giacinto Gigli, *Diario romano (1608–1670)* (Rome: Tumminelli, 1958), p. 72. Similar views are also presented at p. 48.

should not be surprising because that city feeds itself mostly on novelties [deaths of popes] and everybody hopes to improve their fortune through the frequent mutation of government and prince. But,
in this case, such hope is vanished by the youth and good health of
His Beatitude. . . .[34]

Given these situations, it is not surprising to see that the reports of the
pope's doctors on his illnesses were often perceived as oracles of future
joy.[35] A very similar perspective on the Roman court can be found in the
memoirs of Cardinal Guido Bentivoglio, a former student of Galileo's at
Padua and later one of the cardinals who signed his condemnation in 1633.
He recalled that, when he arrived in Rome as a young courtier during the
pontificate of Clement VIII, he found himself in a blind alley "because
that pontificate was lasting too long" and decided that "at the first new
pontificate, I would become a prelate and follow the usual career."[36] In
1635, another Venetian ambassador could still say that "the true sustenance
of the court consists only in the frequent turnover of popes."[37]

An English visitor was truly puzzled by the peculiar mentality of the
Roman courtiers. In 1620 he wrote to Lord Arundel that "It [is] a strange
and unnaturall thing that in that place, contrary to all others, the long life
of a Prince is sayd to be the ruyne of the people; whose wealth consists in
speedy revolutions. . . ."[38]

Unnatural or not, such a mentality dominated Roman court culture
and politics. When, after having listened to Ciampoli's ritual oration *De
Pontefice eligendo,*[39] the cardinals processed into the conclave on the morning of July 14, 1623, an unidentified courtier shouted: "I hope you'll all

34. Amedeo Pellegrini, ed., *Relazioni inedite di ambasciatori lucchesi alle corti di Roma (sec.
XVI-XVII)* (Rome: Tipografia Poliglotta, 1901), p. 26. A similar view is contained in a later
report dating 1667 (right after the election of Clement IX): "Because one has not seen many
mutations at all during this Pontificate and offices are for the most still held by the same
ministers as before, Rome (which desires novelty) remains little satisfied . . . And the same
prelates, each of them hoping that with the next movement of the wheel [of fortune] they
will have their chance to move up, have no right to complain because, at least, they do not
lack that which is the most common food of the court" (ibid., pp. 56–57). Thanks to Carl
Ipsen for having provided a copy of this source.

35. Richard Palmer, "Medicine at the Papal Court in the Sixteenth Century," in Vivian
Nutton, ed., *Medicine at the Courts of Europe, 1500–1837* (London: Routledge, 1990), pp. 49–
78.

36. "On the next mutation I thought I too would enter the prelature and take the usual
path" (Guido Bentivoglio, *Memorie e lettere*, ed. Costantino Panigada (Bari: Laterza, 1934),
p. 96.

37. Contarini in Barozzi and Berchet, *Relazioni*, 2:353.

38. Quoted in Francis Haskell, *Patrons and Painters* (New Haven: Yale University Press,
1980), p. 3.

39. Giovanni Ciampoli, *Oratio de pontefice maximo eligendo* (Rome: Mascardi, 1623).

become popes within a year!"[40] Apparently, everybody (cardinals included) burst out laughing.

These demographic cycles of the Roman court could be counted on even by people in disgrace. When Ciampoli was brought down by Urban VIII and sent to govern the very undistinguished province of Montalto, Castelli was surprised by the self-control (which he termed a "miracle") with which a usually arrogant courtier like Ciampoli faced such a dramatic shift of fortune.[41] But, if we read Ciampoli's letters written after 1632, we detect a strategy that had escaped Castelli.

At first Ciampoli thought that the pope would eventually calm down and that he would be foolish to relinquish his connection with the Roman court. Then, when he realized that the crisis was more permanent than he had initially expected, he began to revive contacts with other cardinals whom he thought would have a good chance to be elected to the pontificate next time around. To Cardinal Sacchetti he wrote that "I thank Your Excellence most humbly and pray that divine providence will accelerate the time of your supreme exaltation."[42] To Cardinal Aldobrandini he wrote that "I pray God to give you a long life so that you may reach an even greater princedom."[43] And he flattered Cardinal Monti by suggesting that his recent successes "are leading you toward the golden throne."[44] When, in 1641, Montalto, Filomarino, and Orsini reached the cardinalate, Ciampoli congratulated all of them for their promotions, praising one for his "sovereign prerogatives,"[45] telling another that he was "born to become a monarch,"[46] and reminding the third of a "future hope" for a further promotion.[47] Aware of the patronage deadlock he was in, Ciampoli prayed for divine providence to accelerate Urban's death.

As Ciampoli was twenty years younger than Urban, it was a reasonable strategy to cultivate new and old cardinals so as to be lifted out of the backwater as soon as one of them became pope. But, unfortunately for Ciampoli, Urban turned out to be a very healthy man. As Ciampoli knew very well, "political astrology" was no exact science: "We were not born to a century in which we could hope in the new invention of a Political Astrology which would teach the regular periods of the wheel of fortune whose motion often displays most unexpected retrograde motions."[48]

However, there was an upper limit to the period of the wheel of fortune: the length of the pope's life (and of the clients'). Consequently, clients made bets and developed some medium-range strategies and many

40. "Ch'io vi possa veder tutti papa in un anno!" (Quazza, *L'elezione di Urbano VIII*, p. 16, note 2).

41. *GO*, vol. 14, no. 2351, p. 430. Similar remarks about Ciampoli's behavior are found in *GO*, vol. 15, pp. 416, 420, 430, 433.

42. Ciampoli, *Lettere*, p. 35.

43. Ibid., pp. 29–30.

44. Ibid., p. 67.

45. Ibid., p. 94.

46. Ibid., p. 93.

47. Ibid., p. 94.

48. Ibid., p. 52.

more short-term tactics to maximize their chances and reduce their risks. But, most of the time, clients had to adjust or even force their plans into the framework produced by these patronage conjunctures. In some cases (or for some time) some of these strategies and tactics worked. On other occasions, as when Galileo accelerated his attempts to have Copernicanism legitimized after the election of Urban VIII, the forcing of a program to fit the cycles of patronage may have been partially responsible for its eventual failure.

Despite their different degrees of power, all Roman clients played similar games. Ciampoli, Galileo, and Maffeo Barberini himself were all trying to synchronize their moves with the cycles of patronage.[49] All of them had a relatively low starting position and—for different reasons and with different aims—tried to climb to the top in whatever field they had in mind. And Rome was the only place where those high-risk, high-gain bets could be placed; it was the place were the most powerful options of self-fashioning were to be found. Paradoxically, while Rome was the seat of a very conservative institution, it was also a place were change and novelty (within that framework) were the norm. But, for the same reasons, it was also the place in which false steps were most common and dangerous.

The "Fall of the Favorite"

I would now like to move from the analysis of the instability typical of baroque courts (and of the pope's in particular) to the analysis of a specific mechanism responsible for the high turnover among top courtiers and so-called favorites.[50]

Although little has been written about Roman court life at the beginning of the seventeenth century, we are lucky enough to have a contemporary analysis of court culture written by a client of the Barberinis and dedicated to the pope's brother, Cardinal Antonio. Published in 1624, *Che al savio è convenevole il corteggiare* was not the work of an outsider.[51] As a

49. The conjuncture of Roman patronage played an important role in the career of two other friends of Galileo, Giovanni Battista Rinuccini and Monsignor Dini (*GO*, vol. 13, no. 1493, pp. 59–60).

50. The favorite was a figure that (with different degrees of institutionalization) was typical of baroque courts. In Spain, the title was *privado* and represented a specific role within court hierarchy. In France, instead, the favorite was not an official role—although powerful favorites like Concini and Richelieu played a major role at court. For a comparison between Spain and France, see John H. Elliott, *Richelieu and Olivares* (Cambridge: Cambridge University Press, 1984), esp. pp. 34–43. In England it was a less official but nevertheless normal figure, especially during the Stuart and Tudor periods (See Robert Shephard, "Royal Favorites in the Political Discourse of Tudor and Stuart England" [Ph.D. diss., Claremont Graduate School, 1985]). I owe this latter reference to Paula Findlen.

51. Matteo Pellegrini, *Che al savio è convenevole il corteggiare libri IIII* (Bologna: Tebaldini, 1624). It was reissued one year later as *Il savio in corte* (Bologna: Mascheroni, 1625).

member of Cardinal Antonio's entourage, Matteo Pellegrini was well connected and accepted in Roman cultural and political circles. He delivered orations at the academy of Cardinal Savoia (the one run by Mascardi) and became a good friend of Ciampoli and Sforza Pallavicino.[52] *Che al savio è convenevole il corteggiare* went through two French editions by 1639 and set off a literary dispute which led Pellegrini to write *Difesa del savio in corte*.[53] Finally, his later *Delle acutezze* became an internationally known work on eloquence.[54]

Fortuna and the dangers of court intrigues had been privileged tropes of court treatises since the fifteenth century.[55] Such tropes became even more common with the development of the increasingly large, rigidly hierarchical, and professionalized courts typical of the baroque period. By the beginning of the seventeenth century, court treatises no longer presented the court as an ideal place (as in Castiglione's *Book of the Courtier*) but analyzed its power dynamics with surprising lucidity. With the advent of doctrines of *raison d'état* and the awareness that political absolutism was there to stay, courtiers tried to take a more detached, neo-stoic view of their predicament.[56]

Several chapters of Pellegrini's handbook for the literati seeking a career at court are dedicated to specific forms of courtly dangers. Among

52. Matteo Pellegrini, "Che il dir male non è in tutto male," in Agostino Mascardi, ed., *Saggi accademici* (Venice: Baba, 1653), pp. 193–208. The friendship between Pellegrini and Ciampoli—a friendship that continued even during Ciampoli's exile—is also mentioned in Claudio Costantini, *Baliani e i gesuiti* (Florence: Giunti, 1969), pp. 11–13. At the time of his death, Pellegrini was librarian of the Vatican Library—a post he had received from Cardinal Sforza Pallavicino. Basic biographical information on Pellegrini is found in Benedetto Croce, *Problemi di estetica* (Bari: Laterza, 1910), p. 320.

53. Matteo Pellegrini, *Le sage en cour* (Paris: Lancy, 1638; Rocolet, 1639); and idem, *Difesa del savio in corte* (Viterbo: Diotallevi, 1634). The addressee of the *Difesa* was Giovanni Battista Manzini.

54. Matteo Pellegrini, *Delle acutezze* (Genoa: Ferroni, 1939). On Pellegrini's work and its influence, see Croce, *Problemi di estetica*, pp. 322–37; idem, *Poeti e scrittori del pieno e tardo rinascimento*, (Bari: Laterza, 1945), pp. 205–7; Antonio J. Saravia, *O discurso Engenhoso* (São Paulo: Editora Perspectiva, 1980), pp. 91–146; Klaus-Peter Lange, *Theoretiker des Literarischen Manierismus* (Munich: Wilhelm Fink Verlag, 1968), pp. 114–41; Ezio Raimondi, *Trattatisti e narratori del Seicento* (Milan-Naples: Ricciardi, 1960), pp. 109–12; Mario Rosa, "La chiesa e gli stati regionali nell'età dell'assolutismo," in Alberto Asor Rosa, ed., *Il letterato e le istituzioni*, Letteratura Italiana, vol. 1 (Turin: Einaudi, 1982), pp. 324–25.

55. For earlier views on *fortuna*, see Charles Edward Trinkaus, *Adversity's Noblemen* (New York: Columbia University Press, 1940), esp. chap. 5, "The External Conditions of Life," pp. 121–40. A classic fifteenth-century court treatise is Aeneas Sylvius Piccolomini [Pope Pius II], *De curialium miseriis epistola*, ed. Wilfred P. Mustard (Baltimore: Johns Hopkins University Press, 1928).

56. On late sixteenth- and early seventeenth-century *tacitismo* see Gerhard Oestreich, *Neostoicism and the Early Modern State* (Cambridge: Cambridge University Press, 1982).

them we find "The Instability of Favor," "That the Natural Instability of Favor Is in the Interest of the Powerful," "The Danger Caused by the Favorite's Arrogance," "That the Favorite Is at Danger from Envy," "The Dangers of Being the Favorite," and, finally, "The Fall of the Favorite."[57] While Pellegrini treated these issues from a more analytic point of view, other writers dealt with them within fictionalized narratives about the fall of real courtiers. These books emerged all over Europe after the fall of some major favorite like Concini in France or Buckingham in England. Often, these writers dramatized the end of real favorites by going back to historical examples (like that of Emperor Tiberius executing Sejanus) and elaborating on them.[58]

Pellegrini's description of court patronage ties is permeated with the discourse of love and flirtation. For instance, "the favor [*gratia*] of the prince is like Penelope around whom the rival courtiers struggle for the fruits of servitude."[59] This imagery emphasizes that, unlike the modern period in which the pyramid of power is relatively broad at the top, or in which there exist a number of different forms and centers of power, court society had only one source of all types of power: the prince.

Because of this power structure, competition in court society was of a peculiar type. One did not seek success but *gratia,* something that could not be measured in terms of annual income or number of citations in academic journals. *Gratia* was the result of the prince's favor. The further one climbed toward the top, the more one's career was represented as a form of kinship with the prince, to the point of being poetically represented as an exclusive love relationship. As Pellegrini put it, "Two lovers cannot enjoy the pleasures of the loved one at the same time. And there is not space for two on the throne of favor."[60]

Once a courtier had developed patronage ties within a court, his career would be very much tied to that place. Therefore, Pellegrini's repre-

57. Pellegrini, *Che al savio è convenevole il corteggiare,* pp. 56–95.

58. The fall of Concino Concini, Maria de' Medici's main advisor at the French court, provided material for much of this literature. Giovanni Battista Manzini's *Della peripetia di fortuna: Overo sopra la caduta di Seiano* (1620?) is a good example of this literature. It was translated into English as *Observations Upon the Fall of Seianus* (2d ed. London: Harper, 1639). Another example is Pierre Matthieu, *Unhappy Prosperity Expressed in the History of Aelius Seianus, and Philippa the Catanian. With Observations Upon the Fall of Seianus* [a reprint of Manzini's book]. *Lastly Certain Considerations Upon the Life and Services of Monsieur Villeroy* (London: Harper, 1639). See also Ben Jonson, *Seianus and His Fall,* ed. W. F. Bolton (London: Benn, 1966); Francisco de Quevedo y Villegas, "Como ha de ser el privado," *Obras completas,* ed. Felicidad Buendia, vol. 2 (Madrid, 1967); and Mira de Amescua, ed., *La comedia famosa de Ruy Lopez de Avalos (primera parte de Don Alvaro de Luna) como doctrinal de privados y regimiento de principes . . . ,*(Mexico City: Editorial Jus, 1965).

59. Pellegrini, *Che al savio è convenevole il corteggiare,* p. 20.

60. Ibid., p. 63.

sentation of competition among courtiers as a matter of favor or disgrace
(or even of life or death) is not much of an overstatement:

> if the loved one [the prince] caresses one of the lovers he insults the
> other one. What can be more hurting than watching somebody else
> enjoy our lover? Our rage against the one who enjoys our lover and
> hurts us is as great as our craving for her pleasures. But if our lover is
> already taken, what can we hope for, or fear worse? Recapturing her
> is the lover's burden. Seeing [our lover] the prey of somebody else is
> unbearable. The pain of being without her may be consoled by hope
> only. But there are no grounds for hope if somebody else possesses
> her already. The desperate lover cannot do anything but eliminate
> the rival. The cheated courtier sees no way out except in the ruin of
> the Favorite.[61]

The language of desire and jealousy employed by Pellegrini provides a
fitting metaphor for the actual power dynamics at court. The fall of the
favorite was not an accident but a normal process. As indicated by Pellegrini, subordinate courtiers could endure their position and work hard
to increase their favor with the prince only if they were allowed to have
hope of promotion. Routine ruin of favorites provided that hope. As another analysis of the fall of the favorite puts it, "the space you leave needs
to be filled again."[62] Finally, the courtiers welcomed the fall of the favorite
because it washed away their patronage debts toward him.

The uncertainty of favor was the most powerful tool the prince had to
maintain control of the court. Therefore the fall of the favorite was a
mechanism that worked both for the prince and for the upward-moving
courtiers. The fall of the courtier did not harm the prince. Paradoxically,
the prince's power was increased or maintained both by a courtier's becoming the favorite and by his subsequent falling in disgrace. By being
successful, the courtier enhanced the prince's image. If he failed, his fall
would also help the preservation of the prince's power by allowing him to
display it mercilessly and remind the other courtiers of what could happen
to them if they misbehaved. It was a courtly *memento mori*. At the same
time, it would keep their "hope" going, and occasionally (as in the case of
the fall of Ciampoli and Galileo) it would also provide convenient scape-
goats for some of the prince's own political problems. As Giovanni Battista Manzini put it in his *On the Vagaries of Fortune: Or on the Fall of
Sejanus,* "There were some who attributed to Sejanus the responsibility
for all the excesses of Tiberius."[63] Similarly, Sir Walter Ralegh, a favorite
of Queen Elizabeth, wrote that

> Tyrannous princes having incurred the universal hate of [their]
> people, found no means to preserve them from popular fury, as to

61. Ibid., pp. 62–63.   62. Manzini, *Della peripetia di fortuna*, p. 38.   63. Ibid., p. 10.

execute or deliver into their hands their own chief minions [favor-
ites] and intimate counsellors. Example: Tiberius delivered to the
people his favorite Sejanus; Nero, Tigellinus; Henry, King of
Sweden, committed to their fury his best beloved servant George
Preston; Caracalla caused all his flatterers to be slain that had per-
suaded him to kill his brother. The like was done by Caligula,
whereby he escaped himself.[64]

Bacon too commented on the fate of the favorite. Writing to Sir
George Villiers (a favorite) in 1616, Bacon warned him:

Remember then what your true condition is. The King himself is
above the reach of his people, but cannot be above their censures;
and you are his shadow, if either he commits an error and is loath to
avow it, but excuses it upon his Ministers, of which you are the first
in the eye: or you commit the fault, or have willingly permitted it,
and must suffer for it; so perhaps you may be offered as a sacrifice to
appease the multitude.[65]

In short, the fall of the favorite was no accident but rather a routine
process of "seasonal rejuvenation" of the court and of "cleansing" of the
prince's power image. It was almost a ritualized sacrifice that worked both
for the prince and aspiring courtiers.[66] As Manzini put it, by bringing
down Sejanus, Emperor Tiberius was offering a sacrifice to Fortune.[67]

Another crucial feature of the fall of the courtier was its suddenness
and inexorability. Pellegrini noticed that "from the summit of favor one
does not descend through the same steps which lead to the top. Often
nothing stands between one's highest and lowest status."[68] Similarly,
Manzini remarked that "the coming back down from a previous height

64. Sir Walter Ralegh, "Cabinet Council," *The Works of Sir Walter Ralegh* (Oxford: Oxford University Press, 1829), 8:149–50, quoted in Shephard, "Royal Favorites," p. 55.

65. Francis Bacon, "Letter of Advice to Sir George Villiers—the First Version," *The
Works of Francis Bacon*, ed. James Spedding et al. (London: Longmans, 1857–74), 13:14; as
quoted in Shephard, "Royal Favorites," p. 54. In a later version of the same letter, Bacon
remarked that "kings cannot err; that must be discharged upon the shoulders of their ministers" ("Advice to Villiers—Second Version," *Works*, 13:28; quoted in Shepard, "Royal Favorites," p. 54).

66. The tacit complicity between the prince and the courtiers during the fall of the favorite is remarkable. Sometimes, the prince's sending signals of withdrawal from the favorite
created a vacuum around him that facilitated or even produced his fall. At other times, the
rumors spread by courtiers concerning the favorite could be picked up by the prince and
turned into an indictment of the favorite.

67. "Tiberius weaved the plot so judiciously that he honored Sejanus and his son with the
dignity of Priesthood. It was almost as if he were advising him to prepare himself to sacrifice
a victim to Fortune" (Manzini, *Della peripetia di fortuna*, p. 11). The trope of sacrifice also
emerges at p. 22.

68. Pellegrini, *Che al savio è convenevole il corteggiare*, p. 73.

cannot be anything but a sudden jump. Most of the time, one cannot tell the passage from the supreme to the lowest state,"[69] or "He who wants to know what is the greatness one experiences by being close to the prince should make his will, because it is nothing else than a sudden downfall."[70]

The instantaneity of the fall results also from everybody dropping the courtier who seems to be losing the favor of the prince. Fortune herself ceases to accept the tributes of her former protégé: "While [Sejanus] was sacrificing to a statue of Fortune . . . she turned her head not to see or pity him, showing the poor courtier how little one should trust the vanity of Fortune, who cannot be stable even when she is of marble."[71] While the isolation of the falling courtier might have been cruel, it was also rational. Once the prince has decided to drop somebody, nobody could save the victim without joining him in his ruin.[72]

The widespread withdrawal of support from the falling courtier fit the interest of the prince by speeding up the courtier's expulsion. In fact, to be effective, the fall of the favorite had to be quick and inexorable. It was only by being absolute that it would be perceived as a sign of the prince's absolute power in determining the fate of his courtiers. Also, the prince could not under any circumstance admit that he had previously misevaluated the now-disgraced courtier. By doing so, he would present himself as fallible and consequently as not absolutely in control, that is, as not absolutely powerful. As Pelligrini puts it:

> Jurists teach us that unless new evidence emerges, one is not justified in disliking what he had previously liked. Nor can the newly emerged rage of the prince be excused without citing some major misbehavior on the part of the favorite. In fact, one does not condemn the loved one for small things. Because the powerful [prince] tends to dissimulate the small defects of the members of his own family, how could he then punish the same defects when they appear in the one he had elevated to the highest honors? The prince cannot judge the favorite as unworthy of his grace (of which he had previously thought him well worthy) without using the pretext that he had misbehaved in some most serious way.[73]

Literally, the prince represented himself as "sacrificing" his favorite courtier to "justice." Although he was simply dropping his client, the process was ritualistically represented as involving a mixture of rage and sor-

69. Manzini, *Della peripetia di fortuna*, p. 20.   70. Ibid., p. 28.   71. Ibid., p. 14.

72. "The Favorite should not hope to receive support in his ruin. Great falling masses crush those who approach them. It is useless to look for help where everybody desires one's fall . . . He who insults the misfortunes of others applauds the will of the heavens . . . When the tall oak falls, all run to get firewood; when the great Favorite falls, everybody runs to the prey" (Pellegrini, *Che al savio è convenevole il corteggiare*, p. 74).

73. Ibid., p. 75.

row on the part of the prince. By "giving up" his favorite to justice, the prince showed himself to be absolutely just, to the point of being willing to do violence to his own loving feelings for the client. However, as explicitly expressed by Pellegrini, the prince did not need real causes to drop the favorite. The instantaneity of the fall of the courtier and the pretexts the prince might produce in order to represent himself betrayed by the favorite were crucial to represent the sudden extraneity of the prince from the client. Precisely because their relationship had been so close (a closeness that, retrospectively, might seem embarrassing) the ex-favorite had to disappear as soon as possible and the prince had to be represented as completely apart from him and from his "misdeeds."[74]

That Pellegrini was indicating a structural rather than accidental feature of court society in general is confirmed by the many tales of falls of favorites in court literature. Commenting on the fall of James I's favorite, the earl of Somerset, Sir John Holles wrote in 1617 that

This morning I walk to the Tower to visit the afflicted [Somerset], such journeys being necessary sometimes for humiliation, seeing the others' miseries as in a glass, we may behold the misfortune to which all men that live under the will of another be subject, this . . . which modernly is termed reason of state is an arrow which flies over every man's head and no man can escape it without miraculous fortune if he stand in the way; *for some men's ruin are as necessary for Princes' designs, as other men's services.*[75]

The Fall of Galileo

In February 1632 the *Dialogue* was off the press in Florence. Galileo had initially planned to have the book printed in Rome through the Accademia dei Lincei. In the late spring of 1630, he went there with a complete draft of the manuscript to obtain the necessary permits. Although the Master of the Sacred Palace, Father Niccolò Riccardi, issued a provisional imprimatur for the book (so that Galileo could negotiate with printers), he wanted Galileo to introduce a few changes and to write the preface and conclusion in accordance with the pope's instructions.[76] The

74. Manzini had a further interpretation of this predicament: "When the prince has given all he has, to get it back he needs to take away what he had already given. And because it is infamous to take back, most of the time he removes from sight those who make him feel shameful" (Manzini, *Della peripetia di fortuna*, p. 33).

75. Holles to Lord Norris, 1 July 1617; quoted in Linda Levy-Peck, "'For a King not to be bountiful were a fault': Perspectives on Court Patronage in Early Stuart England," *The Journal of British Studies* 25 (1986): 48 (emphasis mine).

76. On Father Riccardi, see Ambrosius Eszer, "Niccolò Riccardi, O.P.—Padre Mostro," *Angelicum* 60 (1983): 428–57.

revised manuscript was to be sent or taken back to Rome in a few months so that Riccardi could give it a final review. Ciampoli would have taken care of the very last changes and, finally, the Lincei would have printed it. However, in August 1630 Prince Cesi died and the Lincei entered a terminal crisis. Moreover, the outbreak of the plague made it difficult to ship the manuscript safely between Florence and Rome.

Delays intervened and Galileo began to press to have the book printed in Florence.[77] Father Riccardi belonged to a Florentine family very close to the Medici and was related to the wife of Niccolini, the Medici ambassador in Rome and one of Galileo's close supporters.[78] Riccardi himself was a friend of Galileo and had previously reviewed the *Assayer* in glowing terms.[79] Galileo used his friendship with Riccardi and Niccolini as well as the power of Medici connections and eventually managed to have the final checking of the manuscript transferred from Rome to the Florentine Inquisitor.[80] Riccardi agreed to send his Florentine colleague specific reviewing instructions and the synopsis of the book's preface and conclusion.[81] The *Dialogue* was reviewed again by the Florentine Inquisitor, approved, and finally printed in February. In April (before the *Dialogue* had arrived in Rome) Ciampoli fell from Urban's grace. He was to leave the city in the following October.

In the summer of 1632 the pope ordered the book to be taken out of circulation and instituted a special commission to investigate Galileo's possible wrongdoings.[82] In the early fall, having considered the report of the special commission, the pope decided to hand the matter over to the Inquisition, which quickly ordered Galileo summoned to Rome. After much delay, Galileo arrived in Rome in February 1633. The process began in April and concluded in June, with Galileo's condemnation to formal imprisonment and to the recitation of the penitential psalms once a week for three years. He was found

> vehemently suspected of heresy, namely for having held and believed a doctrine which is false and contrary to the divine and Holy Scripture: that the sun is the center of the world and does not move from east to west, and the earth moves and is not the center of the world,

77. *GO*, vol. 14, no. 2115, pp. 215–18.

78. As mentioned in chapter 2, note 152, it was in the Palazzo Riccardi that the last Medici apotheosis *cum* Medicean Stars was painted.

79. Galileo had met him in 1624 in Rome, and there is evidence that he then waited for Riccardi to become Master of the Sacred Palace before submitting his manuscript for an imprimatur (*GO*, vol. 14, no. 1984, pp. 77–78).

80. Ibid., no. 2156, pp. 254–55; no. 2162, pp. 258–60.

81. Finocchiaro, *Galileo Affair*, pp. 209–10, 213–14.

82. *GO*, vol. 14, no. 2285, pp. 368–71; no. 2287, p. 372; no. 2289, p. 373; no. 2310, pp. 397–98.

and that one may hold and defend as probable an opinion after it has been declared and defined contrary to the Holy Scripture.[83]

Eventually, Galileo was allowed to go back home to Arcetri in the Florentine hills, where he remained under house arrest until his death in 1642.

The interpretation of the sentence is not as straightforward as it may seem. From other available documents we see that the pope and the Holy Office leveled a number of different accusations at Galileo, and it is difficult to assess which could have legitimately led to Galileo's sentencing and which were little more than juridical pretexts.[84] Galileo had been variously accused of having violated the publication agreements, insulting the pope by having the stupid Simplicio express the doctrine of God's omnipotence, violating the injunction not to hold or defend Copernicus given to him by Bellarmine in 1616, and presenting Copernicus's ideas not hypothetically but absolutely. It also has been difficult to evaluate the role of extrajudicial issues such as personal friendship or hostility toward Galileo by the Pope, in the Congregation of the Holy Office, and among the theologians and mathematicians of the religious orders, or the effect of the political context on Urban's decisions.[85]

Pellegrini's considerations on the downfall of the great courtier may provide a framework in which to contextualize some of these elements. Obviously, I am not claiming that Galileo was Urban's favorite. There were no official favorites in Rome and, although Galileo was well connected at the papal court and visited it every few years, he was not a local courtier.[86] Nevertheless, he did have a special relationship with Urban.[87]

---

83. Finocchiaro, *Galileo Affair*, p. 291.

84. On the problematic juridical dimensions of the trial, see Orio Giacchi, "Considerazioni giuridiche sui due processi contro Galileo," *Nel terzo centenario della morte di Galileo Galilei* (Milan: Società Editrice Vita e Pensiero, 1942), pp. 383–406. Maurice Finocchiaro, in his "The Methodological Background to Galileo's Trial," in William A. Wallace, ed., *Reinterpreting Galileo* (Washington: Catholic University of America Press, 1986), suggests the ambiguous status of the sentence: "Its speaking of 'vehement suspicion of heresy,' and not of heresy per se, is an implicit admission that the available evidence incriminating Galileo was insufficient for a verdict of guilty" (p. 242).

85. For instance, in January 1633 Galileo wrote Diodati saying that the Jesuits were behind his troubles in Rome, an opinion repeated the following April in a letter from Naudé to Gassendi (*GO*, vol. 15, no. 2384, p. 25; no. 2465, p. 88). On the possible role of the Jesuits in the trial, see Westfall, "Galileo and the Jesuits."

86. In his report on the Roman court, Giovanni Ciampoli uses the term *favorito*, but he seems to mean a privileged courtier (such as a secretary or chamberlain) rather than someone who stood out from the taxonomy of usual court roles (Giovanni Ciampoli, "Discorso di Monsignor Ciampoli sopra la corte di Roma," in Guglielminetti and Masoero, "Lettere e prose inedite (o parzialmente edite) di Giovanni Ciampoli," pp. 228, 233. However, the "ruin" of the courtier is an ongoing theme in Ciampoli's "Discorso."

87. Westfall, in "Patronage and the Publication of the *Dialogue*," has stressed the patronage relationship between Urban and Galileo as a key to understanding the publication of the

Through the years, the two had met and communicated regularly and Maffeo's letters always expressed an unconditional admiration for Galileo. Not only did Urban, as Cardinal Barberini, side with him during the dispute on buoyancy at the Florentine court, but in 1620 he wrote for Galileo the *Adulatio perniciosa*—literally an adulatory poem. He signed the letter accompanying the poem *Come fratello* (as your brother), a very unusual title for a cardinal to use with a private gentleman.[88] Then, reporting the pope's reaction to the news that Galileo was planning a trip to Rome at the beginning of 1624, Tommaso Rinuccini wrote to him that

> Three days ago I kissed the feet of Our Holiness, and I swear to Your Lordship that I did not see him so happy with anything as when I mentioned you to him. After having talked a bit about you, and having told him that Your Lordship had a great desire to come to his most holy feet, as soon as your health allowed you to do so, he responded that he would have received great happiness from that, provided that it would not have disturbed you or threatened your health, because one should do whatever possible so that great men like you lived as long as possible.[89]

When Galileo eventually arrived, he was given six audiences with the pope, who presented him with a painting, indulgences, medals, an agnus dei, a promised pension, and a remarkable letter praising Galileo to be shown to the Grand Duke.

The relationship of Galileo and Maffeo Barberini had been one between two independent, noncompeting, and very successful individuals, each of whom, by 1623, had reached the top of his career. Urban was Italy's most important prince and Maecenas, and Galileo was its most conspicuous cultural star. Theirs was a relationship that Maffeo seemed to regard almost as a personal kinship, something that bridged gaps in their different statuses and spheres of activity. This was similar to the relationship between princes and favorites. Favorites did not need to be professional courtiers or to emerge through the political ranks. Also, they did not need to have a distinguished background. And, precisely because their role was so unusual, they were allowed to overlook rules that were laws to others.[90] In a sense, favorites were institutionalized exceptions to court tax-

Dialogue. See also Antonio Favaro, "Oppositori di Galileo VI: Maffeo Barberini," *Atti del Reale Istituto Veneto di Scienze, Lettere ed Arti* 80 (1920–21): 1–46; and Sante Pieralisi, *Urbano VIII e Galileo Galilei* (Rome: Tipografia Poliglotta, 1875).

88. *GO*, vol. 13, no. 1479, pp. 48–49. The only other people who used that title with Galileo were Guidobaldo del Monte (once) and his brother Cardinal Francesco Maria.

89. Ibid., no. 1586, p. 139. This is nearly identical to a letter Maffeo sent Galileo in 1611: "I pray the Lord God to preserve you, because men of great value like you deserve to live a long time to the benefit of the public" (*GO*, vol. 11, no. 591, p. 216).

90. Shephard, "Royal Favorites," pp. 2, 60.

onomies. What mattered was their direct and intimate connection with the prince. Although the connection between Maffeo and Galileo cannot be seen in terms of the typical prince-favorite relationship, it shared some of its features. Galileo was not Urban's political favorite, but he certainly was his intellectual one.

Also, the fall of the courtier as discussed by Pellegrini and others was not an exceptional process but an extreme example of how favor and power normally circulated through networks of princely patronage. Although not all clients fell, the dynamics operating in the fall of the courtier informed the career patterns of less conspicuous clients as well. In particular, the ways in which the image and power of an absolute prince were at stake in his relationship with clients with whom he had developed some sort of kinship did not apply only to the fall of the favorite. While I am not proposing a strict analogy between the fall of the courtier and Galileo's trial, I want to use their similarity as a heuristic device to uncover patronage-laden aspects of the trial that have been left unnoticed by received interpretations.

The analysis will focus on two dimensions of the fall of the favorite. The first is the patron's use of the trope of betrayal to justify getting rid of a formerly close client. The second is how, in order to preserve the image of the prince's power as absolute, the fall of the favorite too must be presented as "absolute," that is, as terrible, inexorable, and overdetermined.

Betrayal as a Trope

Galileo's trial is indirectly connected to Ciampoli's falling from grace with Urban in the spring of 1632, an event that still remains largely unexplained. On April 25 the Medici ambassador in Rome reported to the grand duke that Ciampoli had been dismissed because he had tried to improve on a Latin letter written by the Pope.[91] Ciampoli was well known for his great arrogance (a trait common to favorites) in poetic, literary, and intellectual matters, and it is quite possible that his behavior might have wounded Urban's hypersensitive (political and poetic) ego.[92] The danger associated

91. The diplomatic dispatch (in the Archivio di Stato di Firenze, "Mediceo principato 3351," fol. 324) is reprinted in Antonio Favaro, "Giovanni Ciampoli," in Paolo Galluzzi, ed., *Amici e corrispondenti di Galileo* (Florence: Salimbeni, 1981), 1:167–68, and in Westfall, *Essays on the Trial of Galileo*, 96. This explanation of the fall of Ciampoli is also found in his biography in Giovanni Targioni Tozzetti, *Notizie*, vol. 2, part 1, p. 111. This version (and others) of the story is found in an unpublished biography of Urban VIII cited by Antonio Favaro in "Serie decimottava di scampoli galileiani," *Atti e Memorie della R. Accademia di Scienze, Lettere e Arti in Padova*, new series, 24 (1907–8), pp. 17–19.

92. Ciampoli's arrogance is reported in almost all biographical sources about him, but it is most conspicuous in a satirical piece by Giano Nicio Eritreo, the *Eudemia*. Eudemia is the name of an unknown island where two (ancient) Romans were stranded as a result of a storm. However, Eudemia looks very much like early seventeenth-century Rome, and the

with writing better letters than those written by the prince must have been a common one if a 1629 how-to book for princely secretaries discussed it.[93] Because of his great poetic pride, Urban was particularly sensitive on these matters. Fulvio Testi (a poet himself) was chosen by Duke Francesco d'Este as his ambassador in Rome to take advantage of Urban's soft spots. In one of Testi's letters to the duke we find that,

> Once our discussion was over, I kneeled to depart, but His Holiness made a signal and walked to another room where he sleeps, and, after reaching a small table, he grabbed a bundle of papers and thus, turning to me with a smiling face, he said: *We want Your Lordship to listen to some of our compositions.* And, in fact, he read me two very long Pindaric poems, one in praise of the most holy Virgin, and the other one about Countess Matilde [of Canossa]. I, following the mood, commented on each line with the needed praise, and, after having kissed His Holiness' foot for such an unusual sign of benevolence, left.[94]

Poetic logic paid well. When the duke visited Rome,

> It is impossible to describe the affection with which the pope received him. . . . As the duke is endowed with a quick and vigorous mind, a marvelous eloquence, deep memory, and a good understanding of human letters, he was immediately able to pick and use some of the images included in the pope's recently published poems. Then, he moved from this to reciting verses and entire poems, and the praises they deserve, and the pope was so titillated by hearing

main characters are well recognizable. It is quite likely that Eudemia's Nicorusticus is Rome's Ciampoli. A brilliant mind, Nicorusticus was also "full of a great opinion of himself, he was disdainful of everyone, and, besides himself, he did not think anyone wise and deserving consideration. And this craziness reached such a point that he subjected to the subtlety of his criticism and tried to oust from their ancient thrones those ancient writers who had received so much recognition and appreciation, and whom the consensus and authority of so many centuries had placed at the highest position among poets" (quoted in Luigi Gerboni, *Un umanista nel Seicento* [Città di Castello: Lapi, 1899], p. 128).

93. Panfilo Persico warned his reader of the devastating risks one would take by acting in ways that would have the prince feel threatened by the skills of his secretary (Persico, *Del segretario,* pp. 35–44). In particular: "The wise man tells us that when we serve princes we should take care not to know things. He tells of a Portuguese chevalier who was ordered by his king to write a letter in competition with others. The same king wanted to try his hand, and when the letter written by the chevalier was chosen as the best, [the chevalier] went home and asked to retire [from service], and went to his castles saying that he would not be welcome at court now that the king had realized that he knew more than the king" (p. 84).

94. Quoted in Giovanni de Castro, *Fulvio Testi e le corti italiane nella prima metà del secolo XVII* (Milan: Battezzati, 1875), p. 89. On Urban's poetic activities, see Mario Costanzo, *Critica e poetica del primo Seicento,* vol. 2, *Maffeo e Francesco Barberini, Cesarini, Pallavicino* (Rome: Bulzoni, 1970).

this prince turning panegyrist of his works that he was completely melted by the tenderness of his affection, and was so jubilant that he seemed to lose his mind.[95]

Given Urban's remarkable pride in his poetic skill, it is possible that, by insulting it, Ciampoli may have caused his own fall. However, other explanations were put forward also. Some claimed that he participated in a Spanish conspiracy to overthrow Urban.[96] He was reported as having visited the head of the Spanish party, Cardinal Borgia, at night, under-cover, and riding a mule through a secondary entrance to Borgia's palace, a report that Ciampoli kept rejecting years later not just as false but as plainly silly.[97]

In any event, Ciampoli's fall coincided with Urban's increasingly se-rious political difficulties. His attempt to act as the needle of the scale bet-ween the king of France and the Hapsburgs during the Thirty Years War was far from being effective. In particular, the Spanish and the emperor began to accuse him of favoritism towards the French and weakness to-ward the heretics. In Rome, the climax of the crisis was reached in the so-called scandal of the consistory on March 8. In that meeting, Cardinal Borgia, the Spanish ambassador, read (in front of all the other cardinals present) a harsh protest against the pope's lack of support for his king's military efforts against the Protestants in Germany.[98] Borgia went so far as to intimate that perhaps a council should be convened to assess the pope's will and ability to defend Christianity. Urban and his nephew tried, unsuccessfully, to silence him. Eventually, Urban's brother got up and walked toward Borgia, apparently intending to grab him and take him out of the room, but was physically stopped by Cardinal Sandoval. Finally, Urban was forced to ring the bell and let the guards into the room to con-trol the unrest.[99] Cardinal Pio broke his glasses while Cardinal Spinola tore his hat in anger.[100] Extremely upset at Borgia and desiring immediate revenge, Urban did his best to have him recalled to Madrid. However, fearing that Spain might try to turn this into a pretext for further, more serious challenges (like a possible military invasion from Naples), he had to wait until 1635 for the cardinal's departure.

95. De Castro, *Fulvio Testi e le corti italiane*, p. 90.

96. See Ciampoli's biography in Targioni Tozzetti, *Notizie*, vol. 2, part 1, pp. 110–11.

97. Ciampoli, *Lettere*, pp. 18–20.

98. On this event, see Fernando Gregorovius, *Urbano VIII e la sua opposizione alla Spagna e all'imperatore* (Rome: Fratelli Bocca, 1879), pp. 46–59; and Pastor, *History of the Popes*, 28:287–94. See also Auguste Leman, *Urbain VIII et la rivalité de la France et de la Maison d'Autriche de 1631 à 1635* (Paris: Champion, 1920), pp. 133–45. A number of contemporary dip-lomatic reports about this incident are reprinted in Gregorovius, *Urbano VIII*, pp. 139–51.

99. Gregorovius, *Urbano VIII*, p. 148.

100. Ibid., pp. 142, 148.

In the meantime, Urban turned his anger against Cardinals Ubaldini and Ludovisi, whom he considered accomplices of Borgia. Ludovisi was forced to return to Bologna, while Ubaldini (Urban's order to jail him was rescinded by Urban's nephew) moved out of Rome to Frascati.[101] Ludovisi and Aldobrandini (another cardinal directly involved in the consistory scandal) were two of Ciampoli's close patrons before he had become Urban's *secretario*. By the time Ciampoli was dismissed (perhaps for alleged contacts with the Spanish antipapal faction) Urban was facing a serious and delicate political crisis. He was politically weakened and sensitive to accusations of leniency toward heretics (among whom some people could have later put Galileo). He needed to show he was a firm, decisive, and great papal prince. Finally, he seemed to be undergoing a psychologically difficult period in which he perceived himself as surrounded by enemies.[102] As shown by his patronage of Tommaso Campanella, Urban was a very superstitious man particularly sensitive to negative horoscopes. The difficult political predicament in which he found himself in the spring of 1632 seemed to heighten his paranoia. A May 13 diplomatic dispatch reported that

> From Rome they write that the pope fears poison and has gone to Castel Gandolfo. He has closed himself off there and does not accept visits from anybody who has not been previously searched. The road that goes to Rome is patrolled. He suspects that the [military] maneuvers in the Kingdom of Naples are aimed at him, and that the fleet of the Grand Duke is ready to set sail to Ostia and Civitavecchia. Therefore, he is reinforcing his borders.[103]

Although there seems to be no direct link between Ciampoli's fall and Galileo's trial (except that they both took place during the same difficult political predicament), Urban later used Ciampoli as a useful scapegoat during the Galileo affair,[104] acting like Pellegrini's enraged prince who drops the "betraying" favorite while claiming to regret doing so. On many occasions the Florentine ambassador, Niccolini, reported Urban's rage at Galileo and Ciampoli for having "fraudulently" transgressed their agreements on the publication of the *Dialogue:* "While we were discussing those delicate subjects of the Holy Office, His Holiness exploded into great anger, and suddenly he told me that even our Galileo had dared enter where he should have not have, into the most serious and dangerous sub-

101. Ibid., pp. 59–64.

102. Ibid., p. 61.

103. Ibid., p. 74.

104. Although all the available evidence indicates that the early phase of Ciampoli's fall was not connected to the publication of the *Dialogue,* the same cannot be said for his eventual expulsion from Rome. In fact, Ciampoli left in October, well after the beginning of Galileo's troubles.

jects which could be stirred up at this time."[105] Urban's rages and accusations of betrayal became frequent.[106] On another occasion, the pope claimed that the publication of the *Dialogue* had been a *Ciampolata* (something typical of Ciampoli), hardly an empirical explanation of what had happened.[107] Basically, Urban tried to represent Galileo's alleged betrayal as linked to another one, that of Ciampoli (somebody who was no longer in a position to defend himself). Sometimes, Urban would also accuse the Master of the Sacred Palace of having been duped by Ciampoli or Galileo or, worse, of having participated in their cabal.[108] Interestingly enough, Urban's rage tended to explode whenever the Florentine ambassador would call his attention to the fact that in publishing the *Dialogue* Galileo had followed the instructions he had received from Rome and, indirectly, from the pope himself. In short, the betrayal trope was brought up precisely when Urban's connection to Galileo's "misdeeds" was intimated.

At the same time, in a pattern that accords with Pellegrini's analysis, Urban repeatedly acknowledged the sorrow that all this was causing him because of the closeness and familiarity he had had with Galileo. As he told the Tuscan ambassador, "Galileo had been his friend, they had conversed and dined several times together familiarly, and he was sorry to have to displease him, but one was dealing with the interest of the faith and religion."[109] As in Pellegrini, the prince does not represent himself as dropping his clients for his own personal interests but rather as somebody who is forced to give up his close (but betraying) friend in allegiance to some higher ideal (justice, religion, peace, etc.). Self-interest was turned into purity.

That Urban's accusations of betrayal were not completely believable can be gathered from the way his statements contradict each other and collide with the available evidence. Well into the trial, Urban declared himself completely extraneous to the matter saying that "he was never told anything about it [the imprimatur given by Riccardi] let alone that he ever ordered the license to be granted."[110] However, at an earlier time, Nic-

---

105. *GO*, vol. 14, pp. 383–84, as translated in Finocchiaro, *Galileo Affair*, p. 229. On the same occasion, Urbano stressed his having been betrayed again in the same letter: ". . . and again that his complaint was to have been deceived by Galileo and Ciampoli" (p. 230). A similar papal rage exploded when Niccolini brought up the same argument months later (*GO*, vol. 15, p. 68).

106. *GO*, vol. 14, pp. 383–84, 429; vol. 15, no. 2443, p. 67.

107. *GO*, vol. 15, p. 56.

108. Finocchiaro, *Galileo Affair*, pp. 229, 236, 239, 240, 252.

109. *GO*, vol. 15, pp. 67–68, as translated in Finocchiaro, *Galileo Affair*, p. 247. References to the pope's sadness for being obligated by his duty to Christianity to prosecute a friend are scattered in several other letters of the period. See, for instance, *GO*, vol. 14, no. 2305, p. 392; vol. 15, no. 2443, p. 68.

110. Finocchiaro, *Galileo Affair*, p. 252.

colini pointed out to him that "Signor Galilei had not published without the approval of his ministers and that for that purpose I myself had obtained and sent the prefaces to your city [Florence]."[111] To which the pope "answered with the same outburst of rage, that he had been deceived by Galileo and Ciampoli, that in particular Ciampoli had dared tell him that Signor Galileo was ready to do all His Holiness ordered and that everything [the publication of the *Dialogue*] was fine."[112]

However, it is quite unlikely that Ciampoli could have deceived Urban, because the pope had direct contact with Riccardi about this matter.[113] On June 16, 1630, Raffaello Visconti (a referee for Riccardi) wrote to Galileo that "The Father Master kisses your hands and says that he likes your work and that tomorrow morning he will talk to the pope about the frontispice [preface], and that, as the rest goes, after fixing a few little things like those we fixed together, he will give you the book [back]."[114] Also, in his letters to the Florentine Inquisitor, Riccardi mentioned that he would send them the synopsis of the preface and the conclusion of the book drafted according to the pope's desires.[115]

In that synopsis Galileo was required to include the argument of God's omnipotence, something quite similar to the *Assayer*'s fable of sound. With it, Galileo was expected to stress that whatever phenomenon

111. *GO*, vol. 14, no. 2298, p. 383 as translated in Finocchiaro, *Galileo Affair*, p. 229.

112. Finocchiaro, *Galileo Affair*, p. 229.

113. Therefore, I think that the version of the events given by Giovanfrancesco Buonamici (who wrote it right after the trial in July 1633) should be taken with a grain of salt. In describing a phase of the debate at the Congregation of the Holy Office, Buonamici (without giving any source) claimed: "Then the prosecution turned against Father Monster, who cleared himself first by saying he had received the order to approve the book from His Sanctity himself. But because the Pope denies this and became irritated, Father Monster says that it was ordered him by Secretary Ciampoli under instruction of His Holiness. The Pope replies that he does not believe this. Finally, Father Monster pulls out a note from Ciampoli, who tells him that His Holiness (in whose presence Ciampoli claims to write) orders him to approve the book" (*GO*, vol. 19, p. 410). It seems to me that if Riccardi did have a note from Ciampoli, he would have presented it much earlier than April 1633. Instead, since the summer of 1632, Riccardi had been taking a lot of heat that—had he shown that note—he would have avoided. Ciampoli was ruined anyway, so there would have been no reason for Riccardi to withold that information. Also, I find it extremely improbable that someone as ambitious as Ciampoli would put his career in capital danger in such a stupid (i.e., transparent) way to allow for the publication of Galileo's book. It should be noted that Buonamici's biography of Galileo is factually wrong on several other occasions.

114. *GO*, vol. 14, no. 2032, p. 120.

115. Finocchiaro, *Galileo Affair*, pp. 209, 212, 213–14. Also, in a June 1630 letter to Galileo in Rome, Count Orso d'Elci congratulates Galileo for his success in having Visconti (a reference for Riccardi) negotiate with Urban on his behalf about the inclusion of the tides argument (*GO*, vol. 14, no. 2024, p. 113). On Urban's involvement in the negotiation for the publication, see also Guido Morpurgo-Tagliabue, *I processi di Galileo e l'epistemologia* (Rome: Armando, 1981), pp. 136–39; and Mario d'Addio, "Considerazioni sui processi di Galileo," *Rivista di storia della Chiesa in Italia* 38 (1984): pp. 64–66.

one was investigating could have been produced by God in many ways and that, consequently, any cause we might assume for it cannot be declared the necessary one.[116] In fact, toward the end of the *Dialogue,* Galileo had Simplicio argue that the interpretation of tides (the argument that Galileo saw as closest to a proof of the earth's motion) was not final:

> SIMPLICIO: . . . I confess that your idea seems to me much more in-genious than any others I have heard, but I do not thereby regard it as true and conclusive. Indeed, I always keep before my mind's eye a very firm doctrine, which I once learned from a man of great knowl-edge and eminence [the pope?], and before which one must give pause. From it I know what you would answer if both of you are asked whether God with His infinite power and wisdom could give to the element water the back and forth motion we see in it by some other means other than by moving the containing basin; I say you will answer that He would have the power and the knowledge to do this in many ways, some of them even inconceivable by our intelli-gence. Thus I immediately conclude that in view of this it would be excessively bold if someone should want to limit and compel divine power and wisdom to a particular fancy of his.

> SALVIATI: An admirable and truly angelic doctrine, to which there corresponds very harmoniously another one that is also divine. This is the doctrine which, while it allows us to argue about the constitu-tion of the world, tells us that we are not about to discover how His hands built it (perhaps in order that the exercise of the human mind would not be stopped or destroyed). Thus let this exercise, granted and commanded to us by God, suffice to acknowledge His great-ness; the less we are able to fathom the profound depths of His infi-nite wisdom, the more we shall admire that greatness.[117]

Therefore, Galileo formally complied with the instructions, though to a lesser extent and without the enthusiasm the pope may have desired. Galileo's lukewarm endorsement of Urban's "truly angelic doctrine" was said to enrage the pope, who saw his own ideas in the mouth of the stupid Simplicio at the very end of the book (a reading that may have been sug-gested by some of his adversaries).[118]

---

116. On Urban's theological position about the possibility and nature of scientific knowl-edge, see Oreggi's description of a dialogue between Galileo and Urban on this matter as reproduced in Maurice Finocchiaro, *Galileo and the Art of Reasoning* (Dordrecht: Reidel, 1980), p. 10.

117. Finocchiaro, *Galileo Affair,* pp. 217–18.

118. On Urban's continuing belief about having been "impersonated" by Simplicio, see *GO,* vol. 14, no. 2285, p. 370; no. 2296, p. 379; vol. 16, no. 3227, p. 363; no. 3321, p. 449; no. 3326, p. 455. See also Finocchiaro, *Galileo Affair,* pp. 221, 247. For a brief history of the "Simplicio Affair," see Pieralisi, *Urbano VIII e Galileo Galilei,* pp. 341–87.

These considerations indicate that Urban's accusations of "betrayals" were not necessarily empirical statements but convenient tropes for the prince to use in justifying his decisions while distancing himself from the "culprit."

Inexorable Falls

There is a further point in Pellegrini's analysis that may help us detect an interesting pattern in the later developments of the trial.

Both Pellegrini and Manzini stressed the naiveté of a falling courtier who tries to argue for his innocence and of other courtiers who may try to help him. Once the downfall had begun, it could not be stopped. The prince's power was at stake in the absoluteness of the fall. Any accident would have disrupted the ritual and stained the prince's image. The best strategy was that taken by Ciampoli, that is, to keep quiet and allow himself to be used as a sacrificial lamb to the goddess of Fortune while waiting for the election of a different prince.

Similarly, in October 1632, the Florentine ambassador, Niccolini (somebody who had much more experience than Galileo in Roman courtly matters), wrote to Galileo in Florence that there was no point trying to argue against the claims of the Holy Office,

> Because your claiming to be able to defend and clarify what you wrote will [only] reinforce the thought of condemning the work completely. . . . Regarding then the matter at hand you should realize that it will be necessary for you not to defend those things that the Congregation does not approve, but rather subject yourself to it and retract in the way its cardinals want. Otherwise, you will encounter very great difficulties in the resolution of your case, as it has already happened to many others. Neither can we (speaking Christianly) desire anything different from what they want, *they being a supreme tribunal that cannot err.*[119]

He made the same point a few months later at the height of the trial: "Galileo tries to defend his opinions very strongly; but I exhorted him, in the interest of a quick resolution, not to bother maintaining them and to submit to what he sees they want him to hold or believe about that detail of the earth's motion. He was extremely distressed by this. . . ."[120]

Because of the papal power invested in the trial, the Holy Office expected Galileo to confess, not to argue. As with Pellegrini's disgraced favorite, it was pointless and even harmful for Galileo to try to stop his fall. The proceedings of the trial support this view. Whenever the Holy Office

---

119. *GO*, vol. 14, no. 2223, pp. 417–18 (emphasis mine).

120. Finocchiaro, *Galileo Affair*, p. 249.

needed to negotiate with Galileo (rather than confront him with prear-
ranged questions) it did so under the table and outside the official trial
proceedings. A judicial body representing the power of an absolute prince
could not show itself openly negotiating with a criminal. For instance, as
soon as Galileo arrived in Rome in February 1633, he was privately visited
by consultors of the Congregation of the Holy Office. Although a hopeful
Galileo read these visits as signs of the congregation's benevolence towards him and was pleased to have a chance to discuss his positions infor-
mally with them, I think that these visits had no other purpose than to see
where Galileo stood so that the trial could be directed accordingly.[121] I do
not think that it was accidental that it took two months after Galileo's arrival in Rome for the Holy Office finally to begin to interrogate him. No
surprises were to emerge to stall the unrolling of the pope's rage against
him.[122]

But, unfortunately for the Holy Office, something did emerge in
April. By that time, it seems that the congregation had thought to focus
Galileo's prosecution on the violation of an unsigned injunction that,
according to them, he had received from Cardinal Bellarmine in 1616.
There, Galileo was ordered to "abandon completely the above-mentioned
opinion that the sun stands still at the center of the world and the earth
moves, and henceforth *not to hold, teach, or defend it in any way whatever,
either orally or in writing.*"[123] However, Galileo had an effective self-
defense based on a quite different signed certificate he was given by Bellarmine in 1616, one that granted him the right to discuss the Copernican
doctrine though only in a hypothetical form.[124] This defense may have
created a substantial headache for the inquisitors, who may have seen their
juridical strategies jammed. Again, the Holy Office decided to switch to a
more private and less embarrassing tactic. Father Vincenzo Maculano (the
commissary general of the Holy Office) proposed to negotiate with Galileo privately. As he told Cardinal Francesco Barberini:

> Yesterday, in accordance with the orders of His Holiness, I reported
> on Galileo's case to the Most Eminent Lords of the Holy Congrega-
> tion by briefly relating its current state. Their Lordships approved
> what has been done so far, and *then they considered various difficulties
> in regard to the manner of continuing the case and leading it to a conclu-
> sion;* for in his deposition Galileo denied what can be clearly seen in
> the book he wrote, so that if he were to continue in his negative
> stance it would become necessary to use greater rigor in the admin-

121. *GO,* vol. 15, no. 2413, p. 44; no. 2424, pp. 50–51.

122. On this issue, see also d'Addio, "Considerazioni sui processi di Galileo," p. 91.

123. "Special Injunction (26 February 1616)," as translated in Finocchiaro, *Galileo Affair,*
p. 147 (emphasis mine).

124. "Cardinal Bellarmine's Certificate (26 May 1616)," in ibid., p. 153.

istration of justice and less regard for all the ramifications of this business. Finally, I proposed a plan, namely that the Holy Congregation grant me the authorization to deal extrajudicially with Galileo, in order to make him understand his error and, once having recognized it, to bring him to confess it. The proposal seemed at first bold, and there did not seem to be much hope of accomplishing this goal as long as one followed the road of trying to convince him with reasons; however, after I mentioned the basis on which I proposed this, they gave me the authority.[125]

Although we can only conjecture what specific plan Maculano had in mind, it certainly worked well.[126] Two days later, an excited Maculano reported the good news to Cardinal Barberini, knowing it would please him:

I have not communicated this to anyone else, but I felt obliged to inform Your Eminence immediately, for I hope His Holiness and Your Eminence will be satisfied that in this manner the case is brought to such a point that it may be settled without difficulty. *The Tribunal will maintain its reputation.* . . .[127]

Two days later, Galileo confessed, and the trial headed for a speedy conclusion.[128] As one historian has put it, "the spontaneous confession of Galileo did not save him but rather the judges from a very delicate situation."[129] A few years later, Maculano became a cardinal.

In a sense, the falling favorite could not be given a fair hearing. Galileo's trial was no trial in the modern sense of the word; like most "falls of favorites" it was something resembling a ritual sacrifice. Precisely be-

125. Ibid., p. 276.

126. Morpurgo-Tagliabue, in *I processi di Galileo e l'epistemologia*, pp. 136–40, argues that Maculano seized on a statement uttered by Galileo during the interrogation of 12 April 1633. This was the meeting in which the Inquisition confronted Galileo with defying the injunction—allegedly given him in 1616—not to hold, defend, or teach the Copernican doctrine in any fashion. According to Morpurgo-Tagliabue, Galileo was so surprised and confused by this new accusation that he overdid his defense by arguing that in the *Dialogue* he never tried to defend Copernicus but that he actually tried to refute it. This was definitely a bit too much, for it was in patent contradiction with the content of the book. Maculano may have used this *faux pas* to convince Galileo that this utterance could be easily construed as indicating his bad faith and cunning behavior in the entire matter. Therefore the best thing for him to do was to confess now rather than have the Inquisition go ahead and expose his deceit. On this event, see also d'Addio, "Considerazioni sui processi di Galileo," p. 95.

127. Finocchiaro, *Galileo Affair*, p. 277 (emphasis mine).

128. Niccolini claimed that "Father Commissary [Maculano] seems to be interested in expediting this cause, and wants to impose silence, and if he will succeed at that, it will shorten everything and preserve many from troubles and dangers" (*GO*, vol. 15, no. 2491, pp. 109–10).

129. Morpurgo-Tagliabue, *I processi di Galileo e l'epistemologia*, p. 139.

cause of his closeness to the pope, Galileo was not just another defendant. No discourse could have been developed in which the "humanity" and weakness of the patron (who had allowed such a despicable individual to get so close to him) could be exposed. And there are other ways in which the management of the trial contributed to Galileo's "free fall."

The available evidence indicates that Galileo's earlier attempts to explain his actions to the commission and salvage the salvageable by correcting the book before the matter reached the Inquisition were ignored.[130] When in September 1632 Niccolini mentioned to a representative of the pope that the grand duke would have liked Galileo to have a chance to defend himself *before* any trial was started, he was told that "the Holy Office is not in the habit of hearing defenses."[131] While this response may have reflected the Holy Office's procedures, by the time Niccolini made the request the Holy Office had not yet started its proceedings against Galileo but was still gathering preliminary information through the special commission established by the pope. Two weeks earlier, Niccolini had asked the same question of the pope and received a similar answer: "In these matters of the Holy Office the procedure was simply to arrive at a censure and then call the defendant to recant."[132] When Niccolini asked the pope whether the Holy Office could let Galileo know what kind of charges were being brought against him, Urban answered violently: "I say to your Lordship that the Holy Office does not do these things and does not proceed this way, that these things are never given in advance to anyone, that such is not the custom; besides, he knows very well where the difficulties lie. . . ."[133]

Not only did the Holy Office not want to listen to Galileo's arguments (they were likely to bring Riccardi and Urban himself under the spotlight), it did not want Galileo's version of the story to get out either. Upon his arrival in Rome in February, Galileo was warned by Cardinal Francesco Barberini not to contact, talk to, or accept visits from anybody while at the Medici Palace before the trial (which began only two months later). This unofficial injunction was repeated by the Commissary of the Holy Office a few days later.[134]

The grand duke was the only person who could try to support Galileo's claims that he had done nothing wrong, and that the publication

130. Finocchiaro, *Galileo Affair*, pp. 229–30, 233, 234–35. See also *GO*, vol. 14, no. 2305, p. 392.

131. Finocchiaro, *Galileo Affair*, p. 235.

132. Ibid., p. 229.

133. Ibid., p. 230. See also p. 233; and *GO*, vol. 14, no. 2289, p. 373; no. 2334, p. 419.

134. "Cardinal Barberini has warned him not to socialize and not to bother talking with everyone who comes to visit him, since for various reasons this could cause harm and prejudice" (*GO*, vol. 15, no. 2409, p. 41, as translated in Finocchiaro, *Galileo Affair*, p. 243). For the repetition of the injunction, see *GO*, vol. 15, no. 2414, p. 45.

of the *Dialogue* had been made possible by the approval it received from Riccardi, the Florentine Inquisitor, and indirectly from the pope himself. Consequently, it is not surprising that the pope attempted to prevent the grand duke of Tuscany from getting directly involved in the Galileo affair. In September 1632, Urban repeatedly advised the ambassador, Niccolini, that it would have been wise for the grand duke "to be careful not to get involved . . . because he would not come out of it honorably" and that "His Highness should not get involved but should go slow."[135] A few days later, this warning began to sound more like a threat. On that occasion, the pope told Niccolini that the grand duke "should put aside all respect and affection toward his Mathematician and be glad to contribute himself to shielding Catholicism from any danger."[136] He then repeated

> that one must be careful not to let Signor Galileo spread troublesome and dangerous opinions under the pretext of running a certain school for young people, because he [Urban] has heard something. . . . Furthermore, His Highness should please be careful and have someone be vigilant to ensure some error is not sown throughout the state, which might cause him trouble.[137]

Basically, the pope told the grand duke that he should not let himself be represented as an impious prince supporting a potential heretic.[138] A very similar and not-so-veiled threat was repeated in February 1633.[139] The grand duke's more careful stance during the later part of the trial may indicate that the political implications of this threat must have registered in Florence.[140] That the pope had managed to scare people away from Galileo by presenting him as somebody going down in flames can be seen in the hesitation of many cardinals to accept or respond to the grand

135. Finocchiaro, *Galileo Affair,* p. 230.

136. Ibid., p. 235.

137. Ibid., p. 236.

138. At the same time, the pope tried to keep the grand duke out of the process by pretending that what was being done in Rome was very unusual, something that reflected how much they respected the grand duke and his desires to see Galileo treated well (ibid., pp. 221, 222, 230, 236, 245, 249, 250).

139. *GO,* vol. 15, no. 2428, p. 56. That the grand duke's pressure on behalf of Galileo was only going to hurt him and create serious problems with the pope was also brought up during a conversation between Riccardi and Niccolini (*GO,* vol. 14, no. 2302, p. 388).

140. Ferdinand II's decision not to pay for Galileo's expenses in Rome after the first month may be linked to this. Given the minimal expense involved, the matter was not financial but symbolic: the grand duke did not want to see himself supporting a convicted criminal (*GO,* vol. 15, no. 2509, p. 124). According to Niccolini (who was embarrassed by the grand duke's decision and volunteered to take care of the expenses personally), Galileo's expenses would have been around fifteen scudi per month. The grand duke may have thought that if Galileo remained there for more than a month (the stay he paid for), it probably meant that he was in serious trouble and that a condemnation was likely.

duke's letters on behalf of Galileo. As Niccolini put it: "some Cardinals to whom I delivered the Most Serene letters excused themselves from responding because of the prohibition that exists, and some have even avoided to receiving them for fear of censure."[141]

If we move from considerations of procedure to an analysis of the changing accusations against Galileo, we find a confirmation of the pope's attempt to erase his own involvement with the scandal and to avoid any surprise that might have caused his powerful juridical machine to deviate from its "natural" course. Unlike what Redondi has argued in his *Galileo Heretic* (and what Urban told the Florentine ambassador), I think that the commission the Pope set up in August of 1632 to evaluate Galileo's actions and decide whether the matter should be sent to the Congregation of the Holy Office was not the result of the pope's friendly gesture toward Galileo but was concerned with framing Galileo as carefully as possible so that nobody else would be implicated. In fact, soon after the first commission had convened during the summer of 1632, it became clear that the Holy Office's injunction of 1616 was going to be the crucial piece of evidence against Galileo. As Riccardi told Niccolini in September 1632, Galileo's failure to comply with the orders received in 1616 "was enough to ruin him completely."[142] Similar remarks were collected by Niccolini again in February, a month before the beginning of the hearings.[143]

The report released in September 1632 by the special commission included a brief history of the publication of the *Dialogue* and concluded by listing eight specific items for indictment. Some of them were related to Galileo's alleged transgressions of the publication's agreements while others focused on the content of the text itself. However, all these items were presented as amendable "if the book were judged to have some utility which would warrant such a favor."[144] Not only were these points ineffectual to justify a severe condemnation, they also opened the door to questioning Riccardi's and Urban's involvement in the scandal. Fortunately

141. *GO,* vol. 15, no. 2471, p. 95.

142. In a letter to Cioli dated 11 September 1632, Niccolini reports that Riccardi had told him that "in the files of the Holy Office they have found something which alone is sufficient to ruin Signor Galilei completely; that is, about twelve years ago, when it became known that he held this opinion and was sowing it in Florence, and when on account of this he was called to Rome, he was prohibited from holding this opinion by the Lord Cardinal Bellarmine, in the name of the Pope and the Holy Office. So he says he is not really surprised that His Highness is acting with so much concern, for he has not been told all the circumstances of the business" (*GO,* vol. 14, no. 2302, p. 389, as translated in Finocchiaro, *Galileo Affair,* p. 233).

143. *GO,* vol. 15, no. 2427, p. 55. On that same day Niccolini had an audience with the pope, who confirmed Galileo's violation of Bellarmine's injunction (ibid., p. 56). Finally, the pope brought up the 1616 injunction again with Niccolini a week *before* the publication of the sentence, saying that it had been the violation of that order that was forcing the Holy Office to condemn Galileo (ibid., no. 2518, p. 132).

144. Finocchiaro, *Galileo Affair,* p. 222.

for them, the commission had found a better juridical pretext to nail down Galileo while releasing Urban and his collaborators from their responsibilities.[145] The report ended in this way:

> In 1616 the author had from the Holy Office the injunction that "he abandon completely the above-mentioned opinion that the sun is the center of the world and the earth moves, nor henceforth hold, teach, or defend it in any way whatever, orally or in writing; otherwise the Holy Office would start proceedings against him." He acquiesced in this injunction and promised to obey.[146]

From what we know about the later development of the trial, it seems that it did little more than formalize the accusation identified by the special "friendly" commission.[147]

While the early diplomatic correspondence between Rome and Florence during the summer of 1632 never mentioned the events of 1616 but discussed allegations of Galileo's violation of the conditions for the publication, the juridical outlook changed suddenly in September. With the emergence of the 1616 injunction, the intricate and possibly unclear interaction between Ciampoli, Riccardi, Galileo, and the pope around the publication of the *Dialogue* faded out of the inquisitors' gaze.[148] The text of Galileo's final condemnation glossed over the alleged frauds that Galileo and Ciampoli had committed in the process of obtaining the imprimatur and focused mostly on Galileo's alleged violation of the 1616 injunction. Quite conveniently, Bellarmine—the person whose orders Galileo was claimed to have transgressed—was no longer there to contradict this representation of the events.

Although the legal status of the injunction of 1616 was quite dubious, the pope and the Holy Office were able to use it to represent Galileo as solely responsible for all that happened.[149] Also, not only did Urban and his collaborators come out clean, but, precisely because of his recovered purity, the pope could represent his condemnation of Galileo as abso-

---

145. On this point, see also Favaro, "Oppositori di Galileo VI: Maffeo Barberini," 30, and Morpurgo-Tagliabue, *I processi di Galileo e l'epistemologia*, pp. 135–36.

146. Finocchiaro, *Galileo Affair*, p. 222.

147. However, there were a few voices of dissent within the congregation. On this, see d'Addio, "Considerazioni sui processi di Galileo," pp. 78–80.

148. When, on a few occasions, the Florentine ambassador reminded Urban and Riccardi of the legality of the imprimaturs obtained by Galileo, they displayed their uneasiness by resorting to inconclusive arguments. In September 1618, Urban's secretary Benassi told Niccolini that it was not the first time that a book approved by the Inquisitors was subsequently censored and banned (*GO*, vol. 14, no. 2305, p. 391). Questioned along similar lines, Urban once replied with a joke (ibid., p. 393), while on a different occasion he decided to blame Ciampoli for the entire matter (ibid., no. 2348, pp. 428–29).

149. The injunction bore no signature of either Bellarmine, Galileo, possible witnesses, or notaries.

lutely just. He was not scapegoating Galileo in his own self-interest. Rather, he was sacrificing a former dear friend to prevent the spread of threatening doctrines that might harm the Church. The pope could be represented not as a self-interested human being but as an all-powerful and just papal prince. That was precisely what a good fall of the favorite was supposed to do for the prince.

As noted, the only nontrivial glitch in the trial's otherwise smooth unrolling was taken care of with a little more covert action and some apparently generous offer of leniency.[150] Surprised by the gap between Maculano's promises and the severity of the final sentence, Jerome Langford has hypothesized an internal conflict within the Holy Office between the more lenient position of Maculano and a more rigid, but eventually victorious, stance.[151] I do not think that such a hypothesis is necessary. Seen in the context provided by Pellegrini's analysis, the sincerity of Father Maculano's reassurance of Galileo becomes suspect. In fact, it seems that Maculano (and the other Church official who kept delivering similarly reassuring messages to Galileo and Niccolini) simply tried to keep the victim quiet so that the "sacrifice" would be performed in an orderly way.[152] According to Manzini, Sejanus, on his last day of life, was kept calm by the Roman Senate, which heaped more honors on him so that he would not get nervous and pose any obstacle to Tiberius's plans to get rid of him.[153]

Although the outcome of the trial of 1633 has embarrassed the Catholic Church for centuries, it certainly did not hurt Urban. Princes did not get rid of clients to do themselves harm. On the contrary, the condemnation of Galileo exonerated Urban from a possible scandal, and might have helped him refute insinuations about his weakness against heretics and reduce the political pressure on him. Cardinal Borgia, the Spanish ambassador who had threatened to impeach the pope because of his behavior during the Thirty Years War, was among the ten cardinals in the Congregation of the Holy Office and his name was on the final sentence.[154] Although it is true that, by dropping Galileo, Urban lost one of his most prestigious clients, the slot was soon filled. Writing to Galileo in the spring of 1634, Raffaello Magiotti told him of the arrival in Rome of an

150. *GO,* vol. 15, no. 2486, p. 107.

151. Jerome J. Langford, *Galileo, Science, and the Church* (Ann Arbor: University of Michigan Press, 1971), pp. 142–50, 155.

152. In a sense, Maculano's strategy with Galileo was not formally different from that adopted by the pope with the grand duke. By threatening retaliation for noncooperative behavior while promising leniency and support if collaboration were demonstrated, Maculano and the pope wanted to keep the juridical machinery going without hindrance.

153. Manzini, *Della peripetia di fortuna,* pp. 18–19.

154. However, Cardinal Borgia did not sign the sentence, possibly as a result of his ongoing tensions with the pope.

impressive Jesuit polymath, Athanasius Kircher, the new scientific star who—with his numerous "gems"—was to occupy the Roman stage for many years.[155] In the long run, the Jesuits did manage to become the stars of courtly natural philosophy in Rome.

Although the rest of Urban's long pontificate continued to be politically controversial and financially deadly for the Papal States, the pope held on to his power effectively. He was what we may call a successful tyrant. When he died in July 1644, the diarist Gigli commented that

> He had great happiness because in that time he had accumulated such wealth for his nephews as no other pope has accumulated. All the desirable offices and benefices were vacated during his reign for his nephews, and all those who were his enemies died before him. But in one small thing he did not think of the future needs of his family, that is, he did not keep the friendship and the protection of the princes. For various reasons he had the ill will of the emperor, of the king of Spain, of the Venetians, and of all the rest. . . . The people received great happiness from Urban's death, and they would have done something crazy to his statue had not the conservatori predicted that and called to the Campidoglio a company of soldiers of Constable Colonna armed with pikes and muskets to guard the palace, where there were also pieces of artillery. . . .[156]

Evidently, what Ciampoli had called the "crazy conjunctures and retrogressions" of Rome's wheel of fortune worked very well for Urban and his family, propelling them from obscurity to exceptional power and wealth in a matter of years. However, those same dynamics did not bring good luck to Galileo or Ciampoli (or to all the many other clients who bet and lost their careers at the Roman court).

The Trial and the Structural Constraints of Court Patronage

This refocusing of the trial of Galileo through the lens of the downfall of the favorite has provided a possible framework in which to contextualize the various claims, accusations, and moves that characterized that intricate process. Although cosmological, theological, and juridical arguments were the issues being debated or deployed in the trial, the logic that weaved them together was not that of Aristotle's *Posterior Analytics,* but that of the power image of the absolute prince. In other words, we should not confuse the trigger with the process it sets in motion. At the Roman court, somebody could fall after being accused of having held an opinion contrary to the scriptures or of having revised too heavily a Latin letter

---

155. I owe this point to Steve Harris. Magiotti's letter is in *GO,* vol. 16, no. 2906, p. 65.
156. Gigli, *Diario romano,* pp. 252–54.

written by the pope. At the court of Elizabeth I, a favorite could fall after being accused of having seduced a maid of honor.[157] Although these cases are evidently quite different, they share something in that they all were rooted in the economy of the prince's power image. While a wide range of different events or arguments could trigger somebody's fall at court, the dynamics of the downfall itself displayed much less variability.

It is precisely the economy of the prince's power in its relation with courtly culture and patronage that has been analyzed in this book. While in the earlier chapters I have shown how these dynamics made possible Galileo's self-fashioning as a "new philosopher," this conclusion has suggested how they may have framed his downfall as well. Galileo's career was structured—from beginning to end—by the patronage and culture of a baroque court.

As we have seen, patronage conjunctures played an important role in allowing Galileo to climb the ladder of social status and disciplinary credibility. Not only was Rome the seat of the most important princely court of Italy, but, because of the cyclical changes in the power structure resulting from the frequent turnover of popes, it was a place where patronage conjunctures were both more frequent and, for better or for worse, more powerful than anywhere else. While the peculiar features and power of the Roman court made it very attractive to all ambitious clients, Galileo had additional reasons to seek strong patronage ties with Rome. The pope was the prince whose sacred texts Galileo was trying to reinterpret in order to legitimize his new socioprofessional identity, Copernicanism, and the mathematical analysis of the physical world. Consequently, Galileo could gain much more than other clients from Roman patronage, but for the same reasons he also could lose more.

However, he could not seek patronage through a low-key and more discreet approach because that would have not fitted the patronage codes of a princely Maecenas like Urban. As we have seen, great patrons sought and rewarded conspicuous and consequently controversial clients. Urban's appreciation of the *Assayer* indicates that he subscribed to these cultural codes.[158] Paradoxically, although Galileo had to assume conspi-

157. This was the case with Sir Ralegh. In 1592 rumors circulated that he had seduced and then secretly married Elizabeth Throckmorton, one of the queen's maids of honor. The queen saw this as a great (almost treasonous) offense and "deprived Ralegh of favor for several years and placed him in the Tower for a few months." However, Ralegh was eventually able to recover his standing with the queen (Shepard, "Royal Favorites" p. 222).

158. Also, as analyses of many a courtier's demise tell us, an outsider's quick ascent close to the prince triggered considerable envy among those (like the Jesuits) who, because of their place in the institutional hierarchy, assumed they had priority (ibid., pp. 236–52). This envy was often a major factor in the favorite's subsequent fall. In this regard, we should keep in mind that the Jesuits were quite entitled to think of themselves as the pope's mathematicians. For instance, in 1611, at the time of the certification of Galileo's discoveries, the Jesuit mathematicians were clearly employed as the mathematical experts of the Roman court. Seeing a

CHAPTER SIX

cuous and controversial positions (and time them to patronage conjunctures) to succeed in his courtship of the pope, the reinterpretation of the
Scriptures entailed by his work was a most delicate matter, one that should
have been approached in the most tactful fashion. However, as Galileo
told Cesi in 1623, Urban's pontificate was the last conjuncture he could
possibly expect. In 1630, his disciple Cavalieri repeated that point, adding
that time was running out for Galileo.[159] Consequently, he had to com-
press a delicate and probably lengthy process of social and cognitive legitimation to fit the cycles of the only process of legitimation he had access
to: princely patronage.

Even more paradoxically, given the cosmology and socioprofessional
identity Galileo was trying to legitimize (and their revolutionary implications for the hierarchy among mathematics, philosophy, and theology),
the type of conspicuous and controversial cultural production Galileo
needed in order to court the pope could be developed only at the expense
of the very tradition on which the pope and the Church stood. This was a
dangerous game and even a breeze could upset it. And, as we have seen,
there were all sorts of breezes in Rome. To make things more difficult,
Galileo entered the most delicate phase of the game precisely when his
patronage connections with Rome had declined in number and importance. Cesarini, Cesi, and Cardinal Del Monte had died and Ciampoli
had been expelled. Most of the strong connections with the Roman court
that epitomized the "marvelous conjuncture" of 1623 had disappeared and
Galileo was left with the Medici ambassador as his main (if not only) powerful ally. If the publication of the *Assayer* was in perfect tune with a major
conjuncture, that of the *Dialogue* was not.

As we have seen, the distinction between presenting a claim hypothetically or absolutely was not clear-cut but depended partly on the
reader's outlook.[160] Consequently, Galileo's choice of literary genre for
the dialogue did not provide the intended safety valve against accusations
of writing *ex professo*. For instance, the three reports on the *Dialogue* by
Oreggi, Inchofer, and Pasqualigo seemed to forget that Galileo's claims
were presented by fictitious persons in the context of a fictitious discus-

"foreigner" who was more famous than any of them taking their place may have been quite
disturbing.

159. "I have heard that you are going to Rome at the end of this month, for which I congratulate you, hoping that finally we will see that work that has been awaited by the entire
world. I really think you are doing the right thing because the years pass by and, now that
you still have time and have a good connection with this pope, you will overcome all difficulties" (*GO*, vol. 14, no. 1989, p. 83). That same day Galileo received a similar message from
Ciampoli and Castelli (ibid., no. 1988, p. 82).

160. This ambiguity is the focus of Morpurgo-Tagliabue, *I processi di Galileo e l'epistemologia*.

350

sion.[161] That Galileo's personae sometimes spoke *ex professo* but that they did so in the context of a philosophical comedy (as Campanella saw the *Dialogue*) did not seem to register with the theologians summoned by the Holy Office.[162] They read (or were told to read) a comedy as a treatise. Given the tension between the nominalist discourse (or the papal prince who wanted to enjoy the spectacle from a safe position) and the more realist approach of the new philosopher in search of legitimation, it is not surprising that troubles emerged. And, when they did, it was not difficult for the prince to have the *Dialogue* stripped of its ambiguities and presented as an *ex professo* piece.

From the little evidence available, it does not seem that Urban shared the Aristotelians' epistemological distinction between mathematics and philosophy. Instead, in some sort of Ockhamistic fashion, he seemed to have thought that, because of God's omnipotence, both disciplines could provide only contingent claims. As Urban told Cardinal Hohenzollern in 1624, he really did not think that one could possibly prove the Copernican doctrine as necessarily true.[163] The reason why he liked the *Assayer*'s fable of sound so much and asked Galileo to reproduce that argument in the *Dialogue* reflected this stance. As shown by the dialogue between Urban and Galileo reported by Oreggi, Urban argued that God—being omnipotent—could have produced the same phenomenon in a variety of ways.[164] Consequently, the philosopher's (and the courtier's) pleasure was in discovering the *copia* of nature, the sign of God's immense power and creativity.

Urban was a sophisticated courtier, humanist, and poet, not a Scholastic theologian. He was a courtier-pope (and that is why he appreciated Galileo so much). The notion of God's omnipotence provided Urban with a perfect trope for the knowledge that best fit his predicament as a papal prince: it brought together both his cultural and theological concerns. I would say that, initially, Urban was less concerned with defending the Scriptures from Copernicus than with setting up a discourse which would have left the Scriptures alone while providing a space in which he and other sophisticated courtiers could enjoy brilliant philosophical "gems" produced by authors like Galileo. It was not just safety that Urban had in mind when he wanted Galileo to stress the argument of God's omnipotence: it was also an expression of his good courtly taste. He expected the *Dialogue* to be another virtuoso play of hypotheses like the *Assayer*. Consequently, Urban may have perceived Galileo's unbalanced stance in favor of

---

161. Finocchiaro, *Galileo Affair*, pp. 262–76.

162. *GO*, vol. 14, no. 2284, p. 366.

163. *GO*, vol. 13, no. 1637, p. 182.

164. Finocchiaro, *Galileo and the Art of Reasoning*, p. 10.

Copernicus not simply as theologically (and politically) dangerous; it was also a sign of bad taste—something that may have helped ruin his intellectual kinship with Galileo. By insisting on seeking the final proof of the Copernican doctrine, Galileo was behaving like the man he had poked fun at in the fable of sound for transfixing the cicada. In the very end, Urban's favorite had shown that he had some pedantry left in him.

This volume has presented a study of the interaction between the culture of political absolutism and Galileo's new natural philosophy. Once we see it in this context, Galileo's trial appears as a sign of the structural limits of the type of socioprofessional legitimation offered by court society and political absolutism. Galileo's trial was as much a clash between Aristotelian natural philosophy, Thomistic theology, and modern cosmology as it was a (structurally predictable) clash between the dynamics and tensions of baroque court society and culture.

From Patronage to Academies
A Hypothesis

Galileo & Co.

W HAT CAN THIS "MICROHISTORY" of Galileo's court-based strategies for social and cognitive legitimation tell us about the more general process known as the scientific revolution?

Given the important homologies in the culture of European princely courts, their processes of self-fashioning, and patronage dynamics, some of the perspectives and results presented here may be applicable to the study of other court-based scientific practitioners. Homologies among different courts must have been quite clear to contemporary observers since most court treatises did not deal with specific courts but referred to an institutional type.[1] Specific etiquette rules, dynastic mythologies, titles, and precedence protocols did vary among courts, but their basic structure, the image economy of the prince, and the protocols of honor and patronage were definitely comparable.

Similar considerations apply to the analysis of patronage as the social system of pre-institutionalized science that has been presented here. Because of the quite consistent features of the patronage system throughout Europe (and across many decades), I believe that what has been said about the ways patronage framed one's self-fashioning, ethos, style of argumentation, choice of topics, behavior in disputes, and protocols of legitimation (social and epistemological) is likely to be applicable well outside Galileo's life and career.

While sympathetic readers may grant me this, they may suggest that scientific academies rather than courts were the stages on which the scientific revolution reached its finale. Moreover, they may continue, what can

---

1. Castiglione is the exception, but almost all the other court treatises utilized or mentioned in this study are "generic."

EPILOGUE

the experience of an aggressively individualistic client like Galileo tell us about the emergence of a collective, institutionalized, and experiment-based scientific ethos that emerged in the second half of the seventeenth century and characterized much science thereafter? My response would not be to look for and stress the similarities in scientific style, socioprofessional role, ethos, and institutional setting of Galileo and the members of the Accademia del Cimento, the Royal Society, or the Académie des Sciences. Instead, while granting the important differences, I would argue that an understanding of the process of social and cognitive legitimation provided by the court offers important insights about the genealogy of the later social system, ethos, and style of natural philosophy.[2]

The shift from a social system of science rooted in patronage networks to one centered in scientific institutions was also accompanied by the emergence of new scientific practices. As discussed by Shapin and Schaffer, experiments and collective certification of "matters of fact" became central to the new scientific discourse.[3] With the emergence of experimental philosophy we see a move from a discourse rooted in entertaining disputes to much less contentious forms of knowledge.[4] At the same time, the institutionalization of science that went hand in hand with the establishment of experiments as a fundamental scientific practice was accompanied by a progressive shift from a scientific discourse framed by considerations of "honor" to one centered around the notion of "scientific credibility." Credibility began to be no longer exclusively linked to one's *personal* status or *personal* relationship with a patron. Rather, it became associated to one's membership in scientific *corporations* like the early academies.[5]

With some nontrivial approximation, I would say that in the scientific

2. For a more detailed version of this hypothesis, see Mario Biagioli, "Scientific Revolution, Social Bricolage, and Etiquette," in Roy Porter and Mikulas Teich, eds., *The Scientific Revolution in National Context* (Cambridge: Cambridge University Press, 1992 pp. 11–54), and idem, "Absolutism, The Modern State, and the Development of Scientific Manners," *Critical Inquiry,* forthcoming.

3. Steven Shapin and Simon Schaffer, *Leviathan and the Air Pump* (Princeton: Princeton University Press, 1985), pp. 22–79.

4. Experimental philosophy was perhaps only the most conspicuous example of this trend. In fact, we see that scientific academies tended to adopt "positivistic" approaches to natural phenomena, that is, they tended to avoid searching for final causes and limited themselves to a more descriptive stance. Like experimental philosophy, these approaches reflected the desire for establishing noncontentious and nondogmatic forms of argumentation—discourses that would not disrupt the cohesion of the legitimizing body.

5. Of course, I am not claiming that these changes took place suddenly. Although it is true that early scientific academies routinely extended membership to nobles to enhance their credibility and usually valued the testimony of nobles more heavily than of others, the shift from honor-based to training-based credibility is quite conspicuous once we take a broader chronological perspective.

354

academies the flesh-and-blood patron was eventually replaced by the *persona ficta* of the corporation. For instance, in his *Commentary on the Laws of England,* the eighteenth-century English jurist William Blackstone pointed to the Royal Society and the Royal College of Physicians as examples of corporations.[6] In this new institutional context it was not so much the status and honor of the practitioner (which had been transferred to him from the patron) that bound him to participate in scientific arguments and debates. Instead, it was the client's own membership in a scientific corporation and his ability to operate according to the protocols (the "institutional etiquette") associated with it that presented him as a credible practitioner.[7]

A patron who managed a scientific dispute tended to act not as a judge but as a noncommittal arbiter. Instead, scientific academies passed judgments on "matters of fact," claims that were generally more circumscribed than those we have encountered in disputes involving Galileo. This shift marked the practitioners' fundamental socioprofessional emancipation: their institutions had "internalized" (as corporations) the epistemological legitimacy previously assigned to individual patrons, an emancipation that the practitioners achieved also by presenting their claims within noncontentious discourses like those rooted in "matters of fact."[8]

Although the social distance between patrons and clients (one which prevented the patron from taking sides with one of the clients) was eventually blurred, the epistemological legitimation that went together with that distance was maintained, though in a modified form. Now it was the

6. William Blackstone, *Commentaries on the Laws of England* (London: Strahan, 1800), 1: 471.

7. The later development of scientific curricula, and the extra credibility one obtains by being trained by particularly prestigious academic institutions, fit this broad transition from a notion of credibility based on personal social status to one deriving from institution-based training and the *cursus honorum* one achieves within that framework.

8. We encounter this issue in the first scientific academy for which we have documents. Cesi faced it when he planned to increase the membership of the Lincei. As a result, difficulties may have developed because people of different social statuses and who did not know each other would have entered into private correspondence ("we will encounter frequent occasions to write to many different and unfamiliar members"). Consequently, Cesi wanted to establish some rule about the titles to be used in the Lincean correspondence ("set a norm about letter writing and the titles to be used"). He thought that the received etiquette that was based on social rather than intellectual distinctions should have been dropped and replaced with an internal set of rules. It is interesting that the new philosophical titles were supposed to be used when the members interacted as philosophers and not when they acted as private individuals (*GO,* vol. II, no. 874, p. 507). Cesi's proposal seemed to reflect an actual concern of the Lincei, who were faced with increasingly frequent cross-class interactions among philosophical equals. The Linceans from Naples were the most vocal about the need of introducing appropriate titles (ibid., no. 903, pp. 538–39). We have a case of one of these etiquette misunderstandings (although between two upper-class members), in which Welser scolded Faber for not having informed him of Cesi's proper title (ibid., no. 856, p. 490).

distance between the institution and each of its members that legitimized the practitioners' claims. Although the president of an academy like the Royal Society was not the patron of the institution, he represented its corporate authority.[9] The institution was "above" its members. It was between them and the king—the ultimate source of legitimacy.[10]

Discoveries or critiques were no longer delivered in the form of letters addressed or dedicated to a patron, but were instead sent to or deposited with the secretary of the academy. They became the *Philosophical Transactions* or the Académie des Sciences' *Mémoires*. The intermediate stage of this transition from letters to journals can be seen in the Lincei's "Epistolary Volume" (*Volume epistolico*), a planned edition of their works.[11]

Patrons Out, Experiments In

The relationship between the patrons' honor and the social and cognitive legitimation of science analyzed in Galileo's career might cast some light on the connection between the emergence (in the second half of the seventeenth century) of nondogmatic forms of scientific discourse and the patron-prince's progressive disappearance from the stage.

While clients may have perceived the patron's lack of commitment as a problem, his behavior also reflected a crucially positive feature of the patronage system. As we have seen, the patron's lack of commitment was tied to the very structure of patronage relationships which made these disputes legitimate (though generally irresolvable). Consequently, one could not make them resolvable by rendering them illegitimate (i.e., by getting rid of the hierarchies of social status in which they took place). For instance, experiments (like other forms of noncontentious claims) offered a way out of this deadlock of the patronage system. Experimentally produced

9. I believe the importance of the institution as *persona ficta* in legitimizing its members' claims declined in time. As we see today, the dynamics of credibility cannot be simply framed in terms of scientific institutions and their corporate authority (through these factors still play a role). However, the institution as *persona ficta* played a pivotal role during the early phases of the legitimation of science when radically new socioprofessional roles, a new social system of science, and new scientific practices were being introduced.

10. The transition from an academy with a participating patron to one sponsored by a remote king could be problematic. See David Lux, *Patronage and Royal Science in Seventeenth-Century France* (Ithaca: Cornell University Press, 1989), esp. pp. 81–84.

11. The *volume epistolico* was supposed to be a collection of letters exchanged among the Lincei and other interlocutors on scientific subjects. For instance, Galileo's work on the sunspots was to be published as a *volume epistolico* inclusive not only of Apelle's letter but also of replies and critiques by the other Lincei. Even when—under pressure from Galileo, who was probably concerned with stressing his authorship—the *Istorie* were printed as a separate volume, Cesi printed a number of extra copies to be later bound into the *volume epistolico* to come (*GO*, vol. 11, no. 761, p. 395). See also no. 725, p. 357. On the transition from letters to journals, see Charles Bazerman, *Shaping Written Knowledge* (Madison: University of Wisconsin Press, 1988), pp. 132–33.

"matters of fact" were more circumscribed claims about nature. Such matters were not only theologically safer but—by putting one's honor less on the line—their legitimation involved fewer risks. Then, because their acceptance was inherently linked to collective witnessing, "matters of fact" were the perfect type of claim to be legitimized by a corporation of scientific practitioners rather than by an individual patron. They represented a scientific practice which fit perfectly the new institutional situation of science which, in turn, reflected the practitioners' emancipation from the deadlock of the patronage system. With the introduction of experimental practices we move from spectacular but not necessarily terminable disputes to less (or maybe just differently) spectacular but manageable and terminable debates.

Experiments were not just the most effective way to produce new knowledge, to entertain and attract academicians, to keep clear from accusations of religious or political unorthodoxy, and to produce a platform of collectively acceptable data upon which the academicians' cooperative work and dialogue could be based. *Experiments were also a way out of the deadlock of noncommittal arbitration typical of patronage.* By providing a constructive management of distances in social status well beyond the range of possibilities provided by the patronage system, experimental practices may have been not only an effect but also a cause for the development of scientific institutions.

Expanding on the metaphor of "distance," I would say that in the patronage system the prince's honor could legitimize the scientists and their work only if it was kept in an appropriately removed or sheltered position. If the client tried to get too close, legitimation would not occur because excessive closeness (that is, the client's attempt to secure the patron's direct endorsement of his claims or discoveries) would collide with the prince's concern with maintaining his honor and preventing his power from being put to the test. In a sense, one could get burned by getting too close to the prince and his honor. I think that Galileo's problematic interaction with Pope Urban VIII can be understood also in these terms. Symmetrically, if the client remained too far away—that is, if he could not develop a close patronage relationship—then legitimation would not occur either.

The development of scientific academies and the introduction of experimental practices changed the management of "distance" and legitimation. Not only were "matters of fact" claims which could be legitimized by a corporation of practitioners, but *the prince-patron eventually moved away from the site in which claims were made and debated.* Royal patrons still legitimated these institutions, but they did so through charters rather than through their presence. Louis XIV visited the Académie des Sciences at the Observatory only once, in 1682, in a purely ceremonial event during which all normal activities were suspended. The available evidence suggests that Charles II (the king of England who chartered the Royal Society in 1662) never visited that institution.[12]

We see a transition from the client-patron legitimizing relationship (with all the pros and cons attached to it) to a situation in which the patron leaves the stage but remains a "remote legitimizer" of a scientific corporation which, in turn, legitimizes the claims of its practitioners (provided their claims are cast according to appropriate protocols ensuring the cohesion of that corporation). In a sense, the distance between the client-practitioner and the patron-prince increases, thus allowing for a space in which legitimation can be structured as a more "bureaucratic" process. The patron-prince is still the ultimate source of legitimation, but because his honor is less at stake (as a result of both the distance and the protocols that manage the legitimizing process) the practitioners begin to have access to a different type of authorship. Their physical claims do not need to be represented in hypothetical form, as with Galileo. They become "corporate authors" who can make "positivistic" claims and put forward "matters of fact."

Participating Princes and Disappearing Academicians

Prince Leopold de' Medici's Accademia del Cimento (a scientific academy organized by and around a participating prince) provides a case in which we can observe an intermediate stage of this transition from patronage to institutions.

The Cimento, an informal academy that gathered around Prince Leopold de' Medici between 1657 and 1667 has often been seen as the first academy dedicated to experiments.[13] As noticed by several historians, Leopold never provided his academy with a legal charter. He called it into session or suspended its activity whenever he desired. He set its experi-

12. As can be seen from contemporary reports reproduced by Wolf, the Dauphin's visit to the Académie in 1677 and that of Louis in 1681 were highly ceremonial events, and the royal guests did not participate in any scientific activity (C. Wolf, *Histoire de l'Observatoire de Paris de sa fondation à 1793* [Paris: Gauthier-Villars, 1902], pp. 19–27). See also Alice Stroup's brief but insightful remarks on Le Clerc's engraving of the (imaginary) visit of Louis to the Académie (Alice Stroup, *A Company of Scientists* [Berkeley: University of California Press, 1990], pp. 5–8). On the Royal Society's expectations about Charles II's visit, see Shapin and Schaffer, *Leviathan and the Air Pump*, pp. 31–32; and Simon Schaffer, "Wallification: Thomas Hobbes on School Divinity and Experimental Pneumatics," *Studies in History and Philosophy of Science* 19 (1988): 294–95.

13. The standard sources on the Accademia del Cimento are Giovanni Targioni Tozzetti, *Notizie degli aggrandimenti delle scienze fisiche accaduti in Toscana nel corso di anni LX del secolo XVII* (Florence: Bouchard, 1780; reprint, Bologna: Forni, 1967); and W. E. Knowles Middleton, *The Experimenters* (Baltimore: Johns Hopkins University Press, 1971). See also the very insightful article by Paolo Galluzzi, "L'Accademia del Cimento: 'Gusti' del principe, filosofia e ideologia dell'esperimento," *Quaderni storici* 16 (1981): 788–844; and Michael Segre's recent *In the Wake of Galileo* (New Brunswick: Rutgers University Press, 1991).

mental agenda, paid for the experimental apparatus from his own purse, and drew his academicians from mathematicians and philosophers who were already on the Medici payroll. It seems that the very name "Accademia del Cimento" was a retrospective invention connected to the publication, in 1667, of the *Saggi*—a book presenting a selection of experiments conducted at the academy which, by then, was defunct. Finally, the academy was never formally established or disbanded. It began to meet around 1657, slowed down its activities after 1662, and stopped convening after Leopold became cardinal in 1667 and moved temporarily to Rome. As one academician remarked, the academy was nothing more than an expression of the "prince's whims."[14]

The Cimento's status as an unofficial academy was a direct result of Leopold's participation in it. A prince of Leopold's rank could easily taint his image by working together in an *official* context with his subjects (some of whom were of quite low social background).[15] Leopold controlled the possibilities of status-pollution in various ways. First, he tried to make sure he would be perceived as a princely supervisor rather than an active, hands-on participant. Second, he presented the academy as something belonging to his private sphere.[16] A prince could display himself naked to his servants in the privacy of his bath, but he could not do so in a public space. By the same logic, the participants in the Cimento could not become "academicians" in the sense of being members of an official corporate body; Leopold's status required them to be his "scientific servants."

The same issues of status that made Leopold keep the academy as a fully private enterprise prevented him from entering into scientific disputes. Disputes belonged to people who had an ax to grind, like members of the ignorant and self-interested lower classes.[17] The academy's vocal commitment to the experimental method—one that led to accurate descriptions of experimentally reproduced effects rather than to the explanation of their causes—was not only a result of Leopold's desire to keep clear of possible conflicts with theologians; it reflected the politeness of the philosophical etiquette to which he was bound by his own status.[18] Like his father, Cosimo II, Leopold could have watched Galileo's aggressive performance but could not have adopted that style himself.

14. Galluzzi, "L'Accademia del Cimento," p. 823.

15. For brief biographical sketches of the participants to the Cimento, see Middleton, *Experimenters*, pp. 26–40. On Antonio Oliva—the most "picturesque" of the academicians—see Ugo Baldini, *Un libertino accademico del Cimento: Antonio Uliva* (Florence: Istituto e Museo di Storia della Scienza, 1977).

16. This, in fact, is how Leopold was presented in the preface to the *Saggi* (Giorgio Abetti and Pietro Pagnini, eds., *L'Accademia del Cimento* [Florence: Barbera, 1942], p. 85).

17. Steven Shapin, "The House of Experiment in Seventeenth-Century England," *Isis* 79 (1988): 395–99; and Shapin and Schaffer, *Leviathan and the Air Pump*, pp. 72–76.

18. Abetti and Pagnini, *L'Accademia del Cimento*, pp. 83–87, 124.

By having his "academicians" perform and describe experiments rather than seek their causes, Leopold made sure that the activity of the Cimento would not lead to status-tainting disputes. For analogous reasons, Leopold was exceedingly cautious with having himself invoked as the judge in scientific disputes. When that happened—as with Huygens and Fabri on Saturn's rings—he turned the matter over to his academicians.[19] They were instructed to perform careful experiments on *models* and, without passing any final judgment on the contenders' claims, to report what their experiments *suggested* about the tenability of the contending *hypotheses*.

Similarly, in the *Saggi* Leopold made sure that the academy's activity was represented as having unrolled as smoothly as possible, undisturbed by internal disputes. The frequent strong tensions and explicit disagreements recorded in the academicians' private correspondence were made invisible by the *Saggi*. Moreover, the book was written in a collective voice. No voice of any individual academician (Leopold included) was ever mentioned. Through the *Saggi*'s textual strategies he managed to efface himself sufficiently from the academy's activities to preserve his princely status and yet not enough to delegitimize the academy's results. Unlike the work of Boyle and the Royal Society members, who bound themselves to certify knowledge through "competent" and "open" witnessing managed through a fairly intricate etiquette,[20] the Cimento's results were presented as credible simply on the basis of having been certified by somebody of Leopold's status.[21]

Because of Leopold's effaced but effective presence, the *Saggi* did not need to reproduce the names of the witnesses and experimenters or any other specific circumstantial information about the execution of the experiments. In a sense, Leopold—being present and yet invisible—was the incognito certifier of the academy's work. But, because the *Saggi* did not

19. Albert Van Helden, "The Accademia del Cimento and Saturn's Ring," *Physis* 15 (1973): 237–59.

20. Michael Hunter, *Science and Society in Restoration England* (Cambridge: Cambridge University Press, 1991), p. 36; and Shapin, "House of Experiment," p. 392. As noticed by Norbert Elias, etiquette tended to become more intricate when the risks of status pollution grew higher. Consequently, the Cimento's lack of a cerimonial etiquette reflects its private character—a context which reduced considerably the possibility of status pollution.

21. The diary of the Accademia del Cimento presents a picture of the certification process strikingly similar to that adopted by the Royal Society. On 31 July 1662, "the Academy met at Sig. Lorenzo Magalotti's house, about repeating some experiments that appeared most necessary to the finishing of the work that is to be printed. *All of these, when they have been made easy by practice, have to be done again in the presence of His Highness*" (quoted in Middleton, *Experimenters*, 57 [emphasis mine]). The procedure is very similar to that described by Shapin in "House of Experiment," in which the experiments were tried and perfected at Hooke's apartment/workshop and then re-produced in front of the certifying audience of the Royal Society. The process is identical; what changes is the certifying persona.

mention any academician in particular, the credit for the work of the academy fell by "default" on the prince. Leopold became the author in absentia, the only way in which he could be an author and enhance (rather than jeopardize) his image. The Cimento's unnamed academicians resemble Boyle's technicians studied by Shapin. They were indispensable as *workers* but were not legitimate enough to "make knowledge," that is, to be *authors*.[22] However, unlike Boyle, Leopold did not utilize the academicians' involvement in the experiments to blame them for possible failures. This was not a result of Leopold's good nature but of his high social status. No embarrassing failures could be represented in a princely experimental narrative.[23] To Leopold, any such accident would have been equivalent to an embarrassing etiquette blunder at court.

Unlike Boyle, Leopold had the writer of the *Saggi* give the academicians full credit for having *performed* the experiments. However, despite this apparent difference, there is an underlying similarity between Leopold's and Boyle's textual strategies. In Boyle's case, the assistants were represented as nameless and unable to produce knowledge because it was the patron who had to be presented as the author. It was Boyle who had the status and credibility necessary to "make knowledge." The assistants "collaborated" with him in the sense that they took care of mechanical tasks inappropriate to the patron's status. Leopold's case was different and yet structurally homologous. Having a higher status than Boyle, Leopold was bound to a lower threshold of pollution. Consequently, he could not present himself as participating in scientific activities as much as Boyle could. It was because of this that Leopold's academicians received more credit than Boyle's assistants.

To conclude, I want to point out how the relationship between private and public, or between participation and distance, exemplified by Leopold's involvement in the Cimento embodies the same tensions we have encountered in the patronage environment between the practitioners' desire to have their claims legitimated and the patron's tendency, instead, to assume a noncommittal stance.[24] In the same way that a patron

22. Shapin, "House of Experiment," pp. 373–404; and idem, "The Invisible Technician," *American Scientist* 77 (November-December 1989): 554–63.

23. Similarly—as shown by the lengthy review process of the manuscript—Leopold showed himself extremely worried about the possibility of somebody finding errors in the *Saggi*.

24. Tensions between princes (or patrons) and practitioners were conspicuous in the Cimento as well. For instance, Borelli was frustrated by three features of the academy. First, the academy—being bound by Leopold's concern with avoiding interpretations—did not produce interesting claims. Second, the collective voice with which it presented its results destroyed the authorship of its most able members. Third, the experiments were not part of a specific research program (a program that could not exist, given Leopold's status investment), but wandered in all directions or focused for too long on what Borelli perceived as irrelevant problems.

could legitimate scientific disputes only insofar as they did not require him to assume the role of the judge, a prince like Leopold could participate in scientific activity only if this was presented either as a strictly private enterprise or as one in which he was not explicitly participating. The economy of honor was the same in Galileo's and Leopold's cases. What changed was the way in which these dynamics were articulated through different scientific practices and in institutional settings. More changes along these lines occurred in later scientific academies. While the adoption of experimental practices provided a way out of the deadlocks of patronage, allowing Leopold to participate in and legitimate the academy's findings, his participation resulted in the erasure of his academicians as authors. As in Galileo's preface to the *Sidereus nuncius,* the role of the practitioner was rhetorically effaced while the patron was credited for what the client had discovered. As shown by later scientific academies, for individual authorship to emerge, it was not enough for the prince to stay incognito; he had to leave the stage altogether.

REFERENCES

Manuscript Sources

Archivio linceo (Biblioteca Corsiniana, Rome)
  Manoscritti 2, 3, 4

ASF (Archivio di Stato di Firenze)
  Carte Strozziane, Serie I, 30
  Depositeria generale 389, 396
  Diari di etichetta di guardaroba 1, 2, 4, 5, 6
  Guardaroba medicea 225, 279, 309, 310, 535
  Manoscritti 132, 133, 320, 321
  Mediceo principato 802, 3351, 3761, 5550
  Miscellanea medicea 415, 437, 438, 441, 447, 474, 502
  Tratte 645

ARSI (Archivum Romanum Societatis Iesu)
  ROM 17, 18, 19, 20, 21, 22, 23, 55, 56
  MED 23, 26, 27
  HIST SOC 43

APUG (Archivio della Pontificia Università Gregoriana, Rome)
  Manoscritti 143, 2801

BNCF (Biblioteca Nazionale Centrale di Firenze)
  Galileiani 246
  Fondo Capponi 1

Printed Sources

Abetti, Giorgio, and Pietro Pagnini, eds. *L'Accademia del Cimento.* Florence: Barbèra, 1942.

Accetto, Torquato. *Della dissimulazione onesta.* 1641. Reprinted in S. Caramella and B. Croce, eds. *Politici e moralisti del Seicento.* Bari: Laterza, 1930.

Accolti, Pietro. "Delle lodi di Cosimo II, Granduca di Toscana." In *Raccolta di prose fiorentine,* edited by Carlo Dati, 6:119. Florence: Stamperia di SAR, 1731.

Ago, Renata. *Carriere e clientele nella Roma barocca.* Bari: Laterza, 1990.

Alciati, Andrea. *Emblematum liber.* Augsburg, Steyner 1531.

Alexander, H. G., ed. *The Leibniz-Clarke Correspondence.* Manchester: Manchester University Press, 1956.

Allegri, Ettore, and Alessandro Cecchi. *Palazzo Vecchio e i Medici.* Florence: SPES, 1980.

Altieri Biagi, Maria Luisa. "Il dialogo come genere letterario nella produzione scientifica." In *Giornate lincee indette in occasione del 350 anniversario della pub-*

blicazione del "Dialogo sopra i massimi sistemi" di Galileo Galilei, 143–66. Rome: Accademia dei Lincei, 1983.

———. *Galileo e la terminologia tecnico-scientifica.* Florence: Olschki, 1965.

Amescua, Mira de, ed. *La comedia famosa de Ruy Lopez de Avalos (primera parte de Don Alvaro de Luna) como doctrinal de privados y regimiento de principes* . . . . Mexico City: Editorial Jus, 1965.

Angelozzi, Giancarlo. "Cultura dell'onore, codici di comportamento nobiliari e stato nella Bologna pontificia: Un'ipotesi di lavoro." *Annali dell'Istituto Storico Italo-Germanico in Trento* 8 (1982): 305–324.

Apostolides, Jean-Marie. *Le prince sacrifié.* Paris: Minuit, 1985.

———. *Le roi machine.* Paris: Minuit, 1981.

Archimedes. "On Floating Bodies." In *The Works of Archimedes,* translated by T. L. Heath, 253–300. Cambridge: Cambridge University Press, 1912.

Aretino, Pietro. *Ragionamento delle corti.* Edited by Guido Battelli. Lanciano: Carabba, n.d.

Argomento dell'apoteosi o consagrazione de' Santi Ignatio Loiola e Francesco Saverio rappresentata nel Collegio Romano nelle feste della loro canonizzazione. Rome: Zannetti, 1622.

Ariew, Roger. "Theory of Comets at Paris during the Seventeenth Century." *Journal of the History of Ideas* 53 (1992): 355–72.

Aricò, Denise. "Retorica barocca come comportamento: Buona creanza e civil conversazione." *Intersezioni* 1 (1981): 338–39, 342.

Aristotle. *On the Heavens.* Translated by W. K. C. Guthrie. London: Heinemann, 1939.

———. *Physics III.* Translated by Philip H. Wicksteed and Francis M. Cornford. Vol. 1. London: Heineman, 1980.

Arrighetti, Niccolò. *Delle lodi del Sig. Fillippo Salviati.* Florence: Giunti, 1614.

Asor Rosa, Alberto, ed. *I poeti giocosi dell'età barocca.* Bari: Laterza, 1975.

Ashworth, William. "Divine Reflections and Profane Refractions." In *Gianlorenzo Bernini,* edited by Irving Lavin, 179–95. University Park: Pennsylvania State University Press, 1985.

———. "The Habsburg Circle." In *Patronage and Institutions,* edited by Bruce Moran, 137–67. Rochester, NY: Boydell, 1991.

———. "Iconography of a New Physics." *History and Technology* 4 (1987): 267–97.

———. "Natural History and the Emblematic World View." In *Reappraisals of the Scientific Revolution,* edited by David C. Lindberg and Robert S. Westman, 303–32. Cambridge: Cambridge University Press, 1990.

Avellini, Luisa. "Tra Umoristi e Gelati." *Studi secenteschi* 23 (1982): 109–37.

Bacon, Francis. *Novum organum.* Indianapolis: Bobbs-Merril Company, 1960.

Baldi, Bernardino. "Vita di Federico Commandino." In Filippo Ugolini and Filippo Polidori, eds., *Versi e prose scelte di Bernardino Baldi.* Florence: Le Monnier, 1859, pp. 513–37.

Baldini, Ugo. *Legem Impone Subactis: Studi su filosofia e scienza dei gesuiti in Italia, 1540–1632.* Rome: Bulzoni, 1992.

―――. "Una fonte poco utilizzata per la storia intellettuale: le 'censurae librorum' e 'opinionum' nell'antica Compagnia di Gesù." *Annali dell'Istituto Storico Italo-Germanico in Trento* 11 (1985): 37.

―――. "Additamenta galilaeana I: Galileo, la nuova astronomia e la critica all'aristotelismo nel dialogo epistolare tra Giuseppe Biancani e i revisori romani della Compagnia di Gesù." *Annali dell'Istituto e Museo di Storia della Scienza di Firenze* 9 (1984): 13–43.

―――. "La nova del 1604 e i matematici e filosofi del Collegio Romano." *Annali dell'Istituto e Museo di Storia della Scienza di Firenze* 6 (1981): 63–98.

―――. "La struttura della materia nel pensiero di Galileo." *De homine* 57 (1976): 91–164.

―――. *Un libertino accademico del Cimento: Antonio Uliva*. Florence: Istituto e Museo di Storia della Scienza, 1977.

Baldinucci, Filippo. *Cominciamento e progresso dell'arte dell'intagliare in rame*. Florence: Stecchi, 1767.

Barberi, Ugo. *I Marchesi Bourbon del Monte Santa Maria di Petrella e di Sorbello*. Città di Castello: Tipografia Unione Arti Grafiche, 1943.

Barberini, Maffeo. *Poemata*. Paris, 1625.

Bardi, Giovanni. *Eorum quae vehuntur in aquis experimenta*. Rome: Mascardi, 1614.

Bareggi, Cosimo di Filippo. "In nota alla politica culturale di Cosimo I: L'Accademia Fiorentina." *Quaderni storici* 23 (1973): 527–74.

Bargagli, Girolamo. *Dialogo de' giuochi*. Siena: Bonetti, 1572.

Barker, Peter, "The Optical Theory of Comets from Apian to Kepler," *Physis*. Forthcoming.

Barker, Peter, and Bernard R. Goldstein. "The Role of Comets in the Copernican Revolution." *Studies in History and Philosophy of Science* 19 (1988): 299–319.

Barocchi, Paola. "Introduzione." In Giovanni Maggi, *Bichierografia*. Florence: SPES, 1977.

―――, ed. *Scritti d'arte del Cinquecento*. Vol. 1. Turin: Einaudi, 1977.

Barozzi, Nicolò, and Guglielmo Berchet, eds. *Relazioni degli stati europei lette al Senato dagli ambasciatori veneti del sec. XVII*. Series 3. *Relazioni di Roma*. 10 vols. Bologna, 1856–79.

Barzman, Karen-edis. "Liberal Academicians and the New Social Elite in Grand Ducal Florence." In *World of Art: Themes of Unity and Diversity*, edited by Irving Lavin, 2:459–63. University Park: Pennsylvania State University Press, 1989.

Basile, Bruno. "Galileo e il teologo 'Copernicano' Paolo Antonio Foscarini." *Rivista di letteratura italiana* 1 (1983): 63–96.

Bataille, Georges. *The Accursed Share*. New York: Zone Books, 1988.

Bazerman, Charles. *Shaping Written Knowledge*. Madison: University of Wisconsin Press, 1988.

Becker, Marvin B. *Civility and Society in Western Europe, 1300–1600*. Bloomington: Indiana University Press, 1988.

REFERENCES

Bentivoglio, Guido. *Memorie e lettere*, edited by Costantino Panigada. Bari: Laterza, 1934.

Benzoni, Gino. "Le accademie." In *Storia della cultura veneta*, edited by G. Arnaldi and M. Pastore Stocchi, 4:131–62. Vicenza: Neri Pozza, 1983.

———. *Gli affanni della cultura*. Milan: Feltrinelli, 1978.

Bertelli, Sergio. "Egemonia linguistica come egemonia culturale e politica nella Firenze Cosimiana." *Bibliotheque d'Humanisme et Renaissance* 38 (1976): 249–83.

Bertelli, Sergio, and Giuliano Crifò, eds. *Rituale, cerimoniale, etichetta*. Milan: Bompiani, 1985.

Betti, Benedetto. *Ordine dell'apparato fatto da' Giovani della Compagnia di San Gio. Evangelista*. Florence: Giunti, 1574.

Biagioli, Mario. "Absolutism, the Modern State, and the Development of Scientific Manners." *Critical Inquiry*. Forthcoming.

———. "The Anthropology of Incommensurability." *Studies in History and Philosophy of Science* 21 (1990): 183–209.

———. "Declining Patrons and Pindaric Upstarts: The Lincei." In *Atti del convegno "La concezione del mondo in Europa: Religione, scienza e modernità attraverso l'opera di Alexandre Koyré (1892–1964)"*. Forthcoming.

———. "From Relativism to Contingentism." In *Disunity and Contextualism*, edited by Peter Galison and David Stump. Stanford: Stanford University Press, 1993. Forthcoming.

———. "Galileo the Emblem Maker." *Isis* 81 (1990): 230–58.

———. "Galileo's System of Patronage." *History of Science* 28 (1990): 1–62.

———. "New Documents on Galileo." *Nuncius* 6 (1991): 157–69.

———. "Filippo Salviati: A Baroque Virtuoso." *Nuncius* 7 (1992). In press.

———. "Scientific Revolution, Social Bricolage, and Etiquette." In *The Scientific Revolution in National Context*, edited by Roy Porter and Mikulas Teich, 11–54. Cambridge: Cambridge University Press, 1992.

———. "The Social Status of Italian Mathematicians, 1450–1600." *History of Science* 27 (1989): 41–95.

Bianca, C. "Federico Commandino." In *Dizionario biografico degli italiani*, 26:602–6. Rome: Istituto della Enciclopedia Italiana, 1982.

Biancani, Giuseppe. *Sphaera mundi, seu cosmographia*. Bologna: Bonomi, 1620.

Billacois, François. *The Duel*. New Haven: Yale University Press, 1990.

Bjurstrom, Per. "Baroque Theater and the Jesuits." In *Baroque Art: The Jesuit Contribution*, edited by Rudolf Wittkower and Irma B. Jaffe, 99–110. New York: Fordham University Press, 1972.

Blackstone, William. *Commentaries on the Laws of England*. 4 vols. London: Strahan, 1800.

Blackwell, Richard J. *Galileo, Bellarmine, and the Bible*. Notre Dame: University of Notre Dame Press, 1991.

Bloor, David. "Polyhedra and Abominations of Leviticus: Cognitive Styles in Mathematics." *British Journal of the History of Science* 11 (1978): 245–72.

———. *Wittgenstein: A Social Theory of Knowledge.* New York: Columbia University Press, 1983.

Blumenthal, Arthur R. *Theater Art of The Medici.* Hanover: University Press of New England, 1980.

Boccalini, Traiano. *Ragguagli di Parnaso.* Edited by Luigi Firpo. 3 vols. Bari: Laterza, 1948.

Boissevain, J. *Friends of Friends.* Oxford: Oxford University Press, 1974.

Bortolotti, Ettore. "I cartelli di matematica disfida e la personalità psichica e morale del Cardano." In *Studi e ricerche sulla storia della matematica in Italia nei secoli XVI e XVII.* Bologna: Zanichelli, 1944.

———. "Le matematiche disfide e la importanza che esse ebbero nella storia delle scienze." *Atti della Società Italiana per il Progresso della Scienze* 15 (1927): 163–80.

Bourdieu, Pierre. "Delegation and Political Fetishism." In *Language and Symbolic Power,* 203–19. Cambridge, Mass.: Harvard University Press, 1991.

———. *Distinction: A Social Critique of the Judgement of Taste.* Cambridge, Mass.: Harvard University Press, 1984.

———. *The Logic of Practice.* Stanford: Stanford University Press, 1990.

———. *Outline of a Theory of Practice.* Cambridge: Cambridge University Press, 1977.

———. "The Sentiment of Honour in Kabyle Society." In *Honour and Shame,* edited by J. G. Peristiany, 191–241. Chicago: University of Chicago Press, 1966.

Bourdieu, Pierre, and Jean-Claude Passeron. *La reproduction: Eléments pour une théorie du système d'enseignement.* Paris: Minuit, 1970.

Brahe, Tycho. *Epistolarum astronomicarum libri.* Uraniborg, 1596. Reprinted in *Tychonis Brahe Dani Opera Omnia,* edited by I. L. E. Dreyer. Vols. 6, 7. Amsterdam: Swets and Zeitlinger, 1972.

Brannigan, Augustine. *The Social Basis of Scientific Discoveries.* Cambridge: Cambridge University Press, 1981.

Brecht, Bertolt. *Galileo.* New York: Grove Press, 1966.

Bricarelli, Carlo S. J. "Il P. Orazio Grassi architetto della Chiesa di S. Ignazio in Roma." *Civiltà cattolica* 2 (1922): 13–25.

Brown, Harcourt. *Scientific Organizations in Seventeenth-Century France.* Baltimore: Johns Hopkins University Press, 1934.

Bryson, Frederick R. *The Point of Honor in Sixteenth-Century Italy.* Chicago: University of Chicago Press, 1935.

———. *The Sixteenth-Century Italian Duel.* Chicago: University of Chicago Press, 1938.

Burke, Peter. *The Anthropology of Early Modern Italy.* Cambridge: Cambridge University Press, 1987.

———. *The Italian Renaissance.* Princeton: Princeton University Press, 1986.

———. *Culture and Society in Renaissance Italy 1420–1540.* New York: Scribner's, 1972.

Calligaris, Giacomina. "Viaggiatori illustri ed ambasciatori stranieri alla corte sabauda nella prima metà del Seicento: Ospitalità e regali." *Studi piemontesi* (1975): 151–63.

Campbell, Donald T. "Evolutionary Epistemology." In *The Philosophy of Karl Popper*, edited by Paul A. Schlipp, 1:413–63. La Salle: Open Court, 1974.

Cannadine, David, and Simon Price, eds. *Rituals of Royalty*. Cambridge: Cambridge University Press, 1987.

Carbone, Lodovico. *De pacificatione et dilectione inimicorum* . . . . Florence: Sermartelli, 1583.

Cardi, Agnolo. "La calamita della corte." In *Saggi accademici*, edited by Agostino Mascardi, 242–64. Venice: Baba, 1653.

Carducci, Alessandro. *Il mondo festeggiante, balletto a cavallo fatto nel teatro congiunto al palazzo del Sereniss. Gran Duca per le reali nozze de' Serenissimi Principi Cosimo Terzo di Toscana e Margherita Luisa d'Orleans*. Florence: Stamperia di SAS, 1661.

Carini, Isidoro, ed. "Il conclave di Urbano VIII." In *Spicilegio vaticano di documenti inediti e rari*, 1:345–46. Rome: Loescher, n.d.

Caro, Annibal. *Comedia degli straccioni*. Turin: Einaudi, 1967.

Caroti, Stefano. "Un sostenitore napolitano della mobilità della terra: Il padre Paolo Antonio Foscarini." In *Galileo e Napoli*, edited by Fabrizio Lomonaco and Maurizio Torrini, 81–121. Naples: Guida, 1987.

Carrara, Bellino, S. J. "L' 'Unicuique suum' nella scoperta delle macchie solari." *Memorie della Pontificia Accademia Romana dei Nuovi Lincei* 23 (1905): 191–287; 24 (1906): 47–127.

Carugo, Adriano, and Alistair C. Crombie. "The Jesuits' and Galileo's Ideas of Science and of Nature." *Annali dell'Istituto e Museo di Storia della Scienza di Firenze* 8 (1983): 3–67.

Carugo, Adriano, and Ludovico Geymonat. "Note." In Galileo Galilei, *Discorsi e dimostrazioni matematiche intorno a due nuovo scienze*, 724–26. Turin: Boeringhieri, 1958.

Castelli, Benedetto. *Risposta alle opposizioni del S. Lodovico delle Colombe e del S. Vincenzio di Grazia contro al trattato del Sig. Galileo Galilei* . . . . Florence: Giunti, 1615. Reprinted in *GO*, vol. 4.

Castiglione, Baldassare. *Book of the Courtier*. Translated by Charles Singleton. Garden City, N.Y.: Anchor Books, 1959.

Castro, Giovanni de. *Fulvio Testi e le corti italiane nella prima metà del secolo XVII*. Milano: Battezzati, 1875.

Catalogus universalis pro nundinis Francofurtensibus vernalibus de anno MDCX. Frankfurt: Latomi, 1610.

Cave, Terence. *The Cornucopian Text*. Oxford: Clarendon Press, 1979.

Caverni, Raffaello. *Storia del metodo sperimentale in Italia*. Vol. 4. Florence: 1900. New York: Johnson Reprint, 1972.

Cellini, Benvenuto. *The Autobiography of Benvenuto Cellini*. Trans. John Addington Symonds. New York: Doubleday, 1961.

Cerasoli, F. "Diario di cose romane degli anni 1614, 1615, 1616." *Studi e documenti di storia e diritto* 15 (1894): 280.

Cesarini, Virginio. *Carmina.* Rome: Bernabò, 1658.

Cesi, Federico, "Del natural desiderio del sapere et Institutione de' Lincei per adempimento di esso." In Gilberto Govi, "Intorno alla data di un discorso inedito pronunciato da Federico Cesi fondatore dell' Accademia de' Lincei. *Memorie della R. Accademia dei Lincei,* Classe di Scienze Morali, Storiche e Filologiche, series 3, 5 (1879–80), 249–61.

Chartier, Roger. "Social Figuration and Habitus: Reading Elias." In *Cultural History,* 71–94. Ithaca: Cornell University Press, 1988.

Chiabrera, Gabriello. *La pietà di Cosmo: Dramma musicale rappresentato all'Altezze di Toscana.* Genoa: Pavone, 1622.

———. "Sermone a Gio. Francesco Geri." In *La lirica del Seicento,* edited by Alberto Asor Rosa. Bari: Laterza, 1975.

Ciampoli, Domenico. "Un amico del Galilei: Monsignor Giovanni Ciampoli." In *Nuovi studi letterari e bibliografici,* 3–169. Rocca S. Casciano: Cappelli, 1900.

Ciampoli, Giovanni. *Lettere di Monsignor Giovanni Ciampoli.* Venice, 1676.

———. *Oratio de pontefice maximo eligendo.* Rome: Mascardi, 1623.

———. "Discorso di Monsignor Ciampoli sopra la corte di Roma." In Marziano Guglielminetti and Mariarosa Masoero, "Lettere e prose inedite (o parzialmente edite) di Giovanni Ciampoli." *Studi secenteschi* 19 (1978): 228–37.

Cicognini, Jacopo. *Amor pudico.* Viterbo: Discepolo, 1614.

Clementi, Filippo. *Il Carnevale romano nelle cronache contemporanee.* Rome: Tiberina, 1899. Reprint, Città di Castello: Unione Arti Grafiche, 1939.

Concina, Ennio. *L'Arsenale della Repubblica di Venezia.* Milan: Electa, 1984.

Collins, Harry M. *Changing Order.* London: Sage, 1985.

———. "Public Experiments and Displays of Virtuosity: The Core-Set Revisited." *Social Studies of Science* 18 (1988): 725–48.

———, ed. *Knowledge and Controversy: Studies in Modern Natural Science.* Special issue of *Social Studies of Science* 11 (1981).

Considerazioni di Accademico Ignoto sopra il Discorso del Sig. Galilei. Pisa: Boschetti, 1612. Reprinted in *GO,* vol. 4.

Copernicus, Nicolaus. *On the Revolutions.* In *Complete Works,* translated by Edward Rosen, edited by Jerzy Dobrzycki. Vol. 2. Warsaw-Cracow: Polish Scientific Publishers, 1978.

Coppola, Giovanni Carlo. *Cosmo, ovvero l'Italia trionfante.* Florence: Stamperia di SAS, 1650.

Coresio, Giorgio. *Operetta intorno al galleggiare di corpi solidi.* Florence: Sermartelli, 1612. Reprinted in *GO,* vol. 4.

Costantini, Claudio. *Baliani e i gesuiti.* Florence: Giunti, 1969.

Costanzo, Mario. *Critica e poetica del primo Seicento.* 2 vols. Rome: Bulzoni, 1970.

Covoni, P. F. *Don Antonio de' Medici al Casino di San Marco.* Florence: Tipografia Cooperativa, 1892.

Cox-Rearick, Janet. *Dynasty and Destiny in Medici Art*. Princeton: Princeton University Press, 1984.

Cozzi, Gaetano. *Il Doge Nicolò Contarini*. Rome-Venice: Istituto per la Collaborazione Culturale, 1958.

———. *Paolo Sarpi fra Venezia e l'Europa*. Turin: Einaudi, 1969.

Crane, Thomas Frederick. *Italian Social Customs of the Sixteenth Century*. New Haven: Yale University Press, 1920.

Crapulli, Giovanni. *Mathesis universalis*. Rome: Edizioni dell'Ateneo, 1969.

Croce, Benedetto. *Poeti e scrittori del pieno e tardo rinascimento*. Bari: Laterza, 1945.

———. *Problemi di estetica*. Bari: Laterza, 1910.

Crombie, Alistair C. "Mathematics and Platonism in the Sixteenth-Century Italian Universities and in Jesuit Educational Policy." In *Prismata,* edited by Y. Maeyama and W. G. Saltzer, 63–94. Wiesbaden: Steiner Verlag, 1977.

———. "Sources of Galileo's Early Natural Philosophy." In *Reason, Experiment and Mysticism,* edited by Maria Luisa Righini-Bonelli and William R. Shea, 157–75. New York: Science History Publications, 1975.

D'Addio, Mario. "Considerazioni sui processi di Galileo." *Rivista di storia della Chiesa in Italia* 38 (1984): 64–66.

Dallington, Sir Robert. *Descrizione dello stato del Granduca di Toscana nell'Anno di Nostro Signore 1596*. Florence: All'Insegna del Giglio, 1983. Italian translation of *A Survey of the Great Duke's State of Tuscany. In the Yeare of Our Lord 1596*. London: Blount, 1605.

Daly, Peter M. *Literature in the Light of the Emblem*. Toronto: University of Toronto Press, 1979.

Davies, J., trans. *The Ceremonies of the Vacant See or a True Relation of What Passes at Rome Upon the Pope's Death*. London: H. L. and R. B., 1671.

Davis, James C. *The Decline of the Venetian Nobility as a Ruling Class*. Baltimore: Johns Hopkins University Press, 1962.

Dear, Peter. "Jesuit Mathematical Science and the Reconstitution of Experience in the Early Seventeenth Century." *Studies in History and Philosophy of Science* 18 (1987): 133–75.

———. "*Totius in Verba*: Rhetoric and Authority in the Early Royal Society." *Isis* 76 (1985): 156.

Della Casa, Giovanni. *Galateo*. Venice: 1558. Reprint, Turin: Einaudi, 1975.

Delle Colombe, Ludovico. *Discorso apologetico d'intorno al discorso di Galileo Galilei circa le cose che stanno sull'acqua o che in quella si muovono*. Florence: Pignoni, 1612. Reprinted in *GO,* vol. 4, pp. 313–69.

———. "Contro il moto della terra." Florence, 1610. Reprinted in *GO,* vol. 10, pp. 251–90.

Della Giovanna, Ildebrando. "Agostino Mascardi e il Cardinal Maurizio di Savoia." In *Raccolta di studi critici dedicati a A. Ancona,* 117–26. Florence: Barbèra, 1901.

Del Torre, Maria Assunta. *Studi su Cesare Cremonini*. Padua: Antenore, 1968.

Delumeau, Jean. *Vie économique et sociale de Rome dans la seconde moitié du XVIe siècle*. Paris: De Boccard, 1959.

Descartes, René. *Le monde, ou Traité de la lumière*. Edited and translated by Michael Mahoney. New York: Abaris Books, 1979.

Diaz, Furio. *Il granducato di Toscana: I Medici*. Turin: UTET, 1976.

Dietz-Moss, Janet. "Galileo's 'Letter to Christina': Some Rhetorical Considerations." *Renaissance Quarterly* 36 (1983): 547–76.

———. "The Rhetoric of Proof in Galileo's Writings on the Copernican System." In *Reinterpreting Galileo*, edited by William A. Wallace, 179–204. Washington D.C.: Catholic University of America Press, 1986.

Douglas, Mary. *Cultural Bias*. London: Royal Anthropological Institute, 1978.

———. *Natural Symbols*. New York: Pantheon, 1970.

———. *Purity and Danger*. London: Routledge, 1966.

Drake, Stillman. *Cause, Experiment, and Science*. Chicago: University of Chicago Press, 1981.

———. "The Dispute Over Bodies in Water." In *Galileo Studies*, 166. Ann Arbor: University of Michigan Press, 1970.

———. *Galileo at Work*. Chicago: University of Chicago Press, 1978.

———. "Galileo Gleanings III: A Kind Word for Sizzi." *Isis* 49 (1958): 155–65.

———. "Galileo Gleanings VIII: The Origin of Galileo's Book on Floating Bodies and the Question of the Unknown Academician." *Isis* 51 (1960): 56–63.

———. "Galileo, Kepler, and the Phases of Venus." *Journal of the History of Astronomy* 15 (1984): 198–208.

———. "Galileo's Steps to Full Copernicanism and Back." *Studies in History and Philosophy of Science* 18 (1987): 93–105.

———. *Telescope, Tides, and Tactics*. Chicago: University of Chicago Press, 1983.

———, ed. *Discoveries and Opinions of Galileo*. Garden City, N.J.: Doubleday, 1957.

———, trans. *Galileo Against the Philosophers*. Los Angeles: Zeitlin and Ver-Brugge, 1976.

Drake, Stillman, and C. D. O'Malley, translators. *The Controversy on the Comets of 1618*. Philadelphia: University of Pennsylvania Press, 1960.

Duhem, Pierre. *To Save the Phenomena*. Chicago: University of Chicago Press, 1969.

Durand, Yves, ed. *Hommage à Roland Mousnier: Clientèles et fidélités en Europe a l'époque moderne*. Paris: PUF, 1981.

Ehalt, Hubert Ch. *Ausdrucksformen Absolutischer Herrschaft*. Munich: Oldenbourg, 1980.

Eisenstadt, S. N., and L. Roniger. *Patrons, clients, and friends*. Cambridge: Cambridge University Press, 1984.

Elias, Norbert. "An Essay on Sport and Violence." In *Quest for Excitement*, Norbert Elias and Eric Dunning, 150–74. Oxford: Blackwell, 1986.

———. *The Court Society*. New York: Pantheon, 1983.

————. *The History of Manners*. New York: Pantheon, 1982.

————. *Power and Civility*. New York: Pantheon, 1982.

Elliott, John H. *Richelieu and Olivares*. Cambridge: Cambridge University Press, 1984.

Engelhardt, Tristram H., and Arthur L. Caplan, eds. *Scientific Controversies*. Cambridge: Cambridge University Press, 1987.

Erspamer, Francesco. *La biblioteca di Don Ferrante: Duello e onore nella cultura del Cinquecento*. Rome: Bulzoni, 1982.

Erythraeus, Ianus Nicius. *Pinacotheca*. Leipzig: Gleditschl, 1692.

Eszer, Ambrosius. "Niccolò Riccardi, O.P.—Padre Mostro." *Angelicum* 60 (1983): 428–57.

Evans, R. J. W. "Rantzau and Welser: Aspects of Later German Humanism." *History of European Ideas* 5 (1984): 257–70.

————. *Rudolph II and His World*. Oxford: Oxford University Press, 1973.

Faber, Joannes. *Praescriptiones Lynceae Academiae curante Joanne Fabro Lynceo Bambergensi*. Terni: Guerrero, 1624.

Fagiolo dell'Arco, Maurizio, and Silvia Carandini. *L'effimero barocco: Strutture della festa nella Roma del '600*. 2 vols. Rome: Bulzoni, 1978.

Fagiolo, Marcello, and Maria Luisa Madonna, eds. *Barocco romano e barocco italiano*. Rome: Gangemi, 1985.

Fantoni, Marcello. "Feticci di prestigio: Il dono alla corte medicea." In *Rituale, cerimoniale, etichetta*, edited by Sergio Bertelli and Giuliano Crifò, 141–61. Milan: Bompiani, 1985.

Favaro, Antonio. "Giovanni Ciampoli." In *Amici e corrispondenti di Galileo*, edited by Paolo Galluzzi, 1:135–89. Florence: Salimbeni, 1983.

————. "Giovanfrancesco Sagredo." In *Amici e corrispondenti di Galileo*, edited by Paolo Galluzzi, 1:208–10. Florence: Salimbeni, 1983.

————. "Oppositori di Galileo VI: Maffeo Barberini." *Atti del Reale Istituto Veneto di Scienze, Lettere ed Arti* 80 (1920–21): 1–46.

————. "Adversaria galileiana, serie quarta: Giovanfrancesco Sagredo e Guglielmo Gilbert." *Atti e memorie della R. Accademia di Scienze, Lettere ed Arti in Padova*, new series, 35 (1918–19): 12–15.

————. "Oppositori di Galileo III: Cristoforo Scheiner." *Atti del Reale Istituto Veneto di Scienze, Lettere ed Arti* 78 (1918–19): 1–107.

————. "Di una proposta per fondare in Pisa un Collegio di Lincei (1613)." *Archivio storico italiano* series 5, 42 (1908): 137–42.

————. "Serie decimottava di scampoli galileiani," *Atti e Memorie della R. Accademia di Scienze, Lettere e Arti in Padova*, new series, 24 (1907–8), 5–32.

————. "Serie decimasesta di scampoli galileiani," *Atti e Memorie della R. Accademia di Scienze, Lettere e Arti in Padova*, new series, 22 (1905–6), 5–36.

————. "Un ridotto scientifico in Venezia al tempo di Galileo Galilei." *Nuovo Archivio Veneto* 5 (1893): 199–209.

————. "Intorno ai servigi straordinari prestati da Galileo Galilei alla Repubblica

Veneta," *Atti del Reale Istituto Veneto di Scienze, Lettere, e Arti,* series 7, 1 (1889–90), 91–109.

———. "Galileo Galilei e il P. Orazio Grassi." *Memorie del Reale Istituto Veneto di Scienze, Lettere e Arti* 23 (1887): 203–36.

———, ed. *Carteggio inedito di Ticone Brahe, Giovanni Keplero e di altri astronomi e matematici dei secoli XVI e XVII con Giovanni Antonio Magini.* Bologna: Zanichelli, 1886.

———. "Sulla morte di Marco Velsero e sopra alcuni particolari della vita di Galileo." *Bullettino di bibliografia e storia delle scienze matematiche e fisiche* 17 (1884): 252–70.

———. "La libreria di Galileo Galilei." *Bulletino di bibliografia e storia delle scienze matematiche e fisiche* 19 (1886): 219–93.

———. *Galileo Galilei e lo Studio di Padova.* 2 vols. Florence, 1883. Reprint, Padua: Antenore, 1966.

Favoriti, Augusto. "Virginii Caesarini Vita." In Virginii Caesarini, *Carmina.* Rome: Bernabò, 1658.

Feingold, Mordechai. "Philanthropy, Pomp, and Patronage: Historical Reflections upon the Endowment of Culture." *Daedalus* 116 (1987): 155–78.

Feldhay, Rivka. "Knowledge and Salvation in Jesuit Culture." *Science in Context* 1 (1987): 195–213.

———. "The Discourse of Pious Science." *Science in Context* 3 (1990): 109–42.

———, and Adi Ophir. "Heresy and Hierarchy." *Stanford Humanities Review* 1 (1989): 118–38.

———. "Catholicism and the Emergence of Galilean Science: A Conflict Between Science and Religion." In S. N. Eisenstadt and I. Friedrich Silber, eds. *Knowledge and Society: Studies in the Sociology of Culture Past and Present.* Greenwich, Conn.: JAI Press, 1988, 139–63.

Ferrari, Giovanna. "Public Anatomy Lessons and the Carnival: The Anatomy Theater of Bologna." *Past and Present* 117 (1987): 50–106.

Feyerabend, Paul. *Against Method.* London: Verso, 1975.

———. "Consolations for the Specialists." In *Criticism and the Growth of Knowledge,* edited by Imre Lakatos and Alan Musgrave, 219–29. Cambridge: Cambridge University Press, 1970.

———. "Explanation, Reduction and Empiricism." *Minnesota Studies in the Philosophy of Science* 3 (1962): 28–97.

———. *Farewell to Reason.* London: Verso, 1987.

———. *Science in a Free Society.* London: Verso, 1978.

Findlen, Paula. "The Economy of Scientific Exchange in Early Modern Italy." In *Patronage and Institutions,* edited by Bruce Moran, 5–24. Rochester: Boydell, 1991.

———. *Possessing Nature: Museums, Collecting and Scientific Culture in Early Modern Italy.* Berkeley: University of California Press. Forthcoming.

Finocchiaro, Maurice. *The Galileo Affair.* Berkeley: University of California Press, 1989.

———. *Galileo and the Art Of Reasoning.* Dordrecht: Reidel, 1980.

———. "Galileo's Copernicanism and the Acceptability of Guiding Assumptions." In *Scrutinizing Science,* edited by Arthur Donovan, Larry Laudan, and Rachel Laudan, 49–67. Dordrecht: Kluwer, 1988.

———. "The Methodological Background to Galileo's Trial." In *Reinterpreting Galileo,* edited by William A. Wallace, 241–72. Washington: Catholic University of America Press, 1986.

Fitch Lytle, Guy, and Stephen Orgel, eds. *Patronage in the Renaissance.* Princeton: Princeton University Press, 1981.

Fontenelle, Bernard de. "Eloge de Monsieur Cassini." In *Eloges des académiciens,* 1:287. La Haye: Kloot, 1740.

Foscarini, Paolo Antonio. *Lettera del R.P.M. Paolo Antonio Foscarini Carmelitano sopra l'opinione de' Pittagorici e del Copernico della mobilità della terra e stabilità del sole e del nuovo Pittagorico sistema del mondo.* Naples: Scoriggio, 1615.

Foucault, Michel. *Discipline and Punish.* New York: Vintage, 1979.

———. "Truth and Power." In *Power/Knowledge,* edited by Colin Gordon, 109–33. New York: Pantheon, 1980.

Fragnito, Gigliola. "Parenti e familiari nelle corti cardinalizie del Rinascimento." In *"Familia" del principe e famiglia aristocratica,* edited by Cesare Mozzarelli, 2:565, 570. Rome: Bulzoni, 1988.

Franchini, Dario, et al. *La scienza a corte.* Rome: Bulzoni, 1979.

Fredette, Raymond. "Galileo's 'De Motu Antiquiora.'" *Physis* 14 (1972): 321–48.

Frey, Karl, ed. *Il carteggio di Giorgio Vasari.* Munich: Muller, 1923.

Frommel, Christoph Luitpold. "Caravaggios Frühwerk und der Kardinal Francesco Maria del Monte." *Storia dell'arte* 9–10 (1971): 45, 47.

———. "Papal Policy: The Planning of Rome During the Renaissance." *Journal of Interdisciplinary History* 17 (1986): 39–65.

Fumaroli, Marc. *L'âge de l'éloquence. Rhétorique et 'res literaria' de la Renaissance au seuil de l'époque classique.* Geneva: Droz, 1980.

Fusai, Giuseppe. *Belisario Vinta.* Florence: Seeber, 1905.

Gabrieli, Giuseppe. "Il Carteggio Linceo." *Memorie della R. Accademia Nazionale dei Lincei,* Classe di Scienze morali, storiche e filologiche, series 6; part I, 7 (1938–42), 1–121; part II, section I, 122–536; section II, 537–998; part III, 999–1446.

———. "Il liceo di Napoli." *Rendiconti della Reale Accademia Nazionale dei Lincei,* Classe di Scienze morali, storiche e filologiche, series 6, 14 (1938): 499–564.

———. "Marco Welser linceo augustano." *Rendiconti della Reale Accademia Nazionale dei Lincei,* Classe di Scienze morali, storiche e filologiche, series 6, 14 (1938): 74–99.

———. "L'orizzonte intellettuale e morale di Federico Cesi illustrato da un suo zibaldone inedito." *Rendiconti della Reale Accademia Nazionale dei Lincei,* Classe di Scienze morali, storiche e filologiche, series 6, 14 (1938): 663–725.

———. "Cesi e Caetani." *Rendiconti della Reale Accademia Nazionale dei Lincei,* Classe di Scienze morali, storiche e filologiche, series 6, 13 (1937), 255–69.

————. "Una gara di precedenza accademica nel Seicento fra Umoristi e Lincei." *Rendiconti della Reale Accademia Nazionale dei Lincei,* Classe di Scienze morali, storiche e filologiche, series 6, 11 (1935), 235–57.

————. "Bibliografia Lincea: II, Virginio Cesarini e Giovanni Ciampoli." *Rendiconti della Reale Accademia Nazionale dei Lincei,* Classe di Scienze morali, storiche e filologiche, series 6, 8 (1932), 422–62.

————. "Due prelati lincei in Roma alla corte di Urbano VIII: Virginio Cesarini Giovanni Ciampoli." *Atti dell'Accademia degli Arcadi* 3 (1929–30): 171–200.

————. "Memorie Tiburtino-Cornicolane di Federico Cesi fondatore e principe dei Lincei." *Atti e Memorie della Società Tiburtina di Storia e Arte* 9–10 (1929–30): 230–47.

————. "Vita romana del 600 nel carteggio inedito di un medico tedesco in Roma." In *Atti del Primo Congresso Nazionale di Studi Romani,* 1:813–27. Rome: Istituto di Studi Romani, 1929.

————. "Il Palazzo Cesi a Tivoli." *Atti e memorie della Società Tiburtina di Storia e Arte* 8 (1928): 262–68.

————. "Relazione del Conclave di Gregorio XV." *Archivio della Reale Società Romana di Storia Patria* 50 (1927): 5–32.

————. "Verbali delle adunanze e cronaca della prima Accademia Lincea (1603–1630)." *Memorie della Reale Accademia Nazionale dei Lincei,* Classe di Scienze morali, storiche e filologiche, series 6, 2 (1927): 463–512.

————, ed. "Il carteggio della vecchia accademia di Federico Cesi." *Memorie della Reale Accademia Nazionale dei Lincei,* Classe di scienze morali storiche e filologiche, series 6, 7 (1936–41).

Galassi Paluzzi, Carlo. *Storia segreta dello stile dei gesuiti.* Rome: Mundini, 1951.

Galilei, Galileo. *Dialogo di Cecco da Ronchitti in perpuosito della stella nova.* In *Galileo Against the Philosophers,* translated by S. Drake, 28–32. Los Angeles: Zeitlin and Ver Brugge, 1976.

————. *Dialogue Concerning the Two Chief World Systems.* Translated by Stillman Drake. Berkeley: University of California Press, 1967.

————. *Discourse on Bodies in Water.* Edited by Stillman Drake. Urbana: Illinois University Press, 1960.

————. *Istoria e dimostrazioni intorno alle macchie solari e loro accidenti comprese in tre lettere scritte all'illustrissimo signor Marco Velseri Linceo Duumviro d'Augusta e Consigliero di Sua Maestà Cesarea.* Rome: Mascardi, 1613.

————. *Opere.* Edited by Antonio Favaro. 20 vols. Florence: Barbera, 1890–1909.

————. *Sidereus nuncius.* Translated by Albert Van Helden. Chicago: University of Chicago Press, 1989.

————. *Two New Sciences.* Translated by Stillman Drake, 27–28. Madison: University of Wisconsin Press, 1974.

Galison, Peter, *How Experiments End.* Chicago: University of Chicago Press, 1987.

————. "The Trading Zone: Coordinating Action and Belief." Paper presented at UCLA, November 1989. Forthcoming in *Image and Logic.*

Galluzzi, Paolo. "L'Accademia del Cimento: 'Gusti' del principe, filosofia e ideologia dell'esperimento." *Quaderni storici* 16 (1981): 788–844.

———. "Il mecenatismo mediceo e le scienze." In *Idee, istituzioni, scienza ed arti nella Firenze dei Medici*, edited by Cesare Vasoli, 189–215. Florence: Giunti-Martello, 1980.

———. *Momento*. Rome: Edizioni dell'Ateneo, 1979.

———. "Il Platonismo del tardo Cinquecento e la filosofia di Galileo." In *Ricerche sulla cultura dell'Italia moderna*, edited by P. Zambelli, 39–79. Bari: Laterza, 1973.

Galluzzi, Riguccio. *Istoria del granducato di Toscana sotto il governo della Casa Medici*. Florence: Cambiagi, 1781.

Garin, Eugenio. "Galileo the Philosopher." In *Science and Civic Life in the Italian Renaissance*, 117–44. New York: Anchor, 1969.

Garuffi, G. M. *L'Italia accademica, o sia le accademie aperte a pompa e decoro delle lettere più amene nelle città italiane*. Rimini: Dandi, 1688.

Gassendi, Pierre. *Viri illustris Nicolai Claudii Fabricii de Peiresc, senatoris Aquisextiensis vita*. The Hague: Vlacq, 1655.

Geertz, Clifford. "Deep Play: Notes on the Balienese Cockfight." In *The Interpretation of Cultures*, 412–53. New York: Basic Books, 1973.

Gellner, Ernest, and John Waterbury, eds. *Patrons and Clients in Mediterranean Societies*. London: Duckworth, 1977.

Gerboni, Luigi. *Un umanista nel Seicento*. Città di Castello: Lapi, 1899.

Giacchi, Orio. "Considerazioni giuridiche sui due processi contro Galileo." In *Nel terzo centenario della morte di Galileo Galilei*, 383–406. Milan: Società Editrice Vita e Pensiero, 1942. [A publication of the Università Cattolica del S. Cuore.]

Giacobbe, G. C. "Il *Commentarium de certitudine mathematicarum disciplinarum* di Alessandro Piccolomini." *Physis* 14 (1972): 162–93.

———. "Epigoni del Seicento della *Quaestio de certitudine mathematicarum*: Giuseppe Biancani." *Physis* 18 (1976): 5–40.

———. "Francesco Barozzi e la *Quaestio de certitudine mathematicarum*." *Physis* 14 (1972): 357–74.

———. "La riflessione metamatematica di Pietro Catena." *Physis* 15 (1973): 178–96.

Gigli, Giacinto. *Diario romano (1608–1670)*. Rome: Tumminelli, 1958.

Gilbert, Neal W. "The Early Italian Humanists and Disputation." In *Renaissance Studies in Honor of Hans Baron*, edited by A. Molho and J. Tedeschi, 203–26. Florence: Sansoni, 1971.

Gingerich, Owen, and Robert Westman. "The Wittich Connection: Conflict and Priority in Late Sixteenth-Century Cosmology." *Transactions of the American Philosophical Society* 78 (1988): part 7.

Giordani, Enrico, ed. *I sei cartelli di matematica disfida di Lodovico Ferrari coi sei contro-cartelli in riposta di Nicolò Tartaglia*. Milan: Luigi Ronchi, 1876.

Giovio, Paolo. *Dialogo dell'imprese militari e amorose*. Edited by Maria Luisa Doglio. Rome: Bulzoni, 1978.

References

Giraldi, G. *Delle lodi di D. Ferdinando G. D. di Toscana.* Florence: Giunti, 1609.

Goffman, Erving. "The Nature of Deference and Demeanor." *American Anthropologist* 58 (1956): 481.

Govi, Gilberto. "Intorno alla data di un discorso inedito pronunciato da Federico Cesi fondatore dell'Accademia de' Lincei e da esso intitolato: Del natural desiderio di sapere et Istitutione de' Lincei per adempimento di esso." *Memorie della Reale Accademia Nazionale dei Lincei,* Classe di Scienze morali, storiche e filologiche, series 3, 5 (1879–80): 244–61.

Grant, Edward. "Ways to interpret the Terms 'Aristotelian' and 'Aristotelianism' in Medieval and Renaissance Natural Philosophy." *History of Science* 25 (1987): 336–58.

Grassi, Orazio. *De iride disputatio optica.* Rome: Mascardi, 1617.

———. *De tribus cometis anni MDCXVIII disputatio astronomica.* Rome: Mascardi, 1619. Reprinted in *GO,* vol. 6, pp. 19–35.

———. *Libra astronomica ac philosophica.* Rome: Naccarini, 1919. Reprinted in *GO,* vol. 6, pp. 107–79.

———. *Ratio ponderum librae et simbellae* Paris: Cramoisy, 1626. Reprinted in *GO,* vol. 6, pp. 375–500.

Gravit, Francis W. "The 'Accademia degli Umoristi' and Its French Relationships." *Papers of the Michigan Academy of Science, Arts and Letters* 20 (1935): 505–21.

Grazia, Vincenzo di. *Considerezioni sopra il Discorso di Galileo Galilei.* Florence: Pignoni, 1613. Reprinted in *GO,* vol. 4.

Greenblatt, Stephen. *Renaissance Self-Fashioning.* Chicago: University of Chicago Press, 1980.

Greengrass, Mark. "Noble Affinities in Early Modern France: The Case of Henri I de Montmorency, Constable of France." *European History Quarterly* 16 (1986): 275–311.

Gregorovius, Fernando. *Urbano VIII e la sua opposizione alla Spagna e all'imperatore.* Rome: Fratelli Bocca, 1879.

Guazzo, Stefano. *La civil conversazione.* Brescia: Bozzola, 1574.

———. *Dialoghi piacevoli.* Venice: Bertano, 1585.

Guglieminetti, Marziano, and Mariarosa Masoero. "Lettere e prose inedite (o parzialmente edite) di Giovanni Ciampoli." *Studi secenteschi* 19 (1978): 131–257.

Guiducci, Mario. *Discorso delle comete.* Florence: Cecconcelli, 1619. Reprinted in *GO,* vol. 6.

Gundersheimer, Werner L. "Patronage in the Renaissance: An Exploratory Approach." In *Patronage in the Renaissance,* edited by Guy Fitch Lytle and Stephen Orgel, 3–23. Princeton: Princeton University Press, 1981.

Hagstrom, Warren. "Gift Giving as an Organizing Principle in Science." In *Science in Context,* edited by Barry Barnes and David Edge, 21–34. Cambridge, Mass.: MIT Press, 1982.

———. *The Scientific Community.* New York: Basic Books, 1965.

REFERENCES

Hahlweg, Kai, and C. A. Hooker, eds. *Issues in Evolutionary Epistemology*. Albany: State University of New York Press, 1989.

Hannaway, Owen. "Laboratory Design and the Aim of Science: Andreas Libavius versus Tycho Brahe." *Isis* 77 (1986): 585–610.

Harley, David. "Honour and Property: The Structure of Professional Disputes in Eighteenth-Century English Medicine." In *The Medical Enlightenment of the Eighteen Century,* edited by Andrew Cunnigham and Roger French, 138–64. Cambridge: Cambridge University Press, 1990.

Hartner, Willy. "Galileo's Contribution to Astronomy." In *Galileo, Man of Science,* edited by Ernan McMullin, 185. New York: Basic Books, 1967.

Haskell, Francis. *Patrons and Painters*. New Haven, Conn.: Yale University Press, 1980.

Heikamp, Detlef. "L'antica sistemazione degli strumenti scientifici nelle collezioni fiorentine." *Antichità viva* 9 (1970): 3–25.

Heilbron, John. *Physics at the Royal-Society during Newton's Presidency*. Berkeley: Office for History of Science and Technology, 1983.

Hellman, Doris C. *The Comet of 1577: Its Place in the History of Astronomy*. New York: Columbia University Press, 1944.

Herr, Richard. "Honor Versus Absolutism: Richelieu's Fight Against Dueling." *The Journal of Modern History* 27 (1955): 281–85.

Hesse, Mary. *Structure of Scientific Inference*. Berkeley: University of California Press, 1974.

Hull, David. "A Mechanism and Its Metaphysics: An Evolutionary Account of the Social and Conceptual Developement of Science." *Biology and Philosophy* 3 (1988): 123–55.

———. *Science as a Process*. Chicago: University of Chicago Press, 1988.

Hunter, Michael. *The Royal-Society and Its Fellows 1660–1700: The Morphology of an Early Scientific Institution*. Chalfont St. Giles: British Society for the History of Science, 1985.

———. *Science and Society in Restoration England*. Cambridge: Cambridge University Press, 1991.

Hutchinson, Keith. "Toward a Political Iconology of the Copernican Revolution." In *Astrology, Science and Society,* edited by Patrick Curry, 95–141. Woodbridge, Suffolk: Boydell, 1987.

Iliffe, Rob. "Author-Mongering: The 'Editor' Between Producer and Consumer." Forthcoming.

———. "'In the Warehouse': Privacy, Property and Priority in the Early Royal Society." *History of Science*. 30 (1992): 29–68.

Imbert, Gaetano. *La vita fiorentina nel Seicento*. Florence: Bemporad, 1906.

Jack, Mary Ann. "The Accademia del Disegno in Late Renaissance Florence." *Sixteenth Century Journal* 7 (1976): 3–20.

Jardine, Nicholas. *The Birth of History and Philosophy of Science*. Cambridge: Cambridge University Press, 1984.

———. "Epistemology of the Sciences." In *The Cambridge History of Renaissance*

Philosophy, edited by Charles B. Schmitt and Quentin Skinner, 685–711. Cambridge: Cambridge University Press, 1988.

———. "The Forging of Modern Realism: Clavius and Kepler Against the Sceptics." *Studies in History and Philosophy of Science* 10 (1979): 141–73.

———. "The Significance of the Copernican Orbs." *Journal for History of Astronomy* 13 (1982): 168–94.

Jeanneret, Michel. *A Feast of Words.* Chicago: University of Chicago Press, 1991.

Jones-Schofield, Christine. *Tychonic and Semi-Tychonic World Systems.* New York: Arno, 1981.

Jonson, Ben. *Sejanus and His Fall.* Edited by W. F. Bolton. London: Benn, 1966.

Kantorowicz, Ernst. *The King's Two Bodies.* Princeton: Princeton University Press, 1957.

———. "Mysteries of State: An Absolutist Concept and Its Late Medieval Origins." *The Harvard Theological Review* 48 (1955): 65–91.

Kent, F. W. *Household and Lineage in Renaissance Florence.* Princeton: Princeton University Press, 1977.

Kent, F. W., Patricia Simons, and J. C. Eade, eds. *Patronage, Art and Society in Renaissance Italy.* Oxford: Oxford University Press, 1987.

Kepler, Johannes. *Ad vitellionem paralipomena.* Frankfurt: Marinum, 1604.

———. *On the Six-Cornered Snowflake.* Oxford: Clarendon Press, 1966.

Kettering, Sharon. "Gift-Giving and Patronage in Early Modern France." *French History* 2 (1988): 131–51.

———. "The Historical Development of Political Clientelism." *Journal of Interdisciplinary History* 3 (1988): 419–47.

———. "The Patronage Power of Early Modern French Noblewomen." *The Historical Journal* 4 (1989): 817–41.

———. *Patrons, Brokers, and Clients in Seventeenth-Century France.* Oxford: Oxford University Press, 1986.

Kiernan, V. G. *The Duel in European History.* Oxford: Oxford University Press, 1988.

Klapisch-Zuber, Christiane. "Kin, Friends, and Neighbors." In *Women, Family, and Ritual in Renaissance Italy,* 68–93. Chicago: University of Chicago Press, 1985.

Kuhn, Thomas S. "The Road Since *Structure.*" In A. Fine, M. Forbes, and L. Wessels, eds., *PSA 1990,* vol. 2, pp. 3–13. East Lansing, Mich.: Philosophy of Science Association, 1991.

———. "Possible Worlds in History of Science." In Sture Allen, ed., *Possible Worlds in Humanities, Arts, and Sciences.* Proceedings of Nobel Symposium 65, pp. 9–32. Berlin: Walter de Gruyter, 1989.

———. "The Presence of Past Science." Paper delivered for the Shearman Memorial Lectures, University College, London, 23–25 November 1987.

———. "Scientific Development and Lexical Change." Paper delivered for the Thalheimer Lectures, John Hopkins University, 12–19 November 1984.

———. "What Are Scientific revolutions?" In Lorenz Krüger, Lorraine J. Daston, and Michael Heidelberger, eds., *The Probabilistic Revolution*, vol. 1: 7–22. Cambridge: MIT Press, 1987.

———. "Commensurability, Comparability, Communicability." *PSA 1982*, edited by Peter D. Asquith and Thomas Nickles, 2:669–88. East Lansing, Mich.: Philosophy of Science Association, 1983.

———. "Second Thoughts on Paradigms." In *The Essential Tension*, 293–319. Chicago: University of Chicago Press, 1977.

———. *The Structure of Scientific Revolutions*. Chicago: University of Chicago Press, 1962.

Lakatos, Imre. *Proofs and Refutations*. Cambridge: Cambridge University Press, 1976.

Lange, Klaus-Peter. *Theoretiker des Literarischen Manierismus*. Munich: Wilhelm Fink Verlag, 1968.

Langedijk, Karla. *The Portraits of the Medici*. 3 vols. Florence: SPES, 1980.

Langford, Jerome J. *Galileo, Science, and the Church*. Ann Arbor: University of Michigan Press, 1971.

Latour, Bruno. *Science in Action*. Cambridge, Mass.: Harvard University Press, 1987.

Lefevre, Renato. "Gli sfaccendati." *Studi romani* 8 (1960): 154–65.

Lemaine, Gerard. "Social Differentiation and Social Originality." *European Journal of Social Psychology* 4 (1974): 17–52.

Leman, Auguste. *Urbain VIII et al rivalité de la France et de la Maison d'Autriche de 1631 à 1635*. Lille: Giard, 1920, and Paris: Champion, 1920.

Lévi-Strauss, Claude. *The Elementary Structures of Kinship*. Boston: Beacon Press, 1969.

———. *The Savage Mind*. Chicago: University of Chicago Press, 1966.

———. "Race and Culture," in *A View from Afar*. New York: Basic Books, 1985, pp. 3–24.

Levy-Peck, Linda. "'For a King not to be bountiful were a fault': Perspectives on Court Patronage in Early Stuart England." *The Journal of British Studies* 25 (1986): 48.

Liberati, Francesco. *Il perfetto Maestro di Casa*. Rome: Bernabò, 1658.

Liceti, Fortunio. *Litheosphorous, sive De lapide Bononiensi, lucem in se conceptam ab ambiente claro mox in tenebris mire conservante*. Udine: Schiratti, 1640.

Litchfield, R. Burr. *Emergence of a Bureaucracy: The Florentine Patricians 1530–1790*. Princeton: Princeton University Press, 1986.

Livesey, Steven J. "William of Ockam, the Subalternate Sciences and Aristotle's Theory of 'Metabasis.'" *British Journal for the History of Science* 19 (1985): 127–45.

Lloyd, G. E. R. "Saving the Appearances." *Classical Quarterly* 28 (1978): 202–22.

Locke, John. *An Essay Concerning Human Understanding*. Edited by A. Fraser. 2 vols. New York: Dover, 1959.

Lumbroso, Giacomo. "Notizie sulla vita di Cassino dal Pozzo." *Miscellanea di storia italiana* 15 (1874): 136–43.

References

Lunadoro, Girolamo. *Relatione della corte di Roma.* Rome: Frambotto, 1635.

Lux, David. *Patronage and Royal Science in Seventeenth-Century France.* Ithaca: Cornell University Press, 1989.

Machamer, Peter. "Galileo and the Causes." In *New Perspectives on Galileo,* edited by Robert E. Butts and Joseph C. Pitt, 161–80. Dordrecht: Reidel, 1978.

Maffei, Scipione. *Della scienza chiamata cavalleresca libri tre.* Rome: Gonzaga, 1710.

Malanima, Paolo. "Concini, Cosimo." *Dizionario biografico degli italiani,* vol. 27, pp. 730–31. Rome: Istituto della Enciclopedia Italiana, 1982.

Malinowski, Bronislaw. "Kula: The Circulating Exchange of Valuables in the Archipelagoes of Eastern Guinea." *Man,* series 1, 19–20 (1920): 97–105.

Manetti, Antonio. "Circa il sito, forma e misura dell'Inferno di Dante Alighieri, poeta eccellentissimo." In *Studi sulla Divina Commedia di Galileo Galilei, Vincenzo Borghini ed altri,* edited by Ottavio Gigli, 35–114. Florence: Le Monnier, 1855.

Manni, Paola. "Galileo accademico della Crusca." In *La Crusca nella tradizione letteraria e linguistica italiana,* 119–36. Florence: Accademia della Crusca, 1985.

Mannucci, Francesco Luigi. "La vita e le opere di Agostino Mascardi." *Atti della Società Ligure di Storia Patria* 42 (1908): 139–76.

Manzini, Giovanni Battista. *Della peripetia di fortuna: Overo sopra la caduta di Seiano.* 1620?

———. *Observations Upon the Fall of Seianus.* 2d ed. London: Harper, 1639.

Maravall, Jose Antonio. *Culture of the Baroque.* Minneapolis: University of Minnesota Press, 1986.

Margani, Margherita. "Sull'autenticità di una lettera attribuita a G. Galilei." *Atti della Reale Accademia delle Scienze di Torino* 57 (1921–22): 556–68.

Margolis, Howard. "Tycho's System and Galileo's Dialogue." *Studies in History and Philosophy of Science* 22 (1991): 259–75.

Marin, Louis. *Portrait of the King.* Minneapolis: Minnesota University Press, 1988.

Marino, Giambattista. *L'Adone.* Paris, 1623. Reprint. Turin: Paravia, 1922.

———. *Epistolario seguito da lettere di altri scrittori del Seicento,* edited by Angelo Borzelli and Fausto Nicolini. Bari: Laterza, 1912.

———. *Lettere.* Turin: Einaudi, 1966.

Marius, Simon. *Mundus iovialis.* Nuremberg: Laur, 1614.

Martinori, Edoardo. *I Cesi.* Rome: Tipografia Compagnia Nazionale Pubblicità, 1931.

Mascardi, Agostino. "Che la corte è vera scuola non solamente della prudenza, ma delle virtù morali." In *Prose vulgari,* 349–67. Venice: Baba, 1653.

———. "Discorso o invettiva fatta in una accademia intorno alla iniquità della fortuna." In *Prose vulgari,* 510–17. Venice: Baba, 1653.

———. "Discorso secondo: Che un cortigiano non dee dolersi, perché venga più favorito in corte l'ignorante che 'l dotto; il plebeo, che 'l nobile." In *Prose vulgari.* Venice: Baba, 1653.

———. "Per l'esequie del Signor D. Virginio Cesarino." In *Prose vulgari.* Venice: Baba, 1653.

———. *Le Pompe del Campidoglio*. Rome: Zanetti, 1624.

———. "Sopra un componimento poetico intorno alla cometa. Al Signor Conte Camillo Molza." In *Prose vulgari*, 151–67. Venice: Baba, 1653.

———, ed. *Saggi accademici*. Venice: Baba, 1653.

Matthieu, Pierre. *Unhappy Prosperity Expressed in the History of Aelius Seianus, and Philippa the Catanian. With Observations Upon the Fall of Seianus. Lastly Certain Considerations Upon the Life and Services of Monsieur Villeroy.* London: Harper, 1639.

Mauss, Marcel. *The Gift*. New York: Norton, 1967.

Maylender, Michele. *Storia delle accademie d'Italia.* 5 vols. Bologna: Capelli, 1926–30.

Memorie delle feste fatte in Firenze per le reali nozze de' Serenissimi Sposi Cosimo Principe di Toscana e Margherita Luisa d' Orleans. Florence: Stamperia di SAS, 1662.

Michelangelo Buonarroti il Giovane. *Elogio di Cosimo II*. Florence: 1621.

Middleton, W. Knowles. *The Experimenters*. Baltimore: Johns Hopkins University Press, 1971.

———. "Science in Rome, 1675–1700, and the Accademia Fisicomatematica of Giovanni Giustino Ciampini." *The British Journal for the History of Science* 8 (1975): 140.

Mistruzzi, Carlo. "La nobiltà nello stato pontificio." *Rassegna degli Archivi di Stato* 23 (1963): 206–44.

Molinari, Cesare. *Le nozze degli dei*. Rome: Bulzoni, 1968.

Montagu, Jennifer. "The Painted Enigma and French Seventeenth-Century Art." *Journal of the Warburg and Courtauld Institutes* 31 (1968): 307, 312.

Moore-Bergeron, David. *English Civic Pageantry 1558–1642*. London: Arnold, 1971.

Moran, Bruce. *The Alchemical World of the German Court*. Stuttgart: Franz Steiner Verlag, 1991.

———. "Privilege, Communication and Chemistry: The Hermetic-Alchemical Circle of Moritz of Hesse-Kassel." *Ambix* 32 (1985): 110–26.

———. "Science at the Court of Hesse-Kassel: Informal Communication, Collaboration, and the Role of the Prince Practioner in the Sixteenth Century." Ph.D. diss., University of California, Los Angeles, 1978.

———. "Wilhelm IV of Hesse-Kassel: Informal Communication and the Aristocratic Context of Discovery." In *Scientific Discovery: Case Studies,* edited by Thomas Nickles, 67–96. Dordrecht: Reidel, 1980.

———, ed. *Patronage and Institutions*. Rochester: Boydell Press, 1991.

Morpurgo-Tagliabue, Guido. *I processi di Galileo e l'epistemologia*. Rome: Armando, 1981.

Muir, Edward. *Civic Ritual in Renaissance Venice*. Princeton: Princeton University Press, 1981.

Nagler, Alois Maria. *Theatre Festivals of the Medici 1539–1637*. New Haven, Conn.: Yale University Press, 1964.

Napolitani, Pier Daniele. "La geometrizzazione della realtà fisica: Il peso specifico in Ghetaldi e in Galileo." *Bollettino di storia della scienze matematiche* 8 (1988): 139–237.

Nelli, Giovanni Battista. *Vita e commercio letterario di Galileo Galilei*. 2 vols. Lausanne: 1793.

Neuschel, Kristen B. *Word of Honor*. Ithaca: Cornell University Press, 1989.

Nigro, Salvatore. "Dalla lingua al dialetto: La letteratura popolaresca." In *I poeti giocosi dell'età barocca*, edited by Alberto Asor Rosa. Bari: Laterza, 1975.

Nussdorfer, Laurie. "City Politics in Baroque Rome, 1623–1644." Ph.D. diss., Princeton University, 1985.

———. *Civic Politics in the Rome of Urban VIII*. Princeton: Princeton University Press, 1992.

O'Malley, J. W. *Praise and Blame in Renaissance Rome*. Durham: Duke University Press, 1979.

Odescalchi, Baldassare. *Memorie istorico critiche dell'Accademia de' Lincei e del Principe Federico Cesi*. Rome: Salvioni, 1806.

Oestreich, Gerhard. *Neostoicism and the Early Modern State*. Cambridge: Cambridge University Press, 1982.

Olmi, Giuseppe. "'In essercitio universale di contemplatione e prattica': Federico Cesi e i Lincei." In *Università, accademie e società scientifiche in Italia e in Germania dal Cinquecento al Settecento*, edited by Laetitia Boehm and Ezio Raimondi, 169–236. Bologna: Il Mulino, 1981.

Orbaan, J. A. F. *Documenti sul barocco in Roma*. 2 vols. Rome: Società Romana di Storia Patria, 1920.

Ordine, Nuccio, et al. *Il dialogo filosofico nel '500 europeo*. Milan: Angeli, 1990.

Outram, Dorinda. "The Language of Natural Power: The 'Eloges' of Georges Cuvier and the Public Language of Nineteenth-Century Science." *History of Science* 16 (1978): 153–78.

———. *Georges Cuvier*. Manchester: Manchester University Press, 1984.

Pagano, Sergio. *I documenti del processo di Galileo Galilei*. Vatican City: Archivio Vaticano, 1984.

Palmer, Richard. "Medicine at the Papal Court in the Sixteenth Century." In *Medicine at the Courts of Europe, 1500–1837*, edited by Vivian Nutton, 49–78. London: Routledge, 1990.

Panofsky, Erwin. *Galileo as a Critic of the Arts*. The Hague: Martinus Nijoff, 1954.

Parker, Geoffrey. *Philip II*. London: Hutchinson, 1979.

Parodi, Severina. *Catalogo degli Accademici dalla Fondazione*. Florence: Sansoni, 1983.

Pastor, Ludwig von. *History of the Popes*. St. Louis, Mo.: Herder, 1938.

Patrizi, Giorgio, ed. *Stefano Guazzo e la civil conversazione*. Rome: Bulzoni, 1990.

Paul, Charles B. *Science and Immortality*. Berkeley: University of California Press, 1980.

Pellegrini, Amedeo, ed. *Relazioni inedite di ambasciatori lucchesi alle corti di Firenze, Genova, Milano, Modena, Parma, Torino*. Lucca: Marchi, 1901.

References

————. *Relazioni inedite di ambasciatori lucchesi alle corti di Roma (sec. XVI-XVII)*. Rome: Tipografia Poliglotta, 1901.

Pellegrini, Matteo. *Che al savio è convenevole il corteggiare libri IIII*. Bologna: Tebaldini, 1624.

————. "Che il dir male non è in tutto male." In *Saggi accademici,* edited by Agostino Mascardi, 193–208. Venice: Baba, 1653.

————. *Delle acutezze*. Genoa: Ferroni, 1939.

————. *Difesa del savio in corte*. Viterbo: Diotallevi, 1634.

————. *Le sage en cour*. Paris: Lancy, 1638; Rocolet, 1639.

Peristiany, J. G., and Julian Pitt-Rivers, eds. *Honor and Grace in Anthropology*. Cambridge: Cambridge University Press, 1992.

Persico, Panfilo. *Del segretario libri quattro*. Venice: Damian Zenato, 1629.

Petrioli Tofani, Annamaria. "Contributi allo studio degli apparati e delle feste medicee." In *Firenze e la Toscana nell'Europa del '500,* 2:645–61. Florence: Olschki, 1983.

Petrioli Tofani, Annamaria, and Giovanna Gaeta Bertelà. *Feste e apparati medicei da Cosimo I a Cosimo II*. Florence: Olschki, 1969.

Petruccelli della Gattina. *Histoire diplomatique des conclaves*. Paris: Librairie Internationale, 1865.

Pevsner, Nikolaus. *Academies of Art*. Cambridge: Cambridge University Press, 1940.

Piccolini, Celestino. "Federico II, Principe de' Lincei, Marchese di Monticelli." *Atti e memorie della Società Tiburtina di Storia e Arte* 9–10 (1929–30): 197–207.

————. "Il Palazzo Cesi a Tivoli." *Atti e memorie della Società Tiburtina di Storia e Arte* 8 (1928): 262–68.

————. "Ricevimenti ai feudatari nel Seicento." *Atti e memorie della Società Tiburtina di Storia e Arte* 7 (1927): 217–37.

Piccolomini, Aeneas Sylvius. *De curialium miseriis epistola*. Edited by Wilfred P. Mustard. Baltimore: Johns Hopkins University Press, 1928.

Pieraccini, Gaetano. *La stirpe dei Medici di Cafaggiolo*. Vol. 2. Florence: Nardini, 1986.

Pieralisi, Sante. *Urbano VIII e Galileo Galilei*. Rome: Tipografia Poliglotta, 1875.

Pitt-Rivers, J. *Mediterranean Countrymen*. Paris: Mouton, 1963.

Pomian, Krzysztof. *Collectionneures, amateurs et curieux*. Paris: Gallimard, 1987.

Praz, Mario. *Studies in Seventeenth-Century Imagery*. Rome: Edizioni di Storia e Letteratura, 1964.

Prickard, A. O. "The Mundus Jovialis of Simon Marius." *The Observatory,* 39 (1916): 367–81, 403–12, 443–52, 498–503.

Prodi, Paolo. *The Papal Prince, One Body and Two Souls: The Papal Monarchy in Early Modern Europe*. Cambridge: Cambridge University Press, 1987.

Prosperi, Adriano, ed. *La corte e il "cortegiano": Un modello europeo*. Rome: Bulzoni, 1980.

Ptolemy. *Tetrabiblos.* Trans. F. E. Robbins. Cambridge: Harvard University Press, 1940.

Quazza, Romolo. *L'elezione di Urbano VIII nelle relazioni dei diplomatici mantovani.* Rome: Reale Società Romana di Storia Patria, 1922.

———. "Il periodo italiano della Guerra dei Trent'Anni." *Rivista storica italiana* 50 (1933): 64–89.

Quevedo y Villegas, Francisco de. "Como ha de ser el privado." In *Obras completas,* edited by Felicidad Buendía, 2:592–635. Madrid: 1967.

Quondam, Amedeo. "L'accademia." In *Letteratura italiana,* edited by Alberto Asor Rosa, 1:864. Turin: Einaudi, 1982.

———, ed. *Le "Carte messaggiere."* Rome: Bulzoni, 1981.

Quondam, Amedeo, and Marzio Achille Romani, eds. *Le corti farnesiane di Parma e Piacenza.* 2 vols. Rome: Bulzoni, 1978.

Quine, Willard V. O. *Word and Object.* Cambridge, Mass.: MIT Press, 1960.

Raimondi, Ezio. *Anatomie secentesche.* Pisa: Nistri-Lischi, 1966.

———. *Trattatisti e narratori del Seicento.* Milan-Naples: Ricciardi, 1960.

Rapp, Richard T. *Industry and Economic Decline in Seventeenth-Century Venice.* Cambridge, Mass.: Harvard University Press, 1976.

Redondi, Pietro. *Galileo Heretic.* Princeton: Princeton University Press, 1987.

Relazioni dei Rettori Veneti di Terraferma. Vol. IV, *Podestaria e Capitanato di Padova.* Milan: Giuffré, 1975. [Published for the Istituto di Storia Economica dell'Università di Trieste.]

La Revue de Mauss 12 (1991). Special issue on "Le don perdu et retrouvé."

Ricci Riccardi, Antonio. *Galileo Galilei e Fra Tommaso Caccini.* Florence: Le Monnier, 1902.

Righini Bonelli, Maria Luisa, and William Shea. *Galileo's Florentine Residences.* Florence: Istituto e Museo di Storia della Scienza, n.d.

Ripa, Cesare. *Iconologia.* Rome: Gigliotti, 1593; Lepido Faci, 1603.

Riquius, Iustus. *De vita Virginii Caesarini.* Padua: Thuilii, 1629.

Robinson, Wade L. "Galileo on the Moons of Jupiter." *Annals of Science* 31 (1974): 165–69.

Rosa, Mario. "La Chiesa e gli stati regionali nell'età dell'assolutismo." In *Il letterato e le istituzioni,* edited by Alberto Asor Rosa, 324–25. Turin: Einaudi, 1982.

Rose, Paul L. *The Italian Renaissance of Mathematics.* Geneva: Droz, 1975.

———. "Letters Illustrating the Career of Federico Commandino." *Physis* 15 (1973): 401–10.

Rosen, Edward. "The Authenticity of Galileo's Letter to Landucci." *Modern Language Quarterly* 12 (1975): 473–86.

———. *Three Copernican Treatises.* New York: Dover, 1939.

———. *Three Imperial Mathematicians.* New York: Abaris Books, 1986.

———, ed. and trans. *Kepler's Conversation with Galileo's Sidereal Messenger.* New York: Johnson, 1965.

Rossi, Paolo. *I filosofi e le macchine, 1400–1700*. Milan: Feltrinelli, 1984.

Rudwick, Martin. *The Great Devonian Controversy*. Chicago: University of Chicago Press, 1985.

Ruffner, James Alan. "The Background and Early Developments of Newton's Theory of Comets." Ph.D. diss. Indiana University, 1966.

Russo, Piera. "L'Accademia degli Umoristi, fondazione, strutture e leggi: Il primo decennio di attività." *Esperienze letterarie* 4 (1979): 47–61.

Sahlins, Marshall. *Stone-Age Economics*. New York: Aldine de Gruyter, 1972.

Santi, Venceslao. "La storia nella *Secchia rapita*." *Memorie della Reale Accademia di Scienze, Lettere e Arti in Modena*, series 3, 6 (1906): 310–33; (1910): 247–397.

Sarasohn, Lisa T. "Nicolas-Claude Fabri de Peiresc and the Patronage of New Science in the Seventeenth Century." *Isis* 84 (1993). In press.

Saravia, Antonio J. *O discurso Engenhoso*. Saõ Paulo: Editora Perspectiva, 1980.

Sarpi, Paolo. "Sopra un decreto della congregazione in Roma in stampa presentato per l'Illustrissimo Signor Conte del Zaffo a 5 maggio 1616." In *Opere*, edited by Gaetano Cozzi and Luisa Cozzi. Milan-Naples: Ricciardi, 1969.

Saussure, Ferdinand de. *Cours de linguistique generale*. Paris: Payot, 1986.

Schaffer, Simon. "Scientific Discoveries and the End of Natural Philosophy." *Social Studies of Science* 16 (1986): 387–420.

———. "Wallification: Thomas Hobbes on School Divinity and Experimental Pneumatics." *Studies in History and Philosophy of Science* 19 (1988): 294–95.

[Scheiner, Christopher]. *Tres epistolae de maculis solaribus scriptae ad Marcum Velserum*. Augustae Vindelicorum, 1612. Reprinted in *GO*, vol. 5.

———. *De maculis solaribus et stellis circa lovem errantibus, accuratior disquisitio ad Marcum Velserum*. Reprinted in *GO*, vol. 5.

Schiebinger, Londa. *The Mind Has No Sex?* Cambridge, Mass.: Harvard University Press, 1989.

Schmidt, S. W., L. Guasti, C. H. Lande, and J. C. Scott. *Friends, Followers and Factions*. Berkeley: University of California Press, 1977.

Schmitt, Charles B. *The Aristotelian Tradition and Renaissance Universities*. London: Variorum, 1984.

———. *Aristotle and the Renaissance*. Cambridge, Mass.: Harvard University Press, 1983.

———. *Studies in Renaissance Philosophy and Science*. London: Variorum, 1981.

Segarizzi, Arnaldo, ed. *Relazioni degli ambasciatori veneti al Senato*. Vol. 3. Bari: Laterza, 1916.

Segre, Michael. "Galileo as a Politician." *Sudhoffs Archiv* 72 (1988): 69–82.

———. *In the Wake of Galileo*. New Brunswick: Rutgers University Press, 1991.

Settis, Salvatore. *Giorgione's Tempest*. Chicago: University of Chicago Press, 1990.

Settle, Thomas B. "Egnazio Danti and Mathematical Education in Late Sixteenth-Century Florence." In *New Perspectives on Renaissance Thought*, edited by John Henry and Sarah Hutton, 24–37. London: Duckworth, 1990.

———. "Galilean Science: Essays in the Mechanics and Dynamics of the 'Discorsi.'" Ph.D. diss., Cornell University, 1966.

————. "Ostilio Ricci, a Bridge between Alberti and Galileo." In *Actes du XIIe Congrès International d'Histoire des Sciences. Paris, 1968,* 229–38. Paris: 1971.

Shapin, Steven. "The House of Experiment in Seventeenth-Century England." *Isis* 79 (1988): 373–44.

————. "The Invisible Technician." *American Scientist* 77 (November–December 1989): 554–63.

————. "Of Gods and Kings: Natural Philosophy and Politics in the Leibniz-Clarke Dispute." *Isis* 72 (1981): 187–215.

————. "Pump and Circumstance: Robert Boyle's Literary Technology." *Social Studies of Science* 14 (1984): 481–520.

————. "A Scholar and a Gentleman." *History of Science,* 29 (1991): 279–327.

————. "Who Was Robert Hooke?" In *Robert Hooke: New Studies,* edited by Michael Hunter and Simon Schaffer, 253–85. Woodbridge, Suffolk: Boydell, 1989.

Shapin, Steven, and Simon Schafer. *Leviathan and the Air Pump.* Princeton: Princeton University Press, 1985.

Shea, William. "Descartes as a Critic of Galileo." In *New Perspectives on Galileo,* edited by Robert E. Butts and Joseph C. Pitt, 139–59. Dordrecht: Reidel, 1978.

————. "Galileo, Scheiner, and the Interpretation of Sunspots." *Isis* 61 (1970): 498–519.

————. "Galileo's Atomic Hypothesis." *Ambix* 17 (1970): 13–27.

————. "Galileo's Discourse on Floating Bodies: Archimedean and Aristotelian Elements." In *Actes du XIIe Congrès International d'Histoire des Sciences. Paris, 1968,* 4:149–53. Paris: 1971.

————. *Galileo's Intellectual Revolution.* New York: Science History Publications, 1972.

Shephard, Robert. "Royal Favorites in the Political Discourse of Tudor and Stuart England." Ph.D. diss., Claremont Graduate School, 1985.

Silli, Graziella. *Una corte alla fine del Cinquecento.* Florence: Alinari, 1927.

Soldani, Jacopo. "Contro i Peripatetici." In Nunzio Vaccalluzzo, *Galileo Galilei nella poesia del suo secolo.* Milan: Sandron, 1910.

Solerti, Angelo. *Musica, ballo e drammatica alla corte medicea dal 1600 al 1637.* Florence: Bemporad, 1905.

Solinas, Francesco, ed. *Cassiano dal Pozzo.* Naples: De Luca, 1989.

Solnon, Jean-François. *La Cour de France.* Paris: Fayard, 1987.

Spezzaferro, Luigi. "La cultura del cardinal del Monte e il primo tempo del Caravaggio." *Storia dell'arte* 9–10 (1971): 76.

Spini, Giorgio, ed. *Architettura e politica da Cosimo I a Ferdinando I.* Florence: Olschki, 1976.

Starn, Randolph, and Loren Partridge. *Arts of Power.* Berkeley: University of California Press, 1992.

Starn, Randolph. "Seeing Culture in a Room for a Renaissance Prince." In *The*

New Cultural History, edited by Lynn Hunt, 205–32. Berkeley: University of California Press, 1988.

Stone, Lawrence. *The Crisis of the Aristocracy, 1558–1641.* Oxford: Oxford University Press, 1967.

Strathern, Marilyn. *The Gender of the Gift.* Berkeley: University of California Press, 1988.

Strong, Roy. *Art and Power: Renaissance Festivals 1450–1650.* Berkeley: University of California Press, 1984.

Stroup, Alice. *A Company of Scientists.* Berkeley: University of California Press, 1990.

Stumpo, Enrico. *Il capitale finanziario a Roma fra Cinque e Seicento.* Milan: Giuffré, 1985.

Sylla Dudley, Edith. "Galileo and the Oxford Calculatores." In *Reinterpreting Galileo,* edited by William A. Wallace, 53–108. Washington, D.C.: Catholic University of America Press, 1986.

Targioni Tozzetti, Giovanni. *Notizie degli aggrandimenti delle scienze fisiche accaduti in Toscana nel corso di anni LX del secolo XVII.* Florence: Bouchard, 1780. Reprint. Bologna: Forni, 1967.

Tasso, Torquato. *Il conte, o vero de l'imprese.* 1594. Reprinted in *I dialoghi di Torquato Tasso,* edited by Cesare Guasti, 3:361–444. Florence: Le Monnier, 1901.

———. *Il malpiglio, o vero de la corte.* 1582. Reprinted in *I dialoghi di Torquato Tasso,* edited by Cesare Guasti, 3:3–10, 18. Florence: Le Monnier, 1901.

Tassoni, Alessandro. *La secchia rapita.* Ronciglione, 1624. Reprinted in *I poeti giocosi dell'età barocca,* edited by Alberto Asor Rosa. Bari: Laterza, 1975.

Tenenti, Alberto. *Piracy and the Decline of Venice 1580–1615.* Berkeley: University of California Press, 1967.

Thoren, Victor E. *The Lord of Uraniborg.* Cambridge: Cambridge University Press, 1990.

Tiepolo, Francesca Maria. "Una lettera inedita di Galileo." *La cultura* 17 (1979): 60, 66.

Torrini, Maurizio. "Giovanni Ciampoli filosofo." In *Novità celesti e crisi del sapere,* edited by Paolo Galluzzi, 267–75. Florence: Giunti Barbera, 1984.

Toulmin, Stephen. *Human Understanding.* Princeton: Princeton University Press, 1972.

Trechman, E. G., trans. *The Diary of Montaigne's Journey to Italy.* London: Hogarth Press, 1929.

Trevor-Roper, Hugh. *Princes and Artists.* London: Thames and Hudson, 1976.

Trexler, Richard. *Public Life in Renaissance Florence.* New York: Academic Press, 1980.

Tribby, Jay. "Of Conversational Dispositions and the *Saggi's* Proem." In *Documentary Culture: Florence and Rome from Grand Duke Ferdinand I to Pope Alexander VII,* edited by Elizabeth Cropper. Florence: Olschki. Forthcoming.

———. "Stalking Civility: Conversing and Collecting in Early Modern Europe." *Rhetorica.* Forthcoming.

————. "Cooking (with) Clio and Cleo: Eloquence and Experiment in Seventeenth-Century Florence." *Journal of the History of Ideas* 52 (1991): 417–39.

Trinkaus, Charles Edward. *Adversity's Noblemen.* New York: Columbia University Press, 1940.

Vaccalluzzo, Nunzio. *Galileo Galilei nella poesia del suo tempo.* Milan: Sandron, 1910.

Van Helden, Albert. "The Accademia del Cimento and Saturn's Ring." *Physis* 15 (1973): 237–59.

————. "Eustachio Divini Versus Christiaan Huygens: A Reappraisal." *Physis* 12 (1970): 36–50.

————. "Galileo and the Telescope." In *Novità celesti e crisi del sapere,* edited by Paolo Galluzzi. Florence: Giunti Barbèra, 1984.

————. "The Invention of the Telescope." *Transactions of the American Philosophical Society* 67 (1977): 20–36.

————. "The Telescope and Authority from Galileo to Cassini." *Osiris* 9 (1993). In press.

Van Helden, Albert, and Mary Winkler. "Representing the Heavens: Galileo and Visual Astronomy." *Isis* 83 (1992): 195–217.

Van Kessel, Elisja M. R. "Joannes van Heeck (1579–?), Co-founder of the Accademia dei Lincei in Rome." *Mededelingen van het Nederlands Instituut te Rome* 38 (1976): 109–34.

Van Melsen, Andrew G. *From Atomos to Atom.* Pittsburgh: Duquesne University Press, 1952.

Varchi, Benedetto. *Storia Fiorentina,* edited by Gaetano Milanesi. 3 vols. Florence: La Monnier, 1857–58.

Varey, J. E. "The Audience and the Play at Court Spectacles: The Role of the King." *Bulletin of Hispanic Studies* 61 (1984): 399–406.

Vasari, Giorgio. *Le opere di Giorgio Vasari.* Edited by Gaetano Milanesi. Florence: Sansoni, 1882.

————. *Vita di Michelangelo.* Edited by Paola Barocchi. Milan-Naples: Ricciardi, 1962.

Verzellino, G. V. "Padre Orazio Grassi giesuita matematico eccellentissimo." In *Memorie degli uomini illustri di Savona,* 2:347–51. Savona, 1891.

Viala, Alain. *Naissance de l'écrivain.* Paris: Minuit, 1985.

Vickers, Brian. "Epideiectic Rhetoric in Galileo's *Dialogo.*" *Annali dell'Istituto e Museo di Storia della Scienza di Firenze* 8 (1983): 69–101.

Villari, Rosario. *Elogio della dissimulazione: La lotta politica nel Seicento.* Bari: Laterza, 1987.

Villifranchi, Giovanni. *Descrizione della barriera e della mascherata fatte in Firenze a XVII & a XIX di Febbraio 1613* . . . . Florence: Sermartelli, 1613.

Wallace, William A. *Galileo and His Sources.* Princeton: Princeton University Press, 1984.

Waller, R. D. "Lorenzo Magalotti in England, 1668–69." *Italian Studies* 1 (1937): 60.

Wazbinski, Zygmunt. *L'Accademia medicea del Disegno a Firenze nel Cinquecento.* 2 vols. Florence: Olschki, 1987.

Weber, Max. "The Sociology of Charismatic Authority." In *From Max Weber,* edited by H. H. Gerth and C. Wright Mills, 245–52. New York: Oxford University Press, 1946.

Weisheipl, James A. "The Nature, Scope, and Classification of the Sciences." In *Science in the Middle Ages,* edited by David C. Lindberg, 461–82. Chicago: University of Chicago Press, 1978.

Weissman, Ronald. *Ritual Brotherhood in Renaissance Florence.* New York: Academic Press, 1982.

———. "Taking Patronage Seriously." In *Patronage, Art and Society in Renaissance Italy,* edited by F. W. Kent, Patricia Simons, and J. C. Eade, 33. Oxford: Oxford University Press, 1987.

Westfall, Richard S. *Essays on the Trial of Galileo.* Vatican City: Vatican Observatory, 1989.

———. "Galileo and the Accademia dei Lincei." In *Novità celesti e crisi del sapere,* edited by Paolo Galluzzi, 189–200. Florence: Giunti Barbera, 1984.

———. "Galileo and the Jesuits." *Essays on the Trial of Galileo,* 31–57. Vatican City: Vatican Observatory, 1989.

———. "*Galileo Heretic:* Problems, As They Appear to Me, With Redondi's Thesis." *Essays on the Trial of Galileo,* 84–103. Vatican City: Vatican Observatory, 1989.

———. "Patronage and the Publication of the *Dialogue.*" *Essays on the Trial of Galileo,* 58–83. Vatican City: Vatican Observatory, 1989.

———. "The Problem of Force in Galileo's Physics." In *Galileo Reappraised,* edited by Carlo Golino, 67–95. Berkeley: University of California Press, 1966.

———. "Science and Patronage: Galileo and the Telescope." *Isis* 76 (1985): 11–30.

Westman, Robert S. "The Astronomer's Role in the Sixteenth Century: A Preliminary Study." *History of Science* 18 (1980): 105–47.

———. "The Comet and the Cosmos: Kepler, Mastlin and the Copernican Hypothesis." In *The Reception of Copernicus' Heliocentric Theory,* edited by Jerzy Dobrzycki, 7–30. Dordrecht: Reidel, 1972.

———. "The Copernicans and the Churches." In *God and Nature,* edited by David C. Lindberg and Ronald L. Numbers, 73–113. Berkeley: University of California Press, 1986.

———. "Kepler's Theory of Hypothesis and the 'Realist Dilemma'." *Studies in History and Philosophy of Science* 3 (1972): 233–64.

———. "The Melanchthon Circle, Rheticus, and the Wittenberg Interpretation of the Copernican Theory." *Isis* 66 (1975): 165–93.

———. "Michael Mastlin's Adoption of the Copernican Theory." *Studia copernicana* 13 (1975): 53–63.

———. "Proof, Poetics and Patronage: Copernicus's Preface to *De revolutionibus.*" In *Reappraisals of the Scientific Revolution,* edited by David C. Lindberg and Robert S. Westman, 167–205. Cambridge: Cambridge University Press, 1990.

References

————. "The Reception of Galileo's *Dialogue*." In *Novità celesti e crisi del sapere,* edited by Paolo Galluzzi, 331–35. Florence: Giunti Barbèra, 1984.

Whigham, Frank. *Ambition and Privilege: The Social Tropes of Elizabethan Courtesy Theory.* Berkeley: University of California Press, 1984.

Wilentz, Sean, ed. *Rites of Power.* Philadelphia: University of Pennsylvania Press, 1985.

Wisan, Winifred L. "The New Science of Motion: A Study of Galileo's 'De motu locali.'" *Archive for History of Exact Sciences* 13 (1974): 222–29.

Wish, Barbara, and Susan Scott Munshower. *Art and Pageantry in the Renaissance and Baroque.* 2 Vols. University Park: Pennsylvania State University Press, 1990.

Wolf, C. *Histoire de l'Observatoire de Paris de sa fondation à 1793,* 19–27. Paris: Gauthier-Villars, 1902.

Yates, Frances. "The Italian Academies." In *Collected Essays,* vol. 2. London: Routledge, 1983.

Zaccagnini, Carlo. *Lo scambio dei doni nel Vicino Oriente durante i secoli XVIII–XV.* Rome: Centro per le Antichità e la Storia dell'Arte del Vicino Oriente, 1973.

Zemon-Davis, Natalie. "Beyond the Market: Books as Gifts in Sixteenth-Century France." *Transactions of the Royal Historical Society* 33 (1983): 69–88.

Zilsel, Edgar. "The Genesis of the Concept of Scientific Progress." In *Roots of Scientific Thought,* edited by Philip P. Wiener and Aaron Noland, 251–75. New York: Basic Books, 1957.

————. "Origins of Gilbert's Scientific Method." In *Roots of Scientific Thought,* edited by Philip P. Wiener and Aaron Noland, 219–50. New York: Basic Books, 1957.

Zorzi, Ludovico. "Introduzione." In Ruzante, *L'anconitana,* Turin: Einaudi, 1965.

————. *Il luogo teatrale a Firenze.* Milan: Electa, 1975.

INDEX

productions, 258; in Roman academic life, 271–72; as source of scientific authority, 276–77

Comets, 9, 62–63, 75, 78, 267–311

Commandino, Federico, 62, 318

Communication breakdowns, 214–18, 220

Concini, Concino, 325, 325n.58

Concini, Cosimo, 54, 54n.145

Consistory scandal, 335–36

Contarini, Alvise, 314, 315, 317n.11

Contarini, Cecilia, 251

Contarini, Francesco, 320

Copernican astronomy: anomalies in, 204; Cesi on, 80; and Galileo, 9, 82, 90–99, 127–28, 149, 220, 226–27, 235, 286–87, 341; legitimation of, 6, 149, 219–20; as physical description of the cosmos, 5–6, 223–25, 234–35; and Scripture, 165, 167

Copernicus: and comets, 275, 275n.24, 282; work of as physical representation of the cosmos, 5–6; work put on the index, 9, 250. See also Copernican astronomy

Coresio, Giorgio, 171, 206, 228, 228n.45, 231

Corpuscularism, 194, 196, 197, 198

Cosimo il Vecchio, 122

Cosimo I de' Medici, 106–8, 110, 122

Cosimo II de' Medici: in buoyancy dispute, 178, 180–81, 207, 228–31; Cosimo-as-cosmos theme, 121–22; in court disputes, 164–65; in court spectacle, 155; enthronement of, 33; Galileo as tutor to, 20, 132; Galileo's position after death of, 279n.41; Galileo's strategy to win support of, 22–24; gift-exchange with Galileo, 45–46; Jupiter associated with, 110; in medal commemorating Medicean Stars, 139, 140; and novelties, 305–6; response to dedication of Medicean Stars, 136, 137–38; *Sidereus nuncius* dedicated to, 110, 128–30, 131–32, 133; socioprofessional ennoblement

of Galileo, 88–89, 129, 130; telescope presented to, 130n

Cosimo III de' Medici, 143–49, 153, 155

Cosma, Saint, 122

Court culture: and absolutism, 149–57; and academic culture, 263–65; artisan culture compared to, 307n.125; and *civiltà*, 156; court diaries, 161n.13; court disputes, 162–209; court nominalism, 303, 311; court spectacles, 155; and dynastic mytholgy, 111–12; fable of sound as epitome of, 301–3; and fall of the favorite, 313–14, 323–29, 333–40, 348; gift-exchange in, 38–39; and *gratia*, 49, 51, 126, 325; and the legitimation of science, 1–10, 149–57; lodestone as analogy for, 125–26; and novelty, 305–6; power centers in court society, 325; qualifications for court life, 112–20; reception of visitors, 47; in Rome, 261, 314–23, 349–50; self-referentiality of, 155; and *sprezzatura*, 51, 52

Court disputes, 162–209

Cozzi, Gaetano, 251

Credibility: and corporations, 354–55; and patronage, 15–19; scientific, 59; and status, 58–59

Cremonini, Cesare, 44n.116, 236n.65, 238, 238n.70

Cristina, Grand Duchess: court culture under regime of, 142; in dispute on Copernican astronomy, 165, 165n.29; marriage to Ferdinand, 31; as patron of Galileo, 23, 24, 33, 120

Damiano, Saint, 122

Dante, 117, 117n.50

Darwin, Charles, 213

Dear, Peter, 17

Della Casa, Giovanni, 114

Delle Colombe, Ludovico: in dispute on buoyancy, 64, 171–81, 183, 191, 196n.124, 197n.125; in dispute on lunar surface, 170, 170n.49; life of,

Ubaldini, Cardinal, 336
Universities: mathematics' place in, 7, 105–6; and scientific revolution, 6; university philosophers satirized, 116–17, 302
Urban VIII (Maffeo Barberini): and *Assayer,* 34, 75, 289, 301, 303, 349, 351; in buoyancy dispute, 75, 76, 165, 179; and Ciampoli, 322, 333–37; and consistory scandal, 335–36; and *Dialogue on the Two Chief World Systems,* 310, 311, 336–39, 350–51; election as pope, 314–15, 317, and Galileo, 33–34, 249, 317, 331–33, 336–39, 357; and Galileo's sunspot observations, 76–77; in Galileo's trial, 343–48; and Leo XI, 318–19
Ursus, 60–61

Van Helden, Albert, 98
Vasari, Giogio, 108, 122n.27, 121
Venice, 29, 32, 89, 119, 246, 250

Venier, Sabastiano, 26, 45, 45n.119
Venus, 93, 99
Viala, Alain, 85, 90
Vinta, Belisario: academic name of, 124n.78; on attractive force of virtue, 124; in court diaries, 161n.13; and Galileo's court position, 11, 23–24, 25, 33; salary of, 104, 104n.4
Virtù, 123–24
Visconti, Raffaello, 338

Weber, Max, 237
Welser, Mark: and Galileo, 70–71, 252; and Scheiner, 64, 136n.118; in sunspot debate, 68–69, 76
Westfall, Richard, 19, 105, 282n.49, 313
Westman, Robert, 10, 17

Zaccagnini, Carlo, 39
Zilsel, Edgar, 1
Zugmann, Johannes, 56